BIO-DYNAMIC
AGRICULTURE

BIO-DYNAMIC AGRICULTURE

An Introduction

by

HERBERT H. KOEPF
Professor, Dr. agr.
Emerson College, Forest Row,
Sussex, England

BO D. PETTERSSON
Lic. agr.
Järna, Sweden

WOLFGANG SCHAUMANN
Dr. med. vet.
Bad Vilbel, West Germany

ANTHROPOSOPHIC PRESS

Originally published in German with the title, *Biologische Landwirtschaft*, this English translation has been made possible through the kind permission of Verlag Eugen Ulmer, Stuttgart, West Germany.

Published in the United States by Anthroposophic Press, Bells Pond, Star Route, Hudson, New York 12534.

ISBN 0-88010-155-5

Cover: Graphic form by Rudolf Steiner.

Library of Congress Cataloging-in-Publication Data

Koepf, Herbert H.
   [Biologische Landwirtschaft, English]
   Bio-dynamic agriculture: an introduction / by Herbert H. Koepf,
Bo D. Pettersson, Wolfgang Schaumann,
      p.   cm.
   Translation of: Biologische Landwirtschaft.
   Includes bibliographical references and index.
   ISBN 0-88010-155-5: $19.95
   1. Organic farming. I. Pettersson, Bo D. II. Schaumann,
Wolfgang. III. Title.
   S6055.K6313   1990
   631.5'84--dc20
                                    90-20957
                                    CIP

Printed in the United States of America

Contents

Foreword

A modern textbook on bio-dynamic farming and gardening has been needed for some time. At present the call for true alternatives is loud and clear. This publication may contribute to the ongoing discussion. The German edition under the title *Biologische Landwirtschaft* appeared in 1974. Thanks to the help from the Bio-Dynamic Farming and Gardening Association and the initiative of Mr. Heinz Grotzke, an English version has now been produced with some alterations and additions.

The book gives an introduction to the bio-dynamic method. Some of the knowledge and concepts which are put into practice in any sound plant and animal husbandry are also included. The research accomplished by bio-dynamic workers up to the beginning of the sixties is published in the works of L. and E. Kolisko and E.E. Pfeiffer.

Contributions to the text were made by K. v Heynitz, G. Merckens, K. Hirsch, H. Grotzke, Richard Zinniker, C. Pank, and also H. Heinze. Mrs. B. Bergstroem supplied most of the photographs.

The passages on animal husbandry were contributed by W. Schaumann. B. Pettersson prepared the sections on quality. All the other chapters were contributed by H.H. Koepf who also did the editing. Mrs. U. Koepf has given invaluable help towards the completing of the manuscript. Anne E. Marshall compiled the index.

The authors are indebted to Mrs. J. Collis for making the translation.

The frame of mind underlying the whole project of this book is well described by the following quotation from a lecture given in 1898 by Emil Schlegel who was a doctor in Tuebingen:

Every age represents a link in the chain of development. just as we today look out into the distant future, so also did the people of olden times. We desire above all to eschew that fundamental evil of research that regards all of today's achievements as perfect and final while looking down with contempt upon the attainments of yesterday. Something that today seems commonplace and vulgar may in the earlier history of human knowledge have stood unchallenged as an important discovery. And views which in coming centuries are likely to grow quite outdated may now stand before our mind's eye as unchangeable truths!

Herbert H. Koepf
Bo D. Pettersson
Wolfgang Schaumann

BIO-DYNAMIC
AGRICULTURE

Chapter One

Bio-Dynamic Agriculture Today

1 A New Way Of Thinking Is Gaining Ground

Farming is more than a trade to be followed for gain. This is how we would like to rephrase the sentence written in 1810 by A. von Thaer, a leading agriculturist of his day. What he actually wrote was: "Agriculture is a trade the purpose of which is to make profits or earn money by the production (and in some cases further processing) of vegetable and animal substances." With this statement he laid the foundation for the way all those connected with this trade have thought and acted since. Yet so many instances show us that we need a far more comprehensive picture of agriculture and its involvement with other spheres of life. This realization stands at the beginning of bio-dynamic agriculture. "The course (of lectures) will show us how intimately the interests of agriculture are bound, in all directions, with the widest spheres of life. Indeed, there is scarcely a realm of human life that lies outside our subject. From one aspect or another, all interests of human life belong to agriculture." This is how Rudolf Steiner spoke on 7 June 1924 in his introduction to the course of lectures he was about to give on agriculture.

Thaer's words were written before "agriculture" in the modern sense came into being. In his day the humus theory, which maintained that plants gain their nourishment from humus, was still valid. Since the middle of our present century this view has partly been accepted once again. In between lay 100 years in which the mineral theory reigned supreme. A few decades before Thaer, European arable farming gained new impulses from the introduction, in Norfolk (England) and other places, of the Norfolk four-course system (roots, barley, clover, wheat). For centuries the fields had been cropped in a three-year course: winter crop, summer crop, and fallow. This had exhausted the land in many regions. Now by a biological method the humus and nutrient content of soils and their structure was improved. More fodder was produced enabling more livestock to be kept, which in turn meant there was more manure. Crop rotation and more intensive use of manure became in due course the tools that made it possible for a farm to improve fertility out of its own resources.

From the middle of the nineteenth century onwards Justus von Liebig contributed to the break-through of the mineral theory and thus the use of mineral fertilizers. Some minerals, such as lime, gypsum or marl, had already been in use in the past, but now interest became concentrated on the use of soluble compounds of nitrogen, phosphorus and potassium as fertilizers. The word "artificial" manure (which in some countries was not replaced by "commercial" or "mineral fertilizer" until the thirties and forties—admittedly for the purpose of influencing public opinion—was used by some with pride and by others with concern.

The introduction of mineral fertilizers is rightly regarded as the beginning of a new chapter in the history of agriculture and horticulture. A chemical way of thought was applied in practice to the phenomena of life. This chemical way of thought sees no fundamental difference between chemical processes taking place inside or outside an organism, noting at best that there are degrees of complexity. This step also marks the beginning of a new attitude toward agricultural affairs in general, an attitude that means that based on careful analysis we make practical applications of isolated factors, frequently without relating such applications sufficiently to the *totality* of the living entity in question, be this a plant, an animal, a farm, or our environment. This sounds abstract, yet it is an exact description of the fashioning principle that has been working for over a hundred years. In it the proud successes and the abysmal failures of agriculture and horticulture are rooted.

The mode of thought just described can be seen in numerous different measures that have been developed for plant and animal production. In farm management it appears as specialization. It affects people's understanding for their work and the aims they set themselves. It has consequences that affect production biology, farm organization and the sociological aspects of farming. It has changed the position of agriculture in the overall economy of a country. It has altered the conditions that determine the meaningfulness of people's work, the satisfaction they derive from it, and the kind of responsibility out of which they act.

2 Emancipated Agriculture

To emancipate means to make independent. By controlling nature to some extent, emancipated agriculture frees itself from a number of constraints that "naturally" rule plant and animal life. It hardly need be said that gardening and farming alter the natural community of life. Whether this is done in a constructive or a destructive way is what matters. A few examples will explain what is meant.

In addition to nitrogen, phosphoric acid and potash (NPK), plants also contain a number of other elements that originate in the soil. Magnesium, iron, calcium, manganese, boron, copper, zinc, sulphur, silicon, sodium, lead, arsenic, mercury and chlorine are some of these essential or non-essential elements.

The effectiveness of fertilizers that contain NPK in rather pure forms depends for the quantity and quality of the harvest on whether the soil is able to make up whatever else is missing. Of some of the elements mentioned it is known whether and in what quantities plants need them. Of others it is not known whether they are necessary, useful or unnecessary. The obvious effects of NPK fertilizers on yields were possible because many soils, though by no means all, were indeed able to provide the necessary supplements. What actually happened was that by producing increasing yields while at the same time neglecting proper humus management, the list of elements to be added to the soil grew longer and longer and the necessary amounts larger and larger. A number of researchers have endeavored to discover to what extent the soil-born elements supplement the nutrient requirements of plants and how the quality of products is influenced by this. One of the foremost among these is W. A. Albrecht of Columbia, Missouri (1943, 1952, 1956, 1961). According to his findings, an imbalance caused by leaching of nutrients or one-sided fertilizing does affect the quality of fodder, the health of livestock and, as he demonstrated with greenhouse thrips and potato scab, susceptibility to disease and pest attacks.

Plant and animal breeding have had a greater effect on the life and performance of cultivated crops and livestock than is generally realized. At the beginning of the twentieth century Mendel's laws were rediscovered, and this laid the foundation for a systematic process of breeding to achieve desired characteristics. In plants one would breed for yield, for certain kinds of resistance, flavor, texture, absence of undesirable substances, hairiness of skin on fruit, and many other characteristics. We have learnt how to "piece together" plants and also animals with certain desired characteristics. New strains spread quickly, and this meant the loss of countless regional and local varieties that had been in use for centuries. Because of their complex hereditary factors these were used for a long time as reserve breeding material. Now we have developed artificially induced mutation brought about by chemicals or radiation. The living variety of animal breeds and grain, vegetable and fruit strains has given way to a uniformity aimed mainly at high yield. We see in this development how the natural variety of plants (and one can say the same of animals) brought into being by differences of location, is gradually disappearing, and as the variety dwindles so do the chances of development.

There have always been pathogenic fungi, bacteria and insect pests that attack cultivated crops and domestic animals, and also weeds that threaten to gain the upper hand. In this realm the strengths and weaknesses of new methods have become obvious even more rapidly than with fertilizing and breeding. The first generation of pesticides consisted of available and recognized strong poisons such as lead arsenate, mercury, copper, nicotine, to mention only a few, and also oil and sulphur. Despite the evident dangers especially with arsenic and mercury, these are still in use today. The most confused situations arise. Recently (*Bio-Dynamics*, 94, 1970) a report appeared on the observations of W. Schuphan who has found that in some grape growing districts of West Germany, detrimental effects of arsenic used twenty-five years ago are still noticeable. At the same time E. A. Woolson of the United States Research Center at Beltsville, Maryland, reported before the American Agricultural Society (*Science News*, 29 11 69)that repeated applications of arsenic compounds represented no or only a small danger to subsequent crops so long as the soil contained large amounts of iron and aluminum. "This reduces or eliminates lasting toxic effects on plants and renders them harmless for consumption by man and animal."

The second generation of agricultural chemicals are substances that have been synthesized by man, that is they do not occur in nature; they are highly toxic even when applied in small amounts. Apart from organic phosphates these are above all the chlorinated hydrocarbons, of which DDT is one of the main representatives. This substance is now banned or severely restricted in many countries. In 1966 in the United States alone, 500 million dollars were spent on DDT (G. M. Woodwell 1967).

It was through this substance and others such as heptachlor, dieldrin, aldrin, etc., that we discovered problems that had hitherto gone unnoticed or indeed had not existed. These include:
— residue formation,
— persistance,
— the largely unknown effects of metabolites,
— worldwide distribution of the substances by the carriers of

the biosphere, air and water,
— unforseeable and incalculable chain reactions in the natural communities of sea, air, forests, other natural systems and also cultivated areas,
— increasing concentrations in the food chain.

This is a long list of uncertainties. G. M. Woodwell mentions that 50% of the DDT sprayed from low-flying aircraft remains in the atmosphere. It is now also increasingly realized that many pesticides evaporate from the soil and are then spread over the earth by movement of the air. DDT persists in the soil for more than ten years. In 1966 the average concentration in human fatty tissues was 11 ppm in the USA, 2.8 in Alaska (Eskimos), 2.2 in Great Britain, 2.3 in West Germany, 5.2 in France, 19.2 in Israel, and 12.8 to 31 ppm in India. A further problem raised by these substances is that they have a wide spectrum of effectiveness. They kill both pests and beneficial animals, and upset the balance in the natural community.

So now we are seeking a third generation of substances and methods of pest and disease control, that is, those that will act selectively, for instance, attractants, chemo-sterilants, juvenile hormones, etc. Biological pest control works fundamentally in the same way: predators or other organisms such as *Bacillus thuringiensis* are introduced that attack a specific pest only. Without any doubt this is an improvement on broad-spectrum pesticides.

Even among those who advocate the above methods, realization is growing that they are inadequate and need supplementing. The presence, density, or absence of any pest in an area is an expression of certain environmental conditions such as the availability of the food it needs or the prevalence of its natural predators. Thus, as has been shown particularly clearly in fruit farming, integrated pest control must be founded on improvements of environmental factors including fertilizing and careful permanent observation of the pest population. Similar realizations are growing in connection with weed control. It is now thought that herbicides can do no more than supplement the more traditional methods of crop rotation, cultivation and mechanical weed control.

These three examples will serve to make our point. The literature on all these counts is vast and fills libraries. Much of it is popularized and available to anyone interested. The experts may not be wholeheartedly in favor of this publicity; knowing the full complexity of the problem they are inclined to shy away from generalizations. On the other hand, however, they can also on occasion rightly be accused of professional short-sightedness. The layman's interest in the subject is all too understandable. While he has every reason to be delighted with the advantages of genuine progress he is also directly affected by the negative consequences of false progress.

How, then, do the above points demonstrate a certain mode of thought? What is the link between the three examples? All three are the result of a particular approach in research. Plants and animals are analyzed. Their growth and metabolism are analyzed as well. Numerous substances are found to be building stones while others have regulating functions. In the same way the development and life conditions of pests and also the factors favoring their propagation are examined. Then from the highly complicated and only partly explained processes those moments are isolated during which it will be most effective to interfere for a particular purpose by means of nutrients and active or toxic substances. Historically there has been a continuous development toward ever greater refinement in the measures taken. The main nutrient substances NPK were followed by the trace elements. Straightforward poisoning was followed by interference at particular key points in the life cycle. The mode of thought that led to this kind of action can be described as technological thinking. The technologist works with a man-made mechanism that is fully known to him in all its details and processes, however complicated. If one part breaks, and it need not be the most important one, the machine stops. The same cannot possibly be said of a living organism. Its processes, its relationship with the environment, its associations are known only in fragments. (A more comprehensive understanding of this complexity is nowadays being sought with the help of cybernetic models.) Living organisms and communities do not react in a predictable way to measures from outside. They have some capacity to compensate for unfavorable influences. For awhile they can cope with excessive demands made on their capacity, so it need not immediately become obvious if a measure undertaken is one-sided. Such one-sidedness can only be avoided if we know the higher system and are able to develop the measure we take from this knowledge. So long as we only recognize isolated phenomena of cause and effect in an organism, we shall be unable to predict the ultimate consequences of an isolated measure. In his lectures on agriculture, R. Steiner describes this as "trying out" a measure. He wants to see in place of this a "rational" approach developed from a comprehensive understanding of the total nature of the organism in question. This book contains many examples of the difference between "trying out" and "acting rationally" each time.

3 Organic Movements

Ever since modern, emancipated agriculture came into being there have been many growers and consumers who have felt uneasy about it. This feeling cannot be explained simply as a consequence of the traditionally conservative attitude of farmers. Even among those who took up the new methods there were many who felt them to be artificial or unnatural. Though this was more a feeling than a rationally adopted view, it soon came to be given expression in the so-called organic or biological movement. This movement has many streams and is by no means ideologically uniform. All of them taken together still amount to a minority in comparison with general agriculture.

Those who do not want to join forces with this movement should nevertheless take the existence of its attitudes into account more seriously than they might at first be inclined from their scientific point of view, even though some quackery can be found here and there. The reaction to emancipated agriculture is life that extends beyond what can be clearly understood by intellectual thought. The practical farmer knows that instinct coupled with sober judgment and experience can achieve a great deal, very often more than abstract knowledge. One who also knows this out of his own experience is the research worker who takes experimentation further than mere analysis. On the other hand it must also be recognized that the designations "organic," "natural," "healthy" are often not clearly defined.

The term "organic agriculture" refers in the first place to manuring. A distinction is made between organic fertilizers on the one hand and inorganic, mineral, chemical or soluble fertilizers on the other. However, minerals such as lime, dolomite, rock phosphate, ground igneous rocks, glauconite, and so on are also used on "organic" farms. Chemical fertilizers in the narrow sense are distinguished from these materials by the degree of solubility, one-sided composition, and rapid physiological effectiveness. In addition, NPK fertilizers often have a secondary effect that is not taken sufficiently into account. They allow a certain freedom in crop rotation and the use of farm manure, indeed in farm organization as such, which in the long run can and often does have negative consequences for soil fertility.

From the start the "biologically" oriented farmers displayed strong interest in soil conservation, and the preservation of the natural environment. They were on the whole keener on this than their "emancipated" agricultural colleagues. In the thirties the worthy Soil Conservation Service was founded in the United States. Its concern became also the concern of the organic movement. Quite early on E. E. Pfeiffer (1938, 1947) was pointing in his writings to the implications of soil erosion and soil conservation. A glance at the magazine of the Soil Conservation Service today reveals a close relationship between its ideas and those of the organic movement.

Adherents of the organic movement have quite clearly rejected the use of chemical pest controls. This became particularly pressing when the second-generation chemical pesticides were introduced, namely, the chlorinated hydrocarbons, the organophosphates, etc. We should not underestimate the fact that for years the organic movement published warnings about these substances that went unheeded by official quarters and the general public alike. These warnings have led to the present generally accepted demand that great care must be taken to monitor these substances and the residues they leave. It is also realized now that it is up to each one of us to play his part in coming to grips with the situation. Rachel Carson's book, which appeared in 1962 and gave indeed the initial impulse in the long overdue awakening, owed much of its impact to the activities of organic and bio-dynamic groups.

Having been founded in 1924, the bio-dynamic movement is probably one of the oldest and most united of the groups in the agricultural organic movement. Before discussing its origins and history, a few words should be said about other organic movements, though it will be impossible here to cover the subject fully.

The Soil Association is one of the more prominent organizations and was founded by Lady E. Balfour, author of *The Living Soil* (1943). It spread particularly in Great Britain and other English-speaking countries. Its headquarters are at Haughley, Suffolk (England), which is the home of the well-known Haughley Experiment with three neighboring farms, one organic, one using organic and mineral fertilizers, and one having no livestock and therefore applying only inorganic fertilizers. It would be going too far to discuss their results here. Details of the extensive analytical program carried out on soils and plants can be found in the annual reports of the farms and also in the summary of the years 1952-1965 by J. F. Ward (1965).

The ideas and comprehensive experience of Sir Albert Howard were also absorbed by the Soil Association. From 1925-30 he developed the Indore method of composting. His work *Manufacture of Humus by the Indore Process* appeared in 1935. The essential guidelines that are still valid today for successful composting were made available by him and Lady E. Balfour to a wide public. Sir Albert Howard published a further book, *An Agricultural Testament*, in 1940. His work developed independently and only slightly later than that of the bio-dynamic movement. The aims and methods of both contain a number of parallels, but there are also some fundamental differences. Howard does not share the concepts that lie behind the bio-dynamic ideas nor do his methods include anything like the specific bio-dynamic preparations.* Studies of historical forms of agriculture, for instance that of China, were further contributors to the ideas of the Soil Association.

In Switzerland it was Dr. Müller who founded and led a national farming movement that became well-known in connection with the Swiss "National Farmers' and Consumers' Cooperative." This is a self-help organization providing extension services, training and distribution of products.

In the United States during the Second World War a movement connected with the name of I. J. Rodale came into being. Between the sixties and seventies, when interest in organic methods rose sharply, his magazine *Organic Gardening* reached a distribution figure of over three quarters of a million. The founding of countless gardening clubs in the United States is also connected with Rodale's work. The Rodale Press publishes material on the recycling of refuse, environmental problems, and also numerous books on gardening. This literature reaches a considerable public. Another group in the United States is an organization called Natural Food Associates that caters for consumers as well as producers.

These and other groups represent the acute unease that is widely felt concerning current methods in agriculture. Their merits cannot only be measured by the scientific level they achieve, because, being private ventures, their efforts are often thwarted by lack of funds and personnel. With their criticisms and suggestions they have not only promoted their own aims but also influenced and sometimes even slowed down official research and official recommendations thus stimulating further consideration. We should recognize that in such an important realm of life as agriculture, which concerns every one of us in its biological, economic and sociological implications, the opposition group that keeps the discussion going plays a most important role.

---

*See p. 206 where preparations are discussed in detail.

4 What Do We Mean By "Natural"?

It seems appropriate to think for a moment about what is meant by the "natural." Surely one thing it does not mean is that people should eat only what nature offers them without any interference from man. They did that in the age of the hunters and gatherers. It also does not mean that the products of field and stable should come about by means of cultivation and husbandry, which change the natural plant and animal communities and their environment. From this point of view all food production is "unnatural." There is indeed a somewhat unclear region between "natural" and "unnatural," but looking more closely at this we should be able to find sensible criteria for deciding what is "natural" and what is not.

The plant and animal community of a natural habitat strive for a certain balance, that is, the number of species and individuals present remains fairly constant or changes only slowly. These changes, occuring according to laws, are called a "succession," and this strives toward a climax state. Such communities differ from place to place and depend on the local combination of soil, climate, topography, etc. Catastrophes such as floods or landslides bring about a new beginning for a habitat where pioneer species can enter that are later followed by others. The balance, which allows every species of flora and fauna a certain share of the available space and food and which is influenced by the inhibiting or sustaining effects the species have on each other, is not rigid. Dry or wet years can swing it away from its equilibrium, but the tendency is always for it to return to the original balance. The science and the study of the environmental relationships between flora and fauna, their suitability for certain conditions and their group dynamics belong to the field of ecology. These groups, together with their habitat, are called an ecological system.

When man interferes by clearing, ploughing, cutting, grazing his cattle, manuring, etc., he creates a new ecological system. If he has insight into the coherence of life and into the cycles of substances, then he can develop this new ecological system

12

toward a situation of balance. All too often, however, he creates impoverished plant communities by establishing monocultures, for instance, growing nothing but cereals year in year out. Such a system does not contain the capacity within itself to maintain a balance. It is unnatural. Similar situations are brought about when both pests and beneficial insects are killed by the use of broad-spectrum pesticides. Such a measure is system-destroying and should be replaced by one which is system-building.

Some mineral fertilizers are present in nature or undergo only minor processing, such as rock phosphates, basic slag or some of the potash materials. Mineral oils are also "natural" in origin. It is often ironically asked what is therefore wrong with them. Natural growth takes place in an almost totally closed cycle of substances. Living plant substance is synthesized in the green leaves while leaf litter, crop residues and other dead substances are decomposed in the living soil layer, so that the various substances can continue on their way in the cycle. The cycle is not entirely closed. It receives additions from the atmosphere and from the mineral reserves in the soil, and it suffers losses through leaching. What matters when substances are added to a cycle is how they suit it. Composts, farm manures and organic waste materials are by their very nature part of the cycle and they "fit" into it. With chemical fertilizers, even in small amounts, this is not necessarily the case.

There are other aspects of what is "natural." The most recent generations of agricultural chemicals are much more strictly "unnatural" than those mentioned hitherto. Let us look at an obvious example in the field of growth hormones. Heteroauxin (β-indolylacetic acid) is a natural growth hormone we have known about for a long time. There are other natural substances that have similar effects. Soils and plants maintain a certain level of auxin activity. These substances, which are effective in high dilutions, are controlled by simultaneous synthesis and breakdown within the organism. So the substances have a relationship with the organism that is system-building and that promotes its functioning (Scheffer, Ulrich 1960). Now there are

13

a number of chemical herbicides, for example, MCPA and 2,4-D, to name only the best known of the phenoxyacetic acids, which have effects comparable to those of growth hormones (W. Koch 1970). They function like growth hormones, they disturb the plant's growth hormone household, or they influence its enzyme system. The plant cannot control these substances, that is, it cannot suppress concentrations that are too high. So they are system-destructive. The same can be said of pesticides, particularly of those that are persistent or break down into toxic metabolites.

These examples serve to demonstrate the fundamental point. Whether it is a matter of feeding, tillage, manuring, administering remedies or anything else, the important thing is the effect of a substance or method used on the living context in which it is applied. The limited aim of removing a specific pest or redressing a specific nutrient deficiency is normally achieved, but far too readily, and it frequently happens that undesirable side effects show up. We have pointed out several forms of what is unnatural. These can be graded as follows, though there are sure to be others that we have not included:

1. bringing about imbalances in ecological systems;
2. applying natural or almost natural substances in doses that are system-destructive or that bypass processes inherent in a system;
3. applying active substances that are foreign to a system and cannot be controlled by it.

5   A Short History Of The Bio-Dynamic Movement

The bio-dynamic movement proceeded from a cycle of eight lectures given by Rudolf Steiner at the farm of the Koberwitz estate in Silesia in 1924. Steiner (1861-1925) became known before the turn of the century as a philosopher and editor of Goethe's scientific writings. Subsequently he worked toward a formulation of anthroposophical spiritual science. This provided a foundation for a number of culturally fruitful impulses that are now developing independently in many countries, in the

Waldorf or Steiner education, in medicine, pharmacy, the Christian Community, the social sciences, and in the arts of eurythmy, speech formation, drama, painting, sculpture, etc. (For more about Steiner's biography and work see *Man and World in the Light of Anthroposophy*, Stewart C. Easton, Anthroposophic Press, Spring Valley, N.Y. 1975).

In a report to members of the Anthroposophical Society given at Dornach, Switzerland, on 20th August 1924, Steiner said about his agricultural lectures:

> In the first place, as you know, it is a matter of a number of farmers who are members of the Anthroposophical Society having asked me for a course of lectures dealing with specifically agricultural aspects and matters concerning agriculture. Those who are farmers in our Society did indeed come from far and wide in order most earnestly to learn of aspects of what can come out of anthroposophical research for this area of human endeavor.

In a memorial volume, Ehrenfried E. Pfeiffer (1970) says the following about the motives of these farmers and their early work:

> In 1922-23 Ernst Stegemann and a group of other farmers went to ask Rudolf Steiner's advice about the increasing degeneration they had noticed in seed strains and in many cultivated plants. What can be done to check this decline and to improve the quality of seed and nutrition? That was their question.
>
> They brought to his attention such salient facts as the following. Crops of lucerne used commonly to be grown in the same field for as many as thirty years on end. The thirty years dwindled to nine, then to seven. Then the day came when it was considered quite an achievement to keep this crop growing in the same spot for even four or five years. Farmers used to be able to seed new crops year after year from their own rye, wheat, oats and barley. Now they were finding that they had to resort to new strains of seed every few years. New strains were being produced in bewildering profusion, only to disappear from the scene again in short order.
>
> A second group went to Dr. Steiner concerned with the increase in animal diseases, with problems of sterility and the widespread foot-and-mouth disease high on the list. Among those in this group were the veterinarian Dr. Joseph Werr, the physician Dr. Eugen Kolisko,

15

and members of the staff of the newly established Weleda, the pharmaceutical manufacturing enterprise.

Count Carl von Keyserlingk brought problems from still another quarter. Then Dr. Wachsmuth and the present writer went to Dr. Steiner with questions dealing particularly with the etheric nature of plants, and with formative forces in general. In reply to a question about plant diseases, Dr. Steiner told the writer that plants themselves could never be diseased in a primary sense, "since they are the products of a healthy etheric world." They suffer rather from diseased conditions in their environment, especially in the soil; the causes of so-called plant diseases should be sought there. Ernst Stegemann was given special indications as to the point of view from which a farmer could approach his task, and was shown some first steps in the breeding of new plant types as a first impetus toward the subsequent establishment of the biological-dynamic movement....

Shortly before 1924, Count Keyserlingk set to work in earnest to persuade Dr. Steiner to give an agricultural course.... This was held from June 7 to 16, 1924 in the hospitable home of Count and Countess Keyserlingk at Koberwitz near Breslau....

In the Agricultural Course, which was attended by some sixty persons, Rudolf Steiner set forth the basic new way of thinking about the relationship of earth and soil to the formative forces of the etheric, astral and ego activity of nature. He pointed out particularly how the health of soil, plants and animals depends upon bringing nature into connection again with the cosmic creative, shaping forces. The practical method he gave for treating soil, manure and compost, and especially for making the bio-dynamic compost preparations, was intended above all to serve the purpose of reanimating the natural forces which in nature and in modern agriculture were on the wane. 'This must be achieved in actual practice,' Rudolf Steiner told me....

The bio-dynamic movement developed out of the cooperation of practical workers with the Natural Science Section of the Goetheanum. Before long it had spread to Austria, Switzerland, Italy, England, France, the North-European countries, and the United States. Today no part of the world is without active collaborators in this enterprise....

The questions Pfeiffer mentions as having been put to Steiner now seem of rather minor importance. Compared with the picture presented by agriculture today, the early twenties appear like the good old days. Cropping was mixed, and agricultural mechanization was in its early stages. The stockless farm was a much discussed innovation. It was also the time when the nitrogen industry built up during the World War needed a market for its product. The same occured in the United States after World War II.

On June 11, 1924, while the lecture course was still in progress, the participants formed the Experimental Circle of Anthroposophical Farmers. At first it was agreed to keep the content of the lectures confidential while working together on the farms and experimentally with the Natural Science Section at the Goetheanum in Dornach to elaborate the suggestions made by Dr. Steiner. "Among the first positive observations were the qualitative improvement of vegetables and also the improvement in the palatability of fodder together with an increase in its satisfying characteristics. Soon the health of domestic livestock improved. It also became apparent that to convert a farm in accordance with the guidelines given in Koberwitz it was necessary to apply every measure that would speed up the development of soil life and support the building up of the whole farm organism." (Postscript to the 1963 German edition of the Agricultural Course.)

Outside opinion then and frequently even today sees bio-dynamic methods as no more than a manuring technique. This is too narrow a view. Those involved realized from the start that any changeover would have to encompass the farm as a whole. The aim is a farm organism that is in harmony with its habitat and in balance with the requirements of those who work there and the market it supplies. The values of traditional and modern methods are measured according to this yardstick. In consequence many a good old-fashioned method is reintroduced, while useful techniques from general agriculture are taken on and entirely new ground is covered. Some of the tasks that were tackled with interest are: application of the manure and spray preparations; composting; hedge-growing and other landscaping measures; optimum use of legumes on the farm including mixed growing with cereals; mulching; growing of herbs and their use in fodder; companion plants; bird

17

protection; pioneer plants on poor arable and woodland; catch cropping; applying green manure; housing of livestock; producing concentrates on the farm, partly by means of hot-air drying of young fodder plants; natural silviculture; various techniques for sound crop husbandry and grass management. The farmers also quite often provided the facilities for experiments concerning Dr. Steiner's suggestions on weeds and pests and on the regeneration of worthwhile but threatened strains.

In Germany the Experimental Circle was banned in 1941, even though some of those responsible for agriculture at a national level at the time were in fact interested in biological methods. Before this, during the thirties, under the purposeful guidance of E. Bartsch, it approached the public through regularly held conferences and the publication of its own literature. It was divided into regional working groups which, with the help of their full-time advisers, regularly exchanged experiences, conducted developmental work and undertook to inform the public. The critical situation in Germany before the World War II gave occasion for B. von Heynitz to bring out a report on 55 medium and smaller farms that existed at the time in Saxony. The following data are taken from this report (B. von Heynitz 1936, 1937).

He himself at that time was farming the two estates of Heynitz and Wunschwitz in the Meissen uplands. Situated mainly on diluvial loam, these two farms comprised together an area of 352 hectares (870 acres) and had soils rated at between 57 and 76 points (best soils in the country = 100). Between 1930 and 1933 they were converted to bio-dynamic methods that were used until 1945. The 1937 report mentions 13 farms of less than 10 ha (25 acres), 31 of 11-50 ha (26-125 acres), 4 of 51-125 ha (126-312 acres) and 7 larger than 125 ha (312 acres). The following list is arranged according to increasing soil fertility and it shows that the farms were distributed fairly evenly over the less and more fertile areas.

The general milk average for Holstein cattle at the time was 3000 kg (6600 lbs). The farms surveyed had a total of 776 cattle and achieved an annual herd average of 3222 kg (7088 lbs) with 3.36% fat. The following table shows the average production figures:

Growing Zone	III	rye zone on the diluvial soils of North Saxony	5 farms
Growing Zone	VII	rye zone on the North Saxon uplands	18 farms
Growing Zone	IV	East Saxon (wheat) and rye zone	2 farms
Growing Zone	V	(rye) and wheat zone of East Saxony	9 farms
Growing Zone	VI	wheat zone of central Saxony	21 farms

	Bio-dynamic farms		State average		Heynitz and Wunschwitz 1933-1937	
	t/ha; bu/t/acre		t/ha; bu/t/acre		t/ha; bu/t/acre	
winter wheat	2.87	40.5 bu	2.57	36.3 bu	3.04	45.0 bu
winter rye	2.58	39.4 "	2.05	31.5 "	2.89	44.4 "
winter barley	2.93	-	2.56	-	-	-
summer barley	2.20	34.5 "	2.14	33.5 "	3.15	49.5 "
oats	2.84	59.5 "	2.34	49.0 "	2.99	62.6 "
oats with beans	2.87	-	-	-	-	-
potatoes	25.6	10.4 t	21.6	8.74 t	24.5	9.91 t
sugar beet	36.7	14.8 "	35.8	14.5 "	29.5	11.9 "
fodder beet	78.4	31.7 "	62.6	25.3 "	-	-
carrots (4 farms)	34.8	14.1 "	-	-	-	-
shelled corn (4 farms)	3.58	1.44 "	-	-	-	-
large-seeded legumes (8 farms)	2.58	1.03 "	-	-	-	-
lucerne hay (1 farm)	7.3	2.95 "	-	-	-	-

Wheat yield was at first not satisfactory at Heynitz because of take-all and eyespot, an after-effect of earlier intensive wheat cropping. The records of the war years are lost, but later on wheat and winter barley yields were usually 4 t/ha (55.9 bu/acre).

Even though the above figures cover a large number of farms, they are nevertheless above the average for the region. Today, 35 years later, yields in general agriculture are higher, but meanwhile bio-dynamic farms are also in some cases achieving wheat yields of between 5 and 6 t/ha (70.5 to 85 bu/acre).

In order to meet the interest of consumers in bio-dynamic products, the Demeter Association was founded at that time on the initiative of the producers. It registered and distributed bio-dynamically grown products under a registered trade mark. We quote here in an abridged form from the writing of Dr. B. von Heynitz (1950) in order to catch in this survey something of the spirit with which people worked in those decades.

> During the course of this article I have a number of times described experiences we had with the quality of our produce. Thus I have mentioned the particular nutrient value of our fodder and the calculations of the official milk inspectors. The average bushel weight of our wheat was higher than that of other farms. Our potatoes contained a higher proportion of starch and it is particularly noteworthy that this high content remained constant so that in spring the potatoes contained almost the same percentage of starch as in the autumn. The sugar content of our sugar beets was 1 to 1½% higher than the refinery average. When dried, the sugar beet leaves gave a better yield than those from beets fertilized with mineral fertilizers. And our herbs, both medicinal and seasonings, contained more essential oils and stronger aromas.
>
> Bio-dynamic products were marketed under the brand name of "Demeter" by the Demeter Association. Marketing centers were established regionally in order to decentralize the work. We had hardly started bio-dynamic cultivation when we began to receive orders for Demeter potatoes. In order that we producers and the consumers might become better acquainted and also so that they should be able to see for themselves what we were in a position to grow and market in Heynitz, I often invited interested people to come on Sundays to visit our garden and farm. The townspeople were delighted to

do so; they came and saw everything and then gave us their orders. As the result of a joint decision, the Demeter Association and a number of farmers in Saxony founded a limited company in 1934. What were the products marketed? I have already mentioned rye bread and zwieback rusks. In addition there was of course dark bread and wholewheat bread. The factories also supplied us with wheat flakes, oat flakes, oatmeal, and barley coffee. In addition there were vegetables, fruit juices, herbs, tisanes, syrup and honey. In the autumn, potatoes accounted for the bulk of our deliveries to our customers. The regional centers had processing agreements with mills, factories and bakeries that obliged them to process our products without mixing them with others or adding anything. Gardeners and farmers were paid the same fixed price as everybody else. The better prices we received simply arose because our wheat had a higher bushel weight and our vegetables achieved a higher grading, which meant that we received premiums just as anyone else would have done.

Our products were highly approved by housewives. The vegetables were particularly praised because they kept better, were tastier and sweeter and also more wholesome. Clinics and hospitals were particularly eager to obtain Demeter products. One difficulty was the high cost of transporting fresh vegetables over long distances by motor vehicle.

The close attention and interest bestowed on our agricultural work by doctors was particularly noticeable. I have had the pleasure of showing many doctors, both singly and in groups, around my farm. At first I thought that their interest in agriculture as such must be slight, but I soon noticed that these visitors in particular were hoping for thorough information. I was particularly impressed by the fact that I was repeatedly asked by doctors to recommend a gardener to them. This was because, when they planned to open a clinic, they hoped to be able to develop their own garden nearby.

We have already mentioned that soon after it commenced, bio-dynamic work spread to a number of European and non-European countries. E. Pfeiffer, who had become known first at the Goetheanum at Dornach, Switzerland, and then in Berlin with a variety of experimental work, moved to the United States in 1939. There he built up the bio-dynamic organization and developed a great variety of work from the Bio-Chemical Research Laboratory he established at Threefold Farm in Spring Valley, N.Y.

Before World War II, the movement was most prevalent in the eastern and central provinces of Germany, lands that have since fallen under other administrations. The few farms left in West Germany after the war continued to work. There, after a few of those who had been responsible for the venture before the war had met each other again, the Experimental Circle for Bio-Dynamic Horticulture and Agriculture was founded. Old contacts were remade, while month-long introductory courses held once or twice a year won the interest of new people. The advisory service was established. Contacts were taken up again with farmers in Switzerland, Holland, Denmark, Sweden, Norway, and then Finland, Britain, France and Austria. National "Experimental Circles" with their own magazines, literature, conferences, etc., came into being in these countries. Advisory services were established in the different regions. The annual Winter Conference held at the Goetheanum in Switzerland, the Free School for Spiritual Science, united a growing number of colleagues.

Research was stimulated by the founding of the Institutes for Bio-Dynamic Research. In addition the facilities at the Goetheanum were available as well as the Institute in Darmstadt, West Germany, since 1950, the experimental and training farm at Rodbjerggaard, Denmark, from 1946-55, and the Institute at Järna, Sweden. Questions dealt with included composting; silage making; the relations between habitat, manuring and the quality of produce; and the bio-dynamic preparations—all this quite apart from the rapidly increasing volume of routine quality checks and advisory aid for the Demeter organization.

The Experimental Circle for Bio-Dynamic Agriculture and Horticulture is the proprietor of the "Demeter" trade mark. In 1954 the new Demeter Association was founded in Germany. More will be said about this in Chapter Ten. The range of fresh, processed and preserved Demeter products now available is extensive. It includes nearly all the vegetable and animal foods that can be grown in the country or made from domestic production.

22

The audience at the lectures Steiner gave in 1924 consisted of professional farmers and gardeners who knew their jobs. They were familiar with all the methods then current for plant and animal production in farm and garden, as well as with the mechanical implements available and the latest methods of farm management and organization. The lectures went beyond this to give those present the content of what anthroposophy has to offer to the subject.

Anyone wishing to work bio-dynamically ought to be familiar with everything that is healthy and useful in present-day agriculture and horticulture. This applies to the farmer himself as well as to the scientist and the consultant, though it must be admitted that to a certain degree everyone will tend toward some measure of specialization nowadays. Those wishing to apply bio-dynamic methods will indeed require even more judgment than their colleagues in general agriculture since they will have to apply even more discrimination and discernment. No one can make a success of any form of agriculture without knowing his trade thoroughly and being absolutely in control of daily routine. This should be seriously taken into account in training, both practical and academic. That students are required to accomplish two years of practical training before starting their courses at agricultural colleges shows a healthy instinct for what will be needed in this field of life. This requirement has never been the custom in some countries, while others used to stipulate it in the past and some still do so today. The tendencies described at the beginning of this book as being typical for general agriculture have also crept into agricultural training. If specialists are educated at colleges, there is inevitably some conflict, wasted effort and even mistakes when they come to put their training into practice. It cannot be left to the farmer by himself to integrate the recommendations of various specialists into the total organism of his farm, even if it is a specialized enterprise with one or few production programs. Either the specialist must have an eye for the farm as a biological unit, or, as is also done nowadays, the farm data are stored in a computer whence the farmer can obtain every decision, however minor. It need hardly be said that such a procedure is insufficient to cope with the life of a whole farm and its environment. In bio-dynamic work, much depends on the individual's capabilities, discernment and the entirely personal and cordial relationship he has to the things he deals with. It is the duty of colleges and training centers to contribute to this relationship.

In the Koberwitz lectures, Steiner gave a variety of suggestions for livestock feeding, building landscapes, manuring, and weed and pest control that can be applied as they stand or that require further elaboration. This does not indicate the source from which they sprang. They stem from an understanding of the natural processes that make up a farm. We have already pointed out that knowledge derived from an analytical approach to a living whole can, in isolated cases, and in a number of ways, be transformed into successful measures but that these measures too often damage the organic whole instead of helping it to flourish. This wider system or organism is of a spiritual nature. It is necessary to work one's way into a living understanding of the coherent organizing forces on which the phenomena of life are based. Steiner does this in lectures through his introduction to the earthly and cosmic environment encircling growth:

> You have now seen what is essential in the discovery of spiritual-scientific methods for agriculture, as it is for other spheres of life. Nature and the working of the spirit through nature must be recognized on a large scale, in an all-embracing sphere. Materialistic science has tended more and more to the investigation of minute, restricted spheres. True, this is not quite so bad in agriculture; here they do not always go on at once to the very minute—the microscopically small, with which they are wont to deal in other sciences. Nevertheless, here too they deal with narrow spheres of activity, or rather, with conclusions that they feel able to draw from the investigation of narrow and restricted spheres. But the world in which man and the other earthly creatures live cannot possibly be judged from such restricted aspects. (Agricultural Course, Lecture Four.)

This shows us the direction in which to seek the relationships between the different manifestations of life, which are frequently disregarded by the analytical process. Achieving a living understanding of these relationships is a task we have as yet hardly begun, but Steiner's ideas about the organism of a farm indicate a path that can be followed by the practicing farmer, since it involves the very things with which his daily work confronts him.

Since it is not our intention to make theoretical statements, a few aphoristic remarks about the expression "analytical approach" should be made. This expression refers to a method of research. The first stage of it is the chemical or physical analysis. For instance, the chemical components or the (tensile) strength of a body are determined. The next stage, usually called an experiment, is the function analysis which aims to isolate individual factors or the interaction of a small number of factors. One wants to know, for instance, how increasing applications of nitrogen affect the growth of grass when combined with different amounts of water. The least analytical way of going about things is when one brings parts of systems together in an experiment in the true sense of the word. For instance, out of the knowledge of individual cultivated plants one might carry out one or several crop rotations suited to the habitat while testing the way the productivity of the soil develops.

Thoughtful people always wondered about the secrets of life's phenomena. Many answers, which are of necessity only part answers, have been given. We can observe in the form and growth of living organisms an order and a coordination of the parts that differs from that in inorganic structures. Few people deny this difference and we are not concerned here with the way such an organization arises. What does concern us is the question, on which opinion is really divided, as to whether there is a difference in principle or only one of complexity between living and non-living substance. If the latter is the case, we can simply apply the universal laws of physics and chemistry to living organisms. It is not our intention here to revive the discussions of vitalism, which lost prestige in the last century. The laws of physics and chemistry do apply to living systems. A boulder or a living animal both fall from a cliff and oxygen oxidizes sugar both inside and outside an organism. What we are after here are those additional laws that make on organism what it is. (The attempt to understand complicated functions of life with the help of cybernetic models does not contribute to answering this particular question. We cannot avoid the choice between attributing the web of life processes to known characteristics of matter or introducing specific categories of life phenomena that are spiritually real.)

We shall restrict ourselves to illuminating a few situations. The practical farmer and the scientist will have different attitudes to the questions asked. It is historically interesting that Justus von Liebig, who is regarded today as the founder of agricultural chemistry, had something to say about these questions. He was a person who did not lack one of the most important characteristics necessary for exploring the phenomena of life, namely reverence. In his second letter on chemistry (New Edition 1967) we read:

> Research into nature teaches us to recognize the history of omnipotence, perfection and unfathomable wisdom of the divine being in all his works and deeds; without knowing this history the perfecting of the human spirit is inconceivable; without it an immortal soul cannot attain to a consciousness of its worth nor of the value it represents in the universe.
>
> In forms and in development that proceed in an orderly fashion according to laws, we may recognize a purpose and an idea, yet our senses perceive the architect only in his work. We cannot see the force that subdues reluctant matter into obeying the prescribed form and order. But our reason recognizes that the idea has an originator and that in the living body there exists a cause that governs the chemical and physical forces and gathers them into forms that are never perceived outside the organism.

This idea of the organism cannot be found at the level of discursive reasoning. It is rather a matter of creating a real relationship with life phenomena, a relationship based on the expanded capacities of a person seeking knowledge through inner training. The concept of reason as it appears in the above quotation indicates the direction in which we may reach a recognition of the independent principle that holds sway in the phenomena of life.

A study of the world of form revealed to the eye by living nature can lead us further. In the plant kingdom, and in another way also in the animal kingdom, we find phenomena in which certain formative principles become manifest in the material world in untold variety. The formative principle of all roses is the same, and yet each one is different. The relationship of the parts to the whole form is also not meaningless, because they belong together in their totality. We are not here seeking to understand the genotype as the alleged cause of the form—this lies on another level of consideration—but rather the form as an independent, spiritual reality. A study of the forms and their metamorphoses through exercises leads gradually to a spiritual participation in the creative, formative principle. On this path of study through exercises we learn step by step to read the gestures of nature by becoming spiritual sharers in them.

This is the method described by Goethe in his writings on botany, mineralogy, meteorology and other fields. Steiner was thoroughly familiar with Goethe's method of observation. Between 1881, when he was just twenty, and 1894, he worked in Vienna and Weimar on editions of Goethe's scientific writings. One of the chief roots of Steiner's quest for knowledge can be sought in Goetheanism. A Goetheanistic consideration of life phenomena is a way to complement modern forms of research, which delve in ever greater detail into searching for causes and are constantly in danger of losing sight entirely of the concrete object as such. The consequences become noticeable when the theory resulting from such research is applied in practice. It is the plant itself, together with the plant community as a whole, that produce yields and determine what is wholesome and what is not. What we learn in studying the creation of form proves to be fruitful also as a procedural method when we are dealing with the processes of synthesis and decomposition of living substances. In other words, there are, in the growing and dying of plants, typical gestures of substance transformation.

Goethean thinking opens up a clear and coherent path toward

anthroposophical knowledge. Much of what Steiner describes in his lectures on agriculture can become more comprehensible through a Goethean approach. The lectures, however, are based on anthroposophy, which goes beyond Goetheanism. Steiner himself defined anthroposophy as "a path of knowledge that strives to lead the spiritual in man to the spiritual in the universe." This statement asserts that there is a spiritual world that is not a fabrication but an objective reality. A spirituality of the same kind exists also in man. The spiritual in man and the spiritual in the universe are knowable. In anthroposophy, the method of attaining such knowledge is described. Steiner, who had such knowledge, presented it in thought forms. In these thought forms it can be grasped by healthy human understanding. It is not a question of "believing" in it or not, but to start with, in ordinary day consciousness, it may have the value of a hypothesis. As such it has proved useful in practice and in research. If we work with these thoughts, we find after awhile that they stimulate out own understanding and awaken in us a capacity for creative thinking. (The objection that this is mere suggestion may seem obvious but it is meaningless unless based on personal experience within the field in question.) So in the method of knowledge on which Steiner's statements are based, the spiritual nature of things is "visible." What is seen and perceived is, of necessity, self-evident. For those whose knowledge is not based on seeing and perceiving, and they are the majority, the statements made can be used in the manner described. When dealing with an object, for instance the bio-dynamic method of cultivation, we can at any point be quite clear about the level on which we are operating: 1. judgment based on ordinary thinking; 2. striving for a Goethean view; 3. application of anthroposophical ideas. No one has time to do this constantly, but the possibility is there. Most of what we shall discuss in this book belongs to the first level, but occasionally the other two will apply.

Those who actually do the work in field and garden need not feel that all this is too theoretical. Their situation is not that of some critical spirits who maintain the peculiar contradiction that what cannot be reached with methods of knowledge existing at present must remain either non-existent or unattainable. Life itself contradicts this attitude again and again. The person who actually works on the land has the advantage of direct experience of the more intimate interplay that exists in the growth of plants, the rhythms of the seasons, the thriving of livestock, and in general in the communities of the field or garden. A personal relationship to all these arises out of his daily work. Experience shows him that the methods are justified. It is not his task to prove it; let the scientists do that. He uses his judgment, which he corrects through experience and which contains a strong element of intuitive insight and instinctive certainty. If these qualities are linked with the practical knowledge he must acquire and with sober common sense, then he will find that he has his own direct access to a way of working that harmonizes with life. This gives him also human value such as a respect for life and a sense of responsibility for the things he deals with daily. Since time immemorial the farmer's calling has been based on a certain inner attitude. He who fosters growth, he who increases and maintains the soil entrusted to his care, does not work only for net profit. A onesided commercial way of thinking does not link the human being with the land. We are beginning to realize how greatly human beings have come to be at variance with themselves and with their fellow men and how in turn men and society are at variance with their environment. Vast numbers of young people are turning away from a civilization that gives them plenty of material goods but not enough human values. They are turning to work once again on the land, where they are seeking content and meaning for their lives. These they hope to find in contact with living things within a daily round small enough to be overseeable. This is not the place to discuss the fact that the socio-economic problems of our society often make such attempts difficult.

R. Steiner's basic works and his lectures on agriculture and allied subjects are available in print (see p. 414).

7 A Brief Description of the Bio-Dynamic Method

The details of the method will be described in the following chapters. Meanwhile it seems appropriate to include a brief summary at this point, since a short description of what the bio-dynamic method is all about is often requested. It is just as possible or impossible to fulfil this request as it is to give a concise definition of modern agriculture as such. Either the definition is brief, in which case it cannot contain much that is concrete. Or it contains concrete statements, in which case it cannot be brief. There is also an additional point. The bio-dynamic method is not merely a number of more or less connected measures; it is a method, that ·is, a coherent procedure. It is a procedure that is founded not only on modern knowledge but also on a modern attitude to knowledge that includes the spiritual. Although, therefore, it is difficult to formulate a concise definition of the bio-dynamic method, we shall include the following sentences from the informative literature made available at a Conference that took place in Wisconsin, USA, in the summer of 1973:

> Bio-dynamic farms need to be diversified in relation to the conditions of each habitat. Cropping includes a balance between soil-improving and soil-exhausting crops. Extensive use is made of legumes for soil building. Companion planting in farms and gardens makes use of the more delicate interrelationships between plants. The cropping systems emphasize lasting productivity.
>
> Manuring materials are chiefly farm produced. The recycling of animal manures and other organic waste materials by proper composting is the basis of the manuring program. The effects of manures on the soil life and on plants are improved by the bio-dynamic preparations. These consist of selected plant and animal substances that have undergone a fermentation process lasting for a part or the whole of a year's seasonal cycle. Preparations are also available that are applied by spraying in order to support soil life and stimulate the plants' utilization of light.
>
> A proper stocking rate is important for crop growing. Optimum production and the good health of the stock are valued more highly than peak performances. Bio-dynamic farms are organized as bal-

anced biological units. The methods employed eliminate a number of diseases, pests and other weaknesses.

Farms run along these lines improve the quality of the ecological system of which they are a part. The food value of their products is high. Not only because agricultural chemicals, which might cause residue problems, are left out, but chiefly because the quality is achieved through obtaining an optimal relationship between soil-born growth factors and atmospheric and cosmic influences. The preparations and attention to rhythmical patterns also contribute largely.

There are bio-dynamic farms, and many private and market gardens in many countries and on all continents. Their products, marketed in a number of countries under the trade mark of "Demeter," are increasingly in demand. They are distinguished by their excellent flavor, keeping properties and high food value.

The development of these farming methods started in 1924. A group of farmers at that time asked Rudolf Steiner to help them bring a new spiritual impulse into their work. The bio-dynamic approach is based on an understanding, out of anthroposophy, of the interrelationships of living organisms and the processes that make up the ecological system, embracing forces working within plants, the soil and from the surrounding universe. Further development is needed. The methods can be applied by anyone who is interested. Over and above their actual application they open for the spirit in man new possibilities of achieving a clear and conscious relationship to the world of forces appearing in living organisms. In turn the daily work is given more of a meaning and an aim. Thus a positive contribution is also made toward alleviating the social and human problems of our time.

8 Conflicts Of Interests And Unsolved Questions

In the organization of a bio-dynamic farm, it is recognized that cropping and animal husbandry aimed at lasting productivity and health, that is, a thriving means of production, are of prime importance. This attitude stands at the source of the movement. Its representatives have in the first place concerned themselves with the biology of production, with questions of quality, etc., and have organized their farms to fit these requirements. Farmers, theoreticians, planners and managers in general agriculture share this attitude only partially or not at all. In their view the determining factors for organization are commercial and labor-saving considerations. Yet the function of agriculture in the life of the individual and society as a whole does not consist merely in providing work and incomes. Only if every farm in its specific situation succeeds in achieving a balance between the needs of the production means, that is, soil, livestock and plants, and the justified economic requirements, can it begin to tackle its other functions. These are unsolved problems or unfulfilled requirements such as: maintenance of a lasing, good means of production; wise use of non-renewable resources; the quality of the environment; the quality of foods; the relationships of farmer and gardener to their work. The organization of the farm should not be aimed solely at achieving parity of income. Only narrow-minded specialization or misunderstood pragmatism should disagree. It is not a romantic illusion to want to attach a decisive importance to these problems. Obviously they raise difficult questions. It is difficult enough to calculate economic projections for the coming five to ten years, but it is even harder to make headway with bio-economic problems. Yet they cannot be avoided.

Interrupted during World War II for about ten years, but never brought to a standstill, agriculture in European and other industrialized countries is involved in a continuing process of restructuring and adaptation. This process was intensified by the economic boom of the fifties and sixties, and the European economic community is not making the changes any easier. Surpluses in some sectors, disparity between agricultural and other incomes and prices, varying national production and price structures, and more recently shortages and soaring prices for imported items, are the essential problems. Energy, raw material, and financial crises that have developed since 1973 have further exacerbated these problems.

Planning authorities and governments favor the depopulation of rural districts in the interests of income parity and the mechanization of farm work. G. Weinschenk (1971) assumes that of 1.1 million farms or holdings of over 2 hectares (5 acres) now existing in West Germany, only 700,000 will be left in 1980. Of these, 200,000 to 300,000 will be full time. This would correspond to about 500,000 male workers working full time on the land. We can deduce from this projection that the majority of these farms will be part-time or just family farms; the shrinkage is undeniable and must be linked with a considerable migration away from the land, though many may only join the army of commuters. Agricultural land is becoming emptied of people because of the increase in farm size and field size. Marginal areas fall into disuse. In some uplands this is expected to account for over 30% of the agricultural land. Thus correspondingly high use of mechanization and chemical fertilizers are needed in the intensive farming areas, while the marginal areas are either reforested or allowed to grow wild, unless enough interested people and usable solutions can be found for extensive hill farming. The essential point is that whether the farming is intensive or extensive, or whether an area is for tourist recreation, there ought to be enough people to care for the land in a healthy way.

E. Reisch pointed out in 1971 how the fight for parity of income with industry has caused individual farms not only to utilize all the possibilities of plant and animal feeding, etc., but also to progress along the path of mechanical and organizational rationalization.

Impressive practical examples of this concentration of cultivation can already be found.... But there is no answer as yet to the decisive question of how the yield/input ratio will develop after ten years or more. We still lack basic information on the effects of substituting artificial and technical measures to increase yields for conditions for lasting soil fertility that resulted from carefully planned land use and crop rotations. What is already certain today is that in many sections of traditional "crop-rotation philosophy," technology has "won" against nature: deep sub-soiling with strong tractors has replaced deep-rooting crops (which can be poor in yield); herbicides have surpassed measures of cropping and mechanical weed control; insecticides have surpassed the long periods for pests to subside; chemical fertilizers have surpassed symbiotic bacteria in supplying nitrogen;

33

possibly some special fungicides will soon have surpassed soil hygiene with regard to prevention of take-all.

The specialization achieved by these and other means is now advanced in agriculture and horticulture. In animal husbandry whole sections can now be entirely separated from the farm as a whole. "The concept of a farm as a rounded off, enclosed 'organism' will be replaced by a modern interpretation of the concept of 'organization,' which includes both a well-thought-out production program and the actual organization of production." In other words though each farm may be different from others, each is to specialize.

We can judge the value of this development, which is actually taking place, when we view it in the light of the questions asked above. The passages cited go so far as to admit that the long-term influence on yield is unknown. Negative effects on the size and quality of yields, and on the environment, will occur soonest in regions with unfavorable or extreme natural conditions. One example and the conclusion drawn in official quarters will serve to demonstrate what is meant.

In January 1970 the British Ministry of Agriculture published a report entitled "Modern Farming and the Soil." This report arose as a result of bad experiences during two exceptionally wet years in the mainly cereal growing regions of middle England. The following quotation from the Preface speaks for itself.

Weather apart, decisions on cropping systems, the realism or otherwise of business management advice, crazes of various kinds, economic pressures, even national agricultural policy and the incentives that accompany it, all affect the raw material of farming, which is the soil. It has not always had too happy a time. There could be a remarkable response from the soil of this country if all forces combined to treat it with respect.

The investigating committee led by Nigel Strutt found that there was no cause for concern with regard to the supply of nutrients for crops. What was worrying, however, was the structure of the soil. In soils with a weak structure the effect of organic substances is of supreme importance. Some soils show a dangerous lack of such substances and cannot be expected to withstand the use or cropping they are subjected to. In addition to the wrong use of heavy equipment, this also stems from crop rotations being too short in medium and high precipitation regions. More drainage is needed. There is an alarming increase of grassweeds, particularly wild oats (*Avena fatua*). On some soils couch grass (*Agropyron repens*) is kept at bay by cultivation and herbicides, but on other soils these methods are ineffective as well as being costly. Grass weeds cannot be entirely controlled by chemical means. It is stressed that the investigating committee realizes the difficulties of combining modern methods (in brackets it adds: "Which are of course inevitable"!) with soil health. Yet healthy soil leads to higher production and thus higher profits. So present agricultural techniques should be reviewed. The farmer should think carefully before changing his system of cropping. The Advisory Service should familiarize itself more thoroughly with the medium and long-term effects of changing a farming system. Soil scientists, agricultural engineers and agricultural advisers should work more closely together. In training programs more emphasis should be placed on the problems of the soil and soil utilization.

Modern methods of plant production comprise rationalization, mechanization, use of agricultural chemicals. Yet here is an official report stating that they are not compatible with the lasting fertility and health of the "means of production." The production capital is being eaten up. Even though the tone of such a report is usually restrained, the need for a harmonizing of economic and biological interests is expressed.

Biologically oriented agriculture does have a contribution to make to this on the basis of the direction in which it is working and on the basis of experience gained hitherto, though it must be admitted that it also still needs further development. The fact that it has a contribution to offer is recognized in some quarters. Recently a motion was made in the Swiss Federal Council for a research institute for biological methods of food growing. Among others P. Matile, a botanist at the Zurich Technical University, was quoted. The following quotation is taken from a lecture he gave on 13 October 1971 (Matile 1972).

Biological agriculture lacks the means necessary for methodical research... and obviously a research institute devoted to this subject and concentrating on developing methods and advising farmers would be extremely useful. Great store must be set by the establishment of such an institute because biological methods in agriculture must be developed as a totality in keeping with agriculture's ecological character. Nothing is simpler than to devise experiments that prove the uselessness of an individual measure taken out of context. An exactly defined total package of coherent measures—composting; special aspects of tillage; more permanent plant cover; renunciation of soluble mineral fertilizers and pesticides; possibly even the choice of varieties—is a mandatory basis for meaningful research. The ecosystems of the whole world have been rendered unstable by the civilizing activities of man, and the realization that we must rethink our use of technology and adapt it to the vital needs of our environment is now widespread. Time also seems to be ripe for consideration of a concept of agriculture that can bring about the stabilizing of our agrarian systems, not via a return to extensive cultivation methods but via the translation of ecological facts into a modern, intensive and, not least, a rational body of methods. Since yields achieved by biological methods are quantitatively quite compatible with the present norm, there would be no question of endangering the world supply of food. On the contrary, the achievements of the extreme targets of agricultural production now existing would increase the problem of overproduction in the industrialized countries while the health of our soils, crops and livestock would be even less guaranteed than is the case at present.

In 1973 the "Swiss Foundation for the Advancement of Biological Farming" was founded and it has started on a program of research (Koepf 1974).

9 Population Growth and Bio-Dynamic Agriculture

The rapidly increasing population of the world is a problem today that must be faced by those who wish to advocate new agricultural methods. Even if we wanted to, it is said, we could not afford to embark on a world-wide system of bio-dynamic or other biological agriculture. As it is, production is not high enough to feed everybody. Even if quality suffers, we must increase our yields still further. Mineral fertilizers are a rapid and relatively cheap means of raising yields in the developing countries. We cannot allow pests to devour a large percentage of the food that should be at the disposal of mankind. The World Food Conference held in Rome in 1974 called for an increased rate in world food production over the next 12 years of 3.6% per annum as against the previous rate of 2.6% per annum.

Populations are increasing in the developing countries more rapidly than anywhere else, as is shown in this table giving figures for the five continents.

World Population Increase (in millions)

	1650	1970	2000 expected
Europe	100	650	860
Asia	330	2050	4000
North America	1	225	310
South America	12	265	590
Africa	100	280	520
World	540	3500	6200

The amount of arable land per head of population is rapidly decreasing.

Hectares of arable land per head of population in the Federal Republic of Germany (1 ha = 2.5 acares)

Year	FRG	in %	Lower Saxony	North-Rhine Westphalia	Baden Württemberg
1939	0.37	100.0	0.64	0.16	0.32
1961	0.30	79.8	0.43	0.11	0.22
1967	0.23	62.5	--	--	--

The decrease of arable land for food production is particularly high in this country and the difference between the more rural and the more industrial States within the country is not great.

The problem of how to feed a growing population is often raised as an argument against bio-dynamic agriculture. But this is not the way to set about it. There is in fact no overall answer. The question must be tackled by detailed regional studies in various parts of the world.

As shown by the figures quoted for the farms in Saxony, the yields achieved by bio-dynamic methods in the thirties were approximately equal to those in ordinary farms. Now that many farms achieve a regular 5-6 t/ha (71-85 bu/acre) of wheat almost as a matter of routine, the odds may have shifted somewhat against bio-dynamic methods, though on good bio-dynamic farms similar yields are reached. The yardstick anyway is the average for the area in question. In rich countries with their wheat, butter, etc., what is produced is as important as how much. As a rule, bio-dynamic farms require a certain amount of livestock. It is not only the disadvantage of the lower conversion rate that must be considered but also the importance of stability that must not be underestimated.

In 1946, when food was scarce in Germany, and indeed in the whole of Europe, F. Scheffer gave a lecture on "Maintenance and Improvement of Soil Fertility." The considerations he voiced then are still of value for organic farms today. In general agriculture at the present time the vision is blurred by motivations of economy and labor, which have become almost entirely dominant. Scheffer started by mentioning the known facts about the balance sheet of nutrients in the soil. A survey conducted by Woermann and Hahn in the thirties had compared the measurable intake by plants of soil nutrients in Germany. The balance sheets for nitrogen and potash were negative. Even when the largest (for those days!) amounts of mineral fertilizers were being applied, the plants still had to draw on the nutrient capital of the soil. Even earlier P. Wagner had drawn up balance sheets for nutrients in field experiments: "On all kinds of soil those for nitrogen and potash were negative when commercial fertilizers were used exclusively.'' Where stable manure was used, with or without commercial fertilizers, that is, where at least some organic manure was applied, the balance was hardly negative or even became positive. In 1946, however, the problem of mass production loomed large, just as it does today in connection with the world supply of food. Scheffer discusses sugar beet, which provides the highest yield in calories per unit area. Root crops are humus-depleting crops. ''Concern that to sandwich in a crop that restores organic matter might reduce the marketing capacity of the farm by reducing the area available for cash crops is unfounded. This is shown clearly by farms that have hitherto included in their rotations a reasonable acreage of lucerne, clover or grass seed. They have thus achieved a balanced humus content in the soil by alternating humus-restoring and humus-depleting crops. The aim of raising the profit margin per unit area is achieved far more easily in the long run since by growing humus-restoring crops we create the conditions for the development of humus-depleting row crops while reducing the risks their production brings.'' These considerations ought to be taken seriously today where less developed forms of agriculture are being converted to more intensive programs.

In 1972/73 in Hesse (West Germany) the negative consequences of onesided grain programs were revealed by an inquiry covering 1,520 fields. Winter wheat sown after clover or beet produced average yields of 4.94 t/ha (69 bu/acre). If the previous year's crop was a cereal, the average yield was 4.76 t/ha (66 bu/acre), and after two years of cereals the yield went down to 4.1 t/ha (57 bu/acre). After several years of cereal crops the yield decreased still further to 3.81 t/ha (53 bu/acre). The same occured with rye, barley and oats, though not quite so drastically. A crop rotation program that includes clover and other fodder crops sustains high yields of cereals (Schwerdt 1974).

Meanwhile in the prosperous countries of the world development has progressed. For economic reasons animal husbandry is now separated from fodder production, and the animal excrement is removed cheaply by labor saving methods instead of being used to best advantage. Indeed, in the specialized production of meat, eggs and dairy products an unknown cost and risk factor that cannot be measured is brought into play and borne by the public at large. It is the loss of soil fertility in the areas where the fodder is grown. Years ago William A. Albrecht pointed out (in conversation) that the sale of alfalfa meal from young soils was the surest way of rapidly exhausting their natural reserves. In *Bio-Dynamics*, No. 80, 1966, the author pointed out a connection between the advance of red clover grown in Wisconsin at the cost of alfalfa, and the insufficient use of farmyard manure. Of course such tendencies when they start to develop are easily doubted. But they show that we must look at food production as a whole and not at individual products alone. The value of record yields should not be overestimated because what counts is the scale of the lasting total output.

In the famine-stricken regions of the world, including the areas where protein shortage is the problem, production increases can only be achieved in small steps via numerous improvements. It is now recognized that imports from rich countries cannot relieve the situation in the long run even if it were possible to overcome the financial and political obstacles hindering such operations. Food shipments can only bring relief after natural disasters or in the very worst situations. What is needed is aid that will help poor regions to help themselves, and such aid includes education and training, creation of local earning possibilities, step-by-step improvement of cultivation methods, only gradual changes of structure investments in improving the infrastructure, simple implements to replace primitive or non-existent tools, and so on. Dr. E. Schumacher, current president of the Soil Association, founded the Intermediate Technology Development Group with whom he has undertaken such tasks and proved that they are workable. (Schumacher 1973).

The objection that not enough organic fertilizer exists to make an overall introduction of organic agriculture possible is also to a large extent misleading. Taken as a whole, the present need is to make existing supplies of organic fertilizers and other soil amendments usable instead of letting them contribute to environmental pollution. Once this has been energetically tackled, it will be time to gauge how much more is needed.

On 5 May 1972 *The Observer* published an interview with G. Bergstrom of Michigan State University and the Nobel Prize winner N. Borlaug who bred the high-yield Mexican dwarf wheat. A number of essential points regarding food production in the future were dealt with and are commented on in the following.

The "Green Revolution" has had some success. In India the wheat harvest rose from 12 million tons in 1965 to 23.3 million tons in 1971. West Pakistan also almost doubled its wheat production during that period. It is true, from 1972 onwards the picture looked much less encouraging. The situation with regard to pesticides is difficult and uncertain. Large-scale growing also requires large amounts of water and mineral fertilizer. In many places water is the limiting factor. In addition, heat equivalent to that from 5 tons of coal is needed to produce one ton of nitrogen fertilizer. The amounts of fertilizer and water theoretically needed by countries like China or India, if they were to produce yields comparable to those of, say, Japan, make it obvious that this would be impossible, quite apart from the resulting degree of environmental pollution that would also have to be taken into account. The idea of breeding wheat varieties with nitrogen-fixing symbiotic bacteria in order to cover nitrogen requirements biologically instead of chemically as suggested by Borlaug becomes understandable. Other attempts are made to build into non-leguminous plants the mechanism some bacteria possess for fixing atmospheric nitrogen (*Nature*, November 29, 1974). Meanwhile, until this has been achieved, it is a matter of recognizing that the total yield of mixed farms should be taken as the yardstick. Yields per area of wheat are not sufficient measure by which to judge yield potential. If the protein deficit could be reduced by growing more large-seeded legumes, of which there are not enough either in tropical or temperate climates, this would at the same time help to raise the yield capacity for wheat on the remaining available acreage.

The energy requirements of agriculture as practiced in the

Western world or in the style of the Green Revolution is high. According to Pimentel et al (1973), the production of enough food for one person in the USA requires the equivalent in energy of 423 litres (112 gallons) of fuel per year. Americans spend only 16.6% of their income on food, which for retail prices is 597 dollars. Primary production costs are a third of this, that is, 199 dollars. On the other hand Americans consume 3110 kcal per day for the production of which, because of the high proportion of animal products, 5280 kcal of plant energy are needed. These cost 38 dollars per 1000 kcal. In India people spend about 23 dollars per year on 2000 kcal per person per day. These 2000 kcal are produced by 2280 kcal of plant energy, which cost 10 dollars per 1000 kcal. For reasons of cost and energy, therefore, new solutions must be found, different from those in the West, to solve the problems of food production in the Third World.

The oceans are by no means inexhaustible sources of food; indeed some of their resources are already overstrained. Seen as a whole, too much fish meal finds its way into the food troughs of pigs and poultry in rich countries instead of going to poor countries. These countries will also in future want to retain more of their oil crops instead of exporting them to the fodder and margarine industries of the West. The obvious conclusion is that the rich countries in temperate zones ought to grow more legumes and oil crops, which means promoting mixed farming instead of subsidizing onesided grain growing.

For the purpose of meat and egg production, too much grain is fed to cattle, pigs and poultry in the rich countries. If a certain change in eating habits could come about, a good deal of grain could be saved. It is a waste, however, that results from a mistaken structural policy of specialization and large-scale cattle rearing in which cattle are fed too much grain. If cattle farming is decentralized, as it used to be, it means that grass, hay and other fodders that are not suitable for human consumption can be utilized. In turn the demand for milk and beef can then be met. Futhermore, animal husbandry on a mixed farm favors the correct use of animal manure, which promotes soil fertility and soil life. The areas used for cash crops produce high and certain grain yields. Saving grain for human consumption is primarily a matter of farm organization and structure. By far too little emphasis is put by scientists and agricultural politics on this aspect of the issue.

China keeps three times as many pigs as the United States. This is made possible by a well-developed system of recycling in which pigs are kept in almost every household and absolutely every scrap of waste is used, including feces. In the specialized beef and poultry factories of the rich countries the least possible use is made of waste matter.

This brings us to a further point. The matter is now being raised of work-intensive agriculture that reverses the concentration of the population in the cities. This type of farming is best carried out in family farms to a limited extent as a part-time job by people whose chief source of income lies elsewhere. This should not be taken as an idea for occupational therapy. It is much more a matter of seeing that it presents a possibility of producing healthy food biologically that contributes to the quality of the environment, produces higher yields and is immune to economic crises. The possibilities include diversified, work-intensive and high-yield crops such as garden and field vegetables and fruit, farm livestock, the utilization and recycling of absolutely all manures and waste products. This is in fact the kind of organization that is typical for bio-dynamic and organic farms: mixed, diversified farm, farm-scale livestock keeping, rational utilization of waste materials. The kind of urbanization that has come about in the industrialized countries has proved impossible for the Third World. The diversified family farm, however, which is so suitable for those countries also, has a future in the richer parts of the world although present agricultural policies are on the whole still working in the opposite direction.

When higher agricultural production is to be achieved in the developing countries, it becomes increasingly urgent to raise the matters of closed cycles of substances; the establishment of lasting soil fertility; prevention of soil erosion; weed and pest control without poisons; and the creation of stable landscapes and plant and animal communities. It could be said that these problems are in fact already as urgent in developing countries as elsewhere. The tasks to be tackled are the same everywhere, though the means may vary from place to place. The abstract question as to what would happen if all agriculture were to be bio-dynamic or organic has little meaning. And the question of how to harmonize quantity, quality and permanence of crops cannot be answered globally but only regionally in the light of the stage of development of the part of the world concerned.

10   What Is The Difference Between Organic And Bio-Dynamic Agriculture?

This question is asked from time to time and a few remarks should be made in answer. The question itself can be somewhat abstract since there are numerous forms of organic agriculture, and it is thus not always clear what is being compared with what.

It has already been pointed out that the bio-dynamic method should not be regarded merely as a manuring method. It covers all aspects of farming and gardening. The aim of achieving a balanced organism determines individual measures and organizational forms. This guiding aim allows for plenty of mobility, which is essential in agriculture, and creates no strict rules as to what must and what must not be done. Since the method is ecologically oriented, strict rules would be out of place anyway. Every farm or garden has its own individual character with regard to its natural, economic and human situation.

Bio-dynamic and organic farms have many methods and also some basic principles in common. It will often be found, however, that with organic farming it is around one particular measure, such as for instance making compost heaps or sheet composting, that a particular method centers.

As with the practical measures, so also with the theoretical basis there are similar areas of common ground. Both share an awareness of ecology that is becoming more widely recognized now. The bio-dynamic movement also has at its disposal the indications given in anthroposophy. It should be remembered that anthroposophy does not claim to have the answer to everything, but it does open up additional possibilities for research and action. One does not have to be an anthroposophist in order to apply the method. Through its application understanding grows and evidence comes to light. This is a field, however, that is special to the bio-dynamic method.

The same applies to the bio-dynamic preparations that are sprayed on to the plants or added to manures. Their value is borne out by application and observation. Their bases are the indications given by R. Steiner about the functions of forces in natural growth.

Chapter Two

Farm Organism and Landscape

General remarks on the term "farm organism": This term appeared early on in bio-dynamic farming. It comprises everything belonging to the farm and living in it—soils, livestock, crops, the people who work there, and then also the wild plants, copses, ponds and streams, wild birds and insects, wild animals, the local climate, the seasons and their rhythms. All these form a living structure of mutual interaction that under the guiding hand of man produces food and fodder. It can be self-sufficient and additions from outside should be kept to a minimum. "Organism" is meant not merely as an image of a complicated totality but is a realistic term indicating an existing reality.

The term "farm organism" does not appear in R. Steiner's lectures on agriculture. But in substance it originates in the following passage:

A farm is true to its essential nature, in the best sense of the word, if it is conceived as a kind of individual entity in itself—a self-contained individuality. Every farm should approximate to this condition. This ideal cannot be absolutely attained, but it should be observed as far as possible. Whatever you need for agricultural production, you should try to possess within the farm itself (including in the "farm," needless to say, the due amount of cattle). Properly speaking, any manures or the like that you bring into the farm from outside should be regarded rather as a remedy for a sick farm. (Second Lecture)

Looking at Steiner's lectures in this light we find that in the concept of the "farm individuality" a basic principle is approached. For an initial understanding three viewpoints can shed further light on what is meant:
1. The cycles of substances and forces;
2. The site or habitat;
3. The farm organization.

1 Cycles Of Substances And Forces

We recognize substances by their perceivable characteristics; a stone has weight, color, solidity, chemical properties, etc. Forces such as those that form a plant or organize a variety of separate processes in their relationship to a total organism, are recognized by what they do. This activity can be described. For instance, the sequence of leaf forms along the stem of a plant are the result of the developmental process. Forces reveal themselves in the creation of form and in processes.

Let us first turn to the cycles of substances and forces in the earth's biosphere. Though the mineral layers may be said to cover one another in a varied sequence in the depths of the earth and though the atmosphere is highly structured through streams of warm and cold air, mist and vapor, yet these realms must seem quite uniform when compared with the wealth of forms and interpenetrations of substances and forces that hold sway in the growth sphere or biosphere. On land this biosphere extends a small distance below the tips of the roots in the ground and above the tops of the trees in the air. It contains the whole profusion of plant forms that come and go with the seasons and all the life processes of the animals above and below the surface of the soil. Some substances can only be produced and only exist in the biosphere (leaving apart water so far as it is penetrated by light). Among these are protein and starch in plants, and humus

47

and clay minerals in the soil. It is in the biosphere that the energy radiated by the sun is transformed into the processes of building up and breaking down that maintain the cycles of substances. Within only a few decimeters considerable qualitative differences occur in the creation of form and substance. The cosmic influences, which we perceive mainly as light and warmth in their rhythmic manifestations, meet with the gaseous substances of the atmosphere and with the solid substances and water in the soil.

Only a few essential aspects of the growth space can be indicated here. Without any claim to completeness they are depicted in the following diagram; some explanatory remarks refer to the diagram from left to right

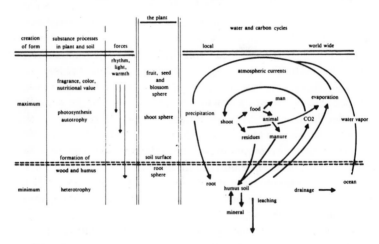

Diagrammatic survey of form creation, processes of substances, cycles of substances, and forces in the growth sphere.

Plant forms. The shaping of plant forms is the most obvious of all the processes. The shoots growing in the light show the greatest variety of forms. The blossoms and in many cases the seeds often excede the green shoots in regularity, constancy (e.g. number of petals), clarity of form. In them the greatest degree of form and regularity is reached. In contrast, the bacteria and fungi and most creatures living in the darkness of the earth lack form. There are only two basic root forms, the tap root, which can thicken out as in carrots or turnips, and adventitious roots, which form a ramified pattern. The formative difference between the spaces above and below the surface of the soil correspond to clear differences in the substance the plant contains.

The processes. In the green leaf sunlight forms sugar and starch out of carbon dioxide in the air and water in the soil. From the sugar and starch arise protein and the whole range of nutrients, physiologically active substances and fibres. The photosynthesis in the green leaf is the fundamental process of all organic substance formation on earth. It takes place in the living cell, chiefly the chloroplast, which consists of chlorophyll and protein.

The roots and also the blossoms, fruit and seeds are "nourished" by the substance formed in the green organs. In contrast to the autotrophy of leaves, the soil microorganisms, the plant organs that are not green, and also animal and man, "nourish" themselves heterotrophically. In the roots the tendency to accumulate fibrous substances is noticeable. This also takes place in wood or in older parts of the plant above ground. Humus is a relatively stable form of organic matter that forms in the soil on the basis of such fibrous substances. The formation of organic substance both below and above ground originates in the leaf.

In the nutritive tissues of seeds and fruit, nutrients are deposited in their highest concentrations; also aromatic substances are formed; the oils and fats, which appear only in this part of the plant, have the highest energy content. Starting from the root and working upwards toward the blossom and seed substances, there is a qualitative intensification of substances in the plant.

The forces. The cosmic growth factors, above all light and warmth in their diurnal and seasonal rhythms, are active at all levels of plant formation and transformation of substance.

Many of the rhythms of light and temperature that determine the development of the plant in its various stages are known.

Cycles of substances. Let us use the carbon cycle as an example since it is the most important in nature. In the cycles section of the diagram, the stream of carbon dioxide is indicated as it streams from the atmosphere to the plant, which is producing nutrients by photosynthesis. Thence, the stream proceeds to animals and man while crop residues and animal manures return to the soil where they are metabolized into carbon dioxide. The carbon dioxide exhaled by the soil and by animals and man closes the circle. A harvest of 4 tons wheat and 6 tons straw contains 4 tons of carbon that need replacing. One hectare (2.5 acres) of land returns 2.5 - 4 tons of carbon to the atmosphere annually in the form of carbon dioxide. This carbon dioxide has an additional important task in the soil; it participates in soil formation and makes nutrients available.

The carbon cycle passes through the atmosphere. The plants draw on this supply, which is always being replenished. Atmospheric currents ensure a thorough mixing over wide areas.

Minerals such as calcium, magnesium, potassium, iron, phosphorus, sulphur and others originating in the ground behave in a different way. They remain at the site and circulate between the leaves of the plant and the humus beneath. The same applies to nitrogen, which, however, originates in the air. As shown by arrows in the diagram, a circulation of nutrients also takes place between the soil minerals and the organic matter, that is, the amount of nutrients circulating in living matter and in humus is increased or replenished from the mineral reserve. But the cycle also suffers losses through leaching. Every farmer knows the progressive loss of calcium from soils in humid areas. On the other hand, nutrients are added out of the atmosphere in small amounts. These are greater near coasts than inland (H. Egner and E. Eriksson 1955). Industry, traffic and ordinary households release substances into the atmosphere that are later deposited on woods and fields in rain. Without the flow of air over vast areas, the carbon dioxide from the burning of fossil fuels in thickly populated areas would soon make them uninhabitable. In the form of sulphuric acid in rainwater, sulphur from burning mineral oils and coal contributes considerably to the leaching of lime from soils.

The nutrient content of the soil within the root zone results from these cycles of substances. There are more nutrients concentrated in the humus layers than in the deeper layers. Since the beginning of soil formation they have been accumulating there. But the total supply of nutrients in the root zone gives only an approximate indication of how much is being supplied to plants. Their availability for the plants is crucial. This depends largely on the microbial life of the soil, which arises as a result of the penetration by humus of the mineral substances. If sufficient humus is formed as a result of the cycles of organic substances, it is then certain that the plants will be able to make full use of the natural supply of nutrients in the soil.

The following table of the nitrogen contents of a number of lighter and heavier soils in Illinois, USA (Millar 1955) gives an example of the biological accumulation of nutrients in the topsoil. Nitrogen is the nutrient element that is not present in the parent rocks of the soil. It accumulates as the result of the organic cycles. We have chosen this example from the American Midwest because the figures stem from 1910 whereas appreciable applications of mineral fertilizers, especially nitrogen, did not commence in this part of the country until after the World War II (US Department of Agriculture 1955). The figures show that heavy soils have a higher nitrogen content than light soils. The lower section is the significant part of the table. The figures for the nitrogen contents refer to layers of the same thickness. The concentration is greatest in the top layer. This is the result of the activity of plants and soil microorganisms during soil formation.

(For the sake of completeness it should be pointed out that these figures, from the transitional region between steppe and originally forested areas, are relatively high. They depict the total nitrogen contents of which the plants receive a small share annually. It might also be added that in the lower layers part of

51

Nitrogen content in various soils at different depths in kg/ha (lbs/acre)

Depth in cm (in inches)		Brown silt loams	Brown loams	Brown sandy loams	Yellow fine sandy loams
0.0-16.7 cm		5540	5190	3380	2390
(0-6 2/3")	total	(5035)	(4720)	(3070)	(2170)
16.7-50 cm		6510	7330	4310	2870
(6 2/3-20)	content	(5920)	(6660)	(3920)	(2610)
50-100 cm		3930	4560	4580	3000
(20-40")	of layer	(3573)	(4145)	(4164)	(2727)
0.0-16.7 cm		5540	5190	3380	2390
(0-6 2/3")	content	(5035)	(4720)	(3070)	(2170)
16.7-50 cm	per 16.7 cm	3250	3660	2150	1430
(6 2/3-20")	(6 2/3")	(2950)	(3330)	(1950)	(1300)
50-100 cm	layer	1310	1520	1530	1000
(20-40")		(1190)	(1380)	(1390)	(910)

the nitrogen is fixed in the clay and therefore not available for plants.)

The following table shows the figures for another element, phosphorus, which originates in the parent rock of the soils (Millar 1955).

Phosphate contents of various soils (% P_2O_5)

Soil group		Number of examples	Layers from the top downwards 1	2	3	4
Podzolic	Sandy	27	0.037	0.030	0.084	-
soils	loams	11	0.079	0.059	0.064	-
Chernozem, silt	loams	6	0.105	0.100	0.083	0.087
Dark brown & brown soils	Sandy loams	3	0.070	0.061	0.051	-

This table shows us a number of things. In the lighter, sandy soils the total amounts are smaller. But the first layer, the topsoil, always has more than the layer beneath as a result of biological accumulation. In podzolic sandy soils the lower layers contain more phosphorus as a result of leaching from the top layer, a phenomenon that does not occur in the loamy soils.

Matters are somewhat more complicated in the case of other elements, for instance, potassium, and also numerous trace elements. In these cases it is the availability of the nutrients for plants that is interesting. In humus-rich, biologically active topsoils this availability is greater than in humus-poor topsoils or in subsoils, which are of course less alive. These examples show that soils can gradually grow "richer" as a result of the cycles of nutrient substances.

Plant varieties differ considerably in their ability to mobilize nutrients from the soil. Those with a high ability enrich the soil with their root and crop residues. Legumes increase the soil's content of nitrogen compounds. They are also rich in calcium, magnesium and boron. Pulses take in more nutrients from the soils than cereals. Lupin, serradella and clover are well-known for their beneficial effects. Deep-rooting alfalfa is particularly efficient at mobilizing nutrients. These plants increase the amount of organic substance circulating in the farm if they are used properly in crop rotations and as fodder and manure. Oats, rye, wheat and barley in that order have decreasing abilities to absorb nutrients from the soil. Some weeds or companion plants are high in nutrient content. Foresters value the capacity of the beech to raise calcium from the lower soil layers to the humus layer of the forest via its leaf litter. Pine stands on poorer sandy soils are improved by interplanting of deciduous trees. Robinia (*Robinia pseudoacacia*) is an important plant for light soils, being a legume that adds nitrogen to the soil. The ability of plants to utilize soil-born nutrients depends on the size and shape of their root systems and on the microorganisms in the soil. Krasilnikov (1961) writes:

Growing near or on the roots, microorganisms, together with the plants, create a special zone, the rhizosphere. Soil in this zone differs in its physical, chemical and biological properties from the soil outside the root zone. Substances present in the soil are subjected to a greater or lesser extent of processing before their absorption by the roots. The role of the rhizosphere microflora reminds one of the digesting organs of animals. The same point of view is held by the American specialist Prof. Clark (1949). He considers that microorganisms living in the rhizosphere perform the same work as do the intestines of animals.

2   The Efficiency Principle In Nature

There is a further viewpoint that extends beyond what has so far been said. The nutrients circulating between soil and plant are also regulated by this cycle. This applies to the amount actually present in the soil, as we have shown in connection with the accumulation in soils of nitrogen and phosphorus, and it also applies to the relative amounts of different nutrients and the seasonal variation of what becomes available to the plants through the activity of the soil microorganisms. Mineral fertilizers tamper with this cycle. They increase the soluble nutrients in the soil and yet for their effectiveness they depend likewise on the way they are regulated by the soil and the cycles mentioned. The soil has to supplement all those nutrients and other growth factors not supplied by fertilizers. The effectiveness of mineral fertilizer thus depends on the degree to which the soil is able to do this. One singular phenomenon has already been mentioned: The nitrogen and potassium balance of the soil-plant system becomes negative when these nutrients are applied as mineral fertilizers. The soil has either to subsidize out of its own store, or else its own store is not utilized to the full. The amount of mineral fertilizer required increases excessively. Hitherto this excess requirement has always been complied with. It offers no problem to a mechanistic way of thinking. But to a biological way of thinking it poses an open question. Plants can absorb a great deal of a nutrient offered in large quantities. In the case of potash they positively indulge in luxury consumption.

The effective use of nutrients by plants is, however, governed by an efficiency principle. Plants manage to produce quantitatively and qualitatively satisfactory yields on relatively smaller amounts of nutrients if they are supplied in the moderate and yet balanced amounts that only a living soil can make available. It must be pointed out here that this efficiency principle operates within the different cycles of substances. The losses arising for instance in the nitrogen cycle or through leaching of minerals brought about by fertilizers can be reduced to a natural minimum when the supply of nutrients to the plant is part of the biological cycle. Kononova et al (1972) recently cited several observations of nitrogen in mineral fertilizer stimulating the mineralization of organic nitrogen in the soil. This can influence leaching and gaseous losses and contribute to a negative nitrogen balance in the soil. This is born out by measurements, made over several years, of water from tile drainage outlets in the farmlands of central Illinois, USA (Koepf 1973). Organic and chemical fertilizers were used on field crops grown in rotation. At regular intervals the nitrate concentrations in the runoff were measured. In two plot pairs the nitrate concentrations after organic fertilizer application averaged 9.7 ppm NO_3 (compared with 46.4 from the chemically fertilized plot) and 7.9 (compared with 49.9). Both organic and chemical fertilizers produced satisfactory yields. It is notable that this effect of organic fertilizers was also observed in crop rotations with a high percentage of grain crops, which leave the soil bare for large parts of the year thus favoring leaching of nitrates. Another example of accelerated nutrient leaching brought about by mineral fertilizers is the poor calcium content of soils observed in a number of countries. Advisory bodies have been concerned with this for some time and have begun to emphasize recommendations for the application of lime. The report of the British Ministry of Agriculture mentioned earlier shows that there are also other growth factors that can become limiting factors.

These indications show quite clearly from another point of view where the differences lie between the orthodox and the biological approaches. In general agricultural practice today the use of chemical fertilizers is accepted without question. This acceptance is characterized by the concept introduced by F. Scheffer of the transformation capacity of soils. By this is meant the fact that soils have differing abilities to transform applications, for instance, of fertilizers into yields. Depending on the soil, a unit of fertilizer will achieve varying yield increases. This concept of transformation capacity overlaps to a large extent with the concept of soil fertility. On fertile soil even relatively small amounts of mineral fertilizer will achieve considerable yield increases. How long such increases can be maintained need not be discussed here. If the soil is very fertile indeed, the use of fertilizer has almost no effect at all, and in soils that have lost much of their fertility through past malpractices even large amounts of mineral fertilizer cannot redress the balance entirely. So, the effectiveness of fertilizers depends on the natural fertility of the soil (see description of the Morrow Plots, p. 95).

Bio-dynamic methods strive as far as is possible to achieve closed cycles of substances. This conforms with the above-mentioned efficiency principle of nutrient utilization, which has full validity in natural locations. The following farm reports give practical examples of farms that of course also "export" their produce.

The closed or almost closed cycles of substances have two effects. Firstly, they increase the effectiveness of the circulating nutrients. A unit of nutrient produces a high yield. Special experiments would be needed to show what conditions achieve maximum results. Meanwhile practice on bio-dynamic farms has shown that this high efficiency is a reality.

Secondly, we can say that the closed cycle of substances maintains systems. With "system" is meant the community of soil, plant and also animal. In this sense a system can be a single garden, a field or a whole landscape. Present-day agricultural methods have a strong tendency to exhaust the means of production more than is necessary. A well-managed bio-dynamic farm steadily increases its production potential. In the discussion on how to produce enough food for a rapidly increasing world population, mineral fertilizers are put forward as a relatively cheap, fast-working means of increasing yields as quickly as possible. This is quite right. But it does not answer the question of lasting soil fertility. Yield is not a measure for soil fertility so long as this can be covered year after year by the application of fertilizers. Concern is growing as to the natural limits of agricultural production and also the limited supplies of raw materials. Among these are phosphates, which are needed not only for agriculture but also in considerable quantities for detergents with which they are washed away into the oceans of the world. Water and energy should also be mentioned. Their efficient use in agriculture depends to a considerable extent on the state of the soil, especially its structure and humus content.

Finally, a third point about the importance of the biological efficiency principle should be brought forward. Fundamentally it has always been a formative force in the development of agriculture, instinctively applied rather than expressed in these words. For thousands of years farmers and gardeners have been returning plant and animal residues, ditch clearings, deposited soil, dead leaves, useless grass from wasteland and many other substances to the organic cycle. Composting is an archetypal agricultural method. Despite deforestation in the catchment basins of the great rivers and the consequent soil erosion, Chinese peasants grow enough food in their small plots to feed themselves, the landowners and an enormous civil service. To fit their circumstances they developed a bio-technology in which the principle of composting in order to return substances to the earth was entirely the right one, though the way in which it is done in practice might be different elsewhere.

We have A. Glyn-Jones to thank for her translation of excerpts from Ibn Al-Awam's book on farming. This Arabian author lived in Seville during the first half of the twelfth century, but it seems that his book is based on much earlier writings of the Nabateans in Chaldea and also possibly some Greek and later works. His sources are partially pre-Christian. The chapter on manure mentions the dung of birds, asses, goats, sheep, cattle and also human excrement. Fresh manure is not recommended as it favors pests. The animal and human excretions are mixed with ash. To counteract the odor, mixing with red earth (probably a clay) is recommended. For growing plants the ash used is even that of other plants of the same species. Sheep dung is the richest. Dung should be mixed with straw and ash and allowed to decompose until the substance "resembles the pharmaceutical preparations in medicine." Liquid manure is made out of a number of ingredients. First-year compost is no good and second-year compost is still too fresh. From the third year on it can be used, but fourth-year compost is the best. The Nabateans, who lived in a hot country (they also knew the value of manure for brackish soils), were obviously familiar with the fact that in their climate mature, that is, stabilized, organic substance should be used. Chinese peasants still practice what is said in this ancient book.

So we see that the principle of closed cycles, or recycling, belongs to the most ancient body of agricultural knowledge. It is a fundamental biological principle that is realized in nature. It is the principle that determines the right biological organization of a farm. Stocking rates in excess of the natural population density; catch crops and hayfields; choice of varieties; planned use of manures; all these help to stimulate the cycles of substances with the result that high yields, including surplus crops that can be sold, are achieved.

3 The Natural Foundation Of The Farm Is Its Habitat

The cycles of substances are maintained by terrestrial and cosmic growth factors. These occur on the earth in a pattern of different combinations. The worldwide green mantle of the continents is structured in accordance with the climatic zones. Starting in the tropical rain forests of the equator we pass through the humid sub-tropics, the savannah, the desert belt, the Mediterranean climate with its wet winters and dry summers, the steppe, the deciduous and coniferous belts, the tundra and finally the Artic zone. Neither the climatic zones nor the growth zones, however, lie in regular belts around the earth. Mountains and plains, the size of continents, and the coastline of land masses bring about the irregularities we know in the weather and thus in the vegetation, preventing the formation of regular zones. A telescoped impression of all this, though with some gaps, can be obtained by climbing from a valley to a mountain top.

Apart from this large-scale structuring, there are also medium and small-scale patterns of the world's vegetative cover. For instance, despite the local variations, sites near the Great Lakes have properties in common by which they are distinguished from the morainic areas some distance away. These in turn are different from the glaciated areas further to the east. Each of these macro-landscapes has its own sub-divisions. In Germany's highlands, for instance, the growth differences on north and south-facing slopes are quite remarkable. In the lowlands where the groundwater table is close to the surface, an altitude difference of only a few hands' breadth often makes a marked difference in the height of grass.

Governed by topography and the underlying geological strata, typical habitats are found close together or more widely separated within a given climatic zone. A habitat in ecology is a combination of the natural environmental growth factors such as climate and soil. For agricultural use, forestry, etc., like habitats lend themselves to like uses. They present the same potential difficulties and advantages with regard to cultivation or any improvements by amelioration. The zonal pattern of the earth's plant cover mentioned earlier is in the main brought about by the climate. For a habitat the local soil properties and the micro-climate are what counts. Both can be improved by man. But the macro-zones have to be taken into account since they favor one branch of farming more than others, for example, crops, dairy farming, fruit farming, special crops, etc.

An important ingredient for the success of a bio-dynamic farmer is a thorough knowledge of the habitats within his farm. He must have a picture of the total natural conditions on which his farm is built. Excessive store has come to be set by fertilizers in the course of the discussion that has been going on for a long time. Yet the local climate and the possibilities of improving it, the planting of trees and shrubs, ameliorations, careful tillage, and not least organizing crop and animal husbandry along lines that are biologically sound, are also factors that play as large a part in successful farming as fertilizers. Habitat research reveals the given situation and the potential that is there. How these are best used will be discussed in all the coming chapters of this book. Meanwhile we shall give a few indications that, it is hoped, will give a clearer picture of what a habitat really is.

Research in many countries has made progress in this field over the last 20-30 years. For instance, there are forest habitat maps for all the States of the Federal Republic of Germany. In addition, numerous specialized maps now exist for fruit farming and viticulture and increasingly also for general agriculture. These are useful not so much for the individual farm or garden as for planning decisions to be made by the authorities. In the face of rapidly increasing populations, industrialization, and transport needs, overall planning is now under way to ensure the existence of recreation areas and the improvement of environmental quality. For this, habitat maps provide information about the natural foundations from which planning must take its start.

In the United States, the "Soil and Water Conservation Plan" is provided to farms by the Soil Conservation Service. There is hardly a farm that does not have this valuable information. Although the primary purpose of the maps, descriptions and management advice given is conservation, these plans provide a good deal of the information that is commonly included in habitat surveys.

In habitat surveys (biogeography) the attempt is made to take into account the totality of growth factors as well as the degree of importance of each factor. As far as is possible, the plant itself is used as a "measuring instrument" and this is backed up by soil analyses. The earlier system of using single plants as indicators has been replaced by vegetation or plant community studies where plant associations are distinguished from one another. This is more exact, and it also requires more experience, since each plant's value as an indicator varies from place to place. The "dryness" of a habitat can be the result of various causes: lack of precipitation; shallow soil over a porous subsoil; low water-holding capacity; shallow rooting in heavy clay. Therefore the information gained from plants should be complemented by the study of the soils and meteorological data. Knowledge of the soil is also important for a further reason: any improvements of a permanent kind have to be applied to the soil itself.

The method of Ellenberg and colleagues (1956, Schreiber 1969) can be quoted as an example of habitat surveys. Each of the following environmental factors is measured or estimated:

temperature; late frost; winter frost; hail;
wind; humidity available for plants; natural
fertility of the soil; its drainage and
aeration; calcium content of the soil.

In addition, the steepness of slopes is reported, which is an important factor in working the land with or without machinery.

Determining the warmth factor involves using not only climatic data but also ecological observations such as noting when certain plants start to sprout in spring, when blossoming begins and when leaf coloration starts in autumn, etc. The severity of winter frosts can be judged by the survival of nut trees. Reports and measurements are used for assessing the dangers from night frost and cold air currents that are particularly threatening to orchards and vineyards. The soil moisture and aeration can be assessed with the help of vegetation studies. The natural fertility of soils depends on the parent material and the soil type. More will be said about this in the third chapter.

The more favorable a habitat is, the less is the farmer limited in his choice for its use. This is shown by the following examples. The actual use that is made of a stretch of land is also determined by custom and traditional forms of land ownership. It should not be forgotten that it is always good to maintain a mixture of woodland, pasture and arable land wherever this is possible. The following list is applicable for semi-arid to humid districts:

Steep; outcrops of rock; dry; poor soil; landslides.	Forest; wasteland.
High temperatures and sun radiation levels favor, while low temperatures and sun radiation levels exclude:	Grape and fruit growing and other special crops.
Wet; steep; heavy soil; regular flooding; high ground-water level; uneven surface; stones; upper regions of hill ranges.	Grazing.
None of the above hindering factors; precipitation usually below 1000 mm/annum (40 in/year).	Crop growing.
Fertile soils mostly in lowlands; mild climate (mostly in districts concentrating on a particular crop).	Commercial vegetable growing.
Theoretically possible everywhere;	Cultivation under glass.

Finally, it is interesting to note which habitat properties can be improved and which cannot. Considerations concerning this always gain in significance when land is changing hands or when a farm is being enlarged. The first and most important decision is where to buy, rent or in any other way take on land for farming or gardening. It happens all too frequently that a city dweller or indeed anybody buying land with the intention of building up a livelihood in the country allows all sorts of motives and desires to play in without really examining the situation and asking whether the resulting work will give him pleasure and satisfaction. The initial decision, if it is to be the right one, should be based on three considerations:

1. How good is the potential of the site, and what are the economically viable improvements that could be made?

2. What is the present transport and market situation and are there plans for development; how well is the farm equipped with buildings and machinery?

3. What do I intend to produce and whom is the farm intended to serve?

The following is a table of indications regarding the first of these three points.

Site conditions that can or cannot be altered at reasonable cost.

---

not	acreage; altitude; slope; aspect; macro- and also to some extent the microclimate; soil texture; soil depth; mineral composition of soil; natural humus content of soil at the site in question.
in some cases	slope (terracing); exposure to wind and frost; water content and aeration of the soil (drainage and irrigation); depth and soil structure (subsoiling); susceptibility to erosion; obstructions such as stones and unevenness; amount and quality of humus.
possible	water content and aeration; susceptibility to erosion; humus and nutrient contents of the soil; structure and biological condition of the soil.

Some of the items appear in two or three of the groups. Every habitat property is related not only to the total landscape but also to the particular farm or garden in question. Amount and quality of the humus in the soil is an obvious example that has to be understood: Humus content and humus quality depend on the macro-climate and on the parent material, that is, on the soil group in question. This naturally arising humus can be improved or impaired by cultivation, which is therefore decisive for the success or failure of the farm or garden. Within meaningful limits, humus can be altered to almost any extent, particularly in small gardens where economic and other limitations of a large farm need not be taken into account.

4 Farm Organization

This theme will be dealt with in detail in the following chapters but here we shall examine the basic principles. In bio-dynamic agriculture, farm organization arises out of the proper application of biological principles. In his lectures on agriculture, R. Steiner spoke of farm organization in so far as it must be based on the needs of plant and animal production. But he made only a few brief remarks on the economic and social structure of a farm. He did, however, express himself on this theme on a number of occasions. In 1957 under the title of "Rudolf Steiner, Land Reform as a Social Challenge Today," Boos collected all the written and verbal statements made by Steiner concerning the role of agriculture, especially ownership and usufruct of land in today's socio-economic context and suggestions for reforms. (This is only available in German). In the first lecture of the Agricultural Course Steiner says:

> For it should go without saying, and every man should recognize the fact, one cannot speak of agriculture, not even of the social forms it should assume, unless one first possesses as a foundation a practical acquaintance with the farming job itself. That is to say, unless one really knows what it means to grow mangolds, potatoes and corn! Without this foundation one cannot even speak of the general economic principles that are involved. Such things must be determined out of the thing itself, not by all manner of theoretic considerations. Nowadays, such a statement seems absurd to those who have heard

university lectures on the economics of agriculture. The whole thing seems to them so well established. But it is not so. No one can judge of agriculture who does not derive his judgment from field and forest and the breeding of cattle.

We see here a difference of principle between bio-dynamic work and agriculture in general. Agrarian structure and the organization of individual farms are in general today almost exclusively determined by the pursuit of increased returns on capital and labor. This is what has maneuvered agriculture and horticulture into their present situations. Many people are now aware of the biological and ecological questionability of this situation. Yet most consider a farm organization built on truly organic principles to be nothing but wishful thinking—which does not, of course, mean that the application of such principles would not be useful or that in fact they are not absolutely essential. Yet the problem is either not seen or its existence is denied. It has certainly neither been thought through nor tackled sufficiently in agrarian circles with all its economic, social and legal consequences.

In contrast, we can point to numerous bio-dynamic and organic farms that have been working for many years in accordance with these principles. The opinion that this way of working is impossible has been superceded by the fact that it is possible. Such farms operate both under ordinary market conditions and under the more favorable circumstances of having their own circle of customers. These remarks apply to individual farms scattered among non-organic operations. For large concentrations of organic farms in one area the problem needs a more detailed discussion.

Ecological farm organization

The general factors influencing this are:

1. the natural conditions, e.g., soils; climate; topography; surface and underground water bodies;
2. variable means of production, i.e.,

65

a) plant and animal species; varieties; strains; breeds; state of development of breeding;

b) actual stock and also the actual technical level of buildings, machinery and management methods on the farm;

3. market and transport situation; short and longer term economic situation including hindrances or help through public incentives;

4. ability and training of the people working on the farm and making decisions.

The purpose of the farm organization is to provide an income and satisfying work for the people mentioned under Point 4. The external influences mentioned under Point 3 must be brought into a harmony with the means of production in Point 2 and the natural base in Point 1, so that a lasting performance, that is, quantity and quality, can be maintained or improved.

First of all we must understand the manner in which a farm differs from a natural habitat uninfluenced by man. A natural habitat houses a plant and animal community and the cycles of substances are almost totally closed. During the course of soil development, the topsoil becomes increasingly enriched with biological substances. The leaves shed in a stretch of woodland or the dying grasses and herbs on a piece of wasteland are the raw materials of humus formation that remain on the spot. Cultivated land is more or less far removed from this natural state, as is shown in the following summary of the three chief forms of land use, forestry, pasture and crops.

1. Forestry	changed tree stands and soil flora; side effects of felling, dragging, path laying, etc.; removal of timber;
2a. Pasture and ranch land	change in vegetation depending on type of livestock; grazing; step of animals; fertilizing; removal of animal products;
2b. Grassland	change in vegetation caused by cutting and grazing; fertilizing; removal of animal products and grass;

| 3. | Crops in field and garden | ground bare at times; monoculture; turning and loosening of topsoil; fertilizing; crops removed once or several times a year. |

The following factors of soil fertility are the chief ones to be affected by these human influences:

1. Contents and nature of organic substance (humus);
2. Amounts and ratios of nutrients made available by the soil;
3. Soil structure and depth of root space (on which depend in in turn the soil's ability to make available rain water, warmth and air to the root system;
4. Occurrence of pathogenic microorganisms including bacteria, fungi and animal pests.

These fertility factors are all interrelated. The amount and type of humus influence not only the nutrition of the plant but also the soil structure and the incidence of diseases and pests. On the other hand, the returns on any current expenditure will be proportionally higher if soil structure and soil life have been improved in previous years. There will be more to say on this in the following chapters. At this point we want to look at the farm as a whole and are concerned mainly with farms in temperate climates ranging from very humid to semi-arid. These comprise Central and Western Europe, Southern and Central Scandinavia, parts of Eastern Europe and the Northern fringe of the Mediterranean climate. Further they include Southeastern Canada, the forest regions and some of the prairie regions of the Northeastern and Eastern parts of the United States. Traditionally diversified farms with livestock, crop growing and varying shares of permanent grasses, and sometimes also woodland, have been familiar in these regions.

Land use in diversified farms is determined by the nature of soil and climate. The proportion of permanent grasses increases in wetter and cooler upland and mountain areas where it almost entirely excludes crop growing. In farms or specialized enterprises in warmer regions the area used for crops such as grapes, other fruit, tobacco, hops, vegetables, etc., increases.

The space not used for permanent grasses or the special crops mentioned then remains for ordinary arable crops. Thus we can distinguish between:

1. Farms using more than 25% of their land for root crops.
2. Farms growing root crops and cereals.
3. Farms growing root crops, cereals and forage crops.
4. Farms growing only forage crops.
5. Farms growing special crops.

This is how the various types of diversified farm are usually listed in Europe. In the North American dairy and corn belt, corn instead of the root crop is the soil-exhausting crop, which also consumes the farm-produced manures. Hayfields, provided the hay is not sold, and manures properly used, restore life and fertility. Permanent grasses are of lesser importance in these production zones.

The division into farm types mentioned above is based on a view of the farm as a whole. Recently this has been abandoned in favor of specialized production programs. On a visit to Davis, California's central agricultural training center, the author was informed that in this State there was no agriculture but there were apple, citrus, tomato, wheat and other industries. Every step toward specialization, most of all specialization in one farm enterprise, destroys connection with the environment that is important, sometimes more, sometimes less, for the thriving of plant and animal. Major plant and animal diseases and pests such as take-all and eyespot in cereals, corn root worm, nematodes in potatoes and cereals, clubroot, infestation, etc., are diseases that are favored by short rotations.

In Germany, wheat grows in a moderately humid climate. Compared with the wheat yields, achieved in regular crop rotations, continuous wheat growing reduces the yields to 50-70% in seasons that, due to weather, favor the spread of take-all and eyespot. It is, therefore, not surprising that those advocating narrow production schemes now call for systemic fungicides (Debruck 1973). Yet the situation is often complicated. For instance, it has been observed for years that on bio-dynamic farms even with more frequent than usual repetition of red clover, there is less "red-clover tiredness" of the soil than on ordinary farms. The depth of the soil and its mobilization of nutrient content can only be fully utilized when plants with different root systems are grown together or in sequence. The special significance of animal manures for soil fertility will be dealt with later. From these few examples we see that crops have important effects on each other. This is the basis of a biologically oriented farm organization.

The clearest indication for the mutual interdependence of different branches of the farm are the cycles of substances. They encompass the whole farm or garden and not an individual field. This is brought about by the manures produced by the farm itself. In addition there is a further difference between a natural habitat and a farm. On a farm the substance cycles are intensified and qualitatively changed, that is, the turnover of organic substances is accelerated and they also contain higher concentrations of nutrients such as nitrogen. Thus, even though the crop is removed from the field, greater fertility and actual production is achieved than is the case in a natural habitat undisturbed by man. This increase is brought about by:

1. Choosing the proper lines of production on the land. This is more or less determined for every farm by the natural conditions. But it is not governed solely by the principle of humus and nutrient replacement. The landscape itself has to be taken into account, that is, the combination of arable land with grasses, woods, wind breaks, fruit trees, etc.

2. The ratio on the arable land of soil-restoring and soil-exhausting crops, that is, the relative areas of root crops, cereals, oil crops, large seeded legumes (pulses), forage and hayfields, catch crops and main crops undersown with clover or grass.

3. The crop rotation, which in hilly or more rainy regions is changed into alternate husbandry; for example, the arable course includes several years of a ley.

4. Feeding livestock on mainly home-grown feeds; proper handling, storage and applicatin of the farm's own manure.

5. Tillage, which influences the depth of the root space, soil life, release of nutrients and turnover of organic matter.

The way in which grasses and arable crops, plant and animal husbandry benefit each other is expressed in the manuring plan that every bio-dynamic farm must work out. How and under what conditions is it possible to achieve the greatest possible self-sufficiency in manures? In production the farm has to be more than self-sufficient since it has to supply the urban population with food as well. To what extent a farm organization can be self-contained depends on whether a high proportion of special crops are grown; whether there is a balance between arable crops and grassland; or whether grassland is the main farm enterprise. Furthermore, one has to look into the number and kind of livestock. Is it possible to meet the nutrient requirements of crops without animal manure? Can closed cycles of substances be maintained with the exclusive use of compost, catch crops, arable hay crops, etc.? Many questions follow from this; for instance, what is it best to buy in: fodder, organic fertilizers, mineral materials, etc.?

Since discussions of this nature cannot be conducted fruitfully in theory, we shall introduce some concrete examples at this point. They show more clearly than theoretical considerations how a real farm organism can be created. B. Pettersson's investigation (1951, 1963, 1964) of a number of Scandinavian farms contains material that is relevant to the matter of farm organization.

The productivity of bio-dynamic farming in Scandinavia

Excerpts from this report will be given here. Its purpose was to determine, through an investigation of a number of bio-dynamic farms in Northern European countries, where and under what conditions the idea of a self-contained farm organism can be realized. The foremost points of interest were yields and the stocking rate. From 1955-58 nineteen farms were inspected of which twelve had been under bio-dynamic cultivation for over ten years. Continued inspection after 1958 showed only minor differences in the results compared with 1955-58.

First, some remarks regarding the method; the farms were divided into groups according to climate and soil:

SS: Southern Scandinavia (Denmark and Southern Sweden). Mixed farming, root crops, cereals and clover grass (1-3 year clover grass). The short growth season hardly allows for any catch crops for green manure.

CS: Central Scandinavia (about the 60th latitude). Primarily cereals and 2-3 year clover grass.

Three soil groups were distinguished: A = best calcium-rich loamy soils; B = medium to poor loamy soils and best sandy soils; C = poor sandy soils, podzolic. The number of farms in each group were: SS A (3); SS B (4); SS C (5); CS B (Sweden and Norway) (7).

It was a matter of registering the movements of crop yields, fodder, buying in and selling of products, livestock, seed. The formula on which this was based is best shown in the following drawing.

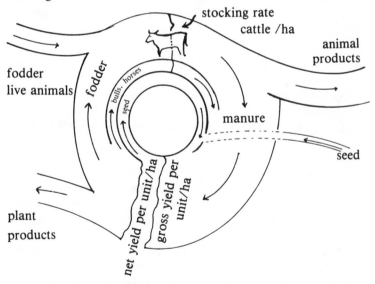

Diagram showing how the farm reports were dealt with. The arrows indicate the direction of product flow.

Limited amounts of organic and inorganic fertilizers had to be taken into account in some cases. In order to be able to calculate the different movements of materials, the investigators converted all the amounts into product units (PU), that is, 1 kg (2.2 lbs) wheat = 1 PU.

Conversion key for Product Units (PU)

1 kg (2.2 lbs) corresponds to	PU	1 kg (2.2 lbs) corresponds to	PU
wheat, rye, barley	1.0	cattle on the hoof	7.0
oats	0.9	beef	4.0
oil cakes	1.2	milk, 4% fat	0.83
hay	0.5	eggs	4.3
potatoes	0.25		
beet, dry matter	0.8	kg N in fertilizers	18.0
white clover seed	14.5	kg P in fertilizers	10.3
rape seed	2.0	kg K in fertilizers	4.8
carrots	0.25		
strawberries	0.5		

Fodder consumption for one cattle unit (CU) was set at 3000 PU per annum. This corresponds to one animal weighing 500 kg (app. 10 cwt or 1100 lbs live weight) and giving 3600 kg (7920 lbs) milk with 4% fat. Animals with a different fodder consumption were given a different PU assessment, using the usual conversion keys.

Using this key, comparisons between yields were possible, namely, gross yield, net yield = gross yield minus seed, and remaining net yield = net yield minus the effect of bought-in fertilizers.

The figures obtained were checked by a second calculation, namely, of net sales as the difference between buying in and selling of products. This second net yield is then: net yield (2) = net sales plus fodder for bulls and horses and plus or minus the changes in livestock numbers and stores. When the books were well kept it was possible to arrive at totals for the two calculations that differed only by 2-3 per cent. In one case the yields were corrected to a result that meant that the differences between the two calculations were never more than 7 per cent on any of the farms during the whole period. *Results.* The main purpose of the investigation was to determine the relation between livestock density and yields. The following table contains the relevant figures. Since there was little alteration in livestock kept, there was no need to take into account the fact that the use of stable manure obtained in one year does not on the whole have an effect until the following year. The table shows the considerable variations in yields from farms in different areas and also the rise in yields running parallel with livestock per hectare or acre.

Investigation period, year of changeover to B-D management, size of farms in the last year of investigation, livestock units per ha (acre) and remaining net yields in PU per ha (acre)*

Group; farm; inspection years	year of change- over	size in ha (acre)	animal units per ha (acre)	remaining net yield, PU/ha (acre)
SS A				
Lejre, 57-61	1957	55.0	0.83	3730
		(135.9)	(0.34)	(1509)
Görlev, 57-61	1953	36.0	1.10	4740
		(89.0)	(0.44)	(1918)
Braade, 57-61	1942	25.5	2.26	5590
		(63.0)	(0.91)	(2262)
SS B				
Mariager, 57-61	1941	28.7	0.91	3420
		(70.9)	(0.37)	(1384)
Östendal, 57-61	1942	11.4	1.76	5620
		(28.2)	(0.71)	(2274)
Gudme, 57-61	1945	5.8	1.36	3970
		(14.3)	(0.55)	(1606)

73

Group; farm; inspection years	year of change-over	size in ha (acre)	animal units per ha (acre)	remaining net yield, PU/ha (acre)
Slude, 57-61	1942	5.5 (13.6)	1.93 (0.78)	4900 (1982)
SS C				
Egtved, 57-61	1957	106.0 (262.0)	1.00 (0.40)	1930 (781)
Herning, 57-61	1941	25.3 (62.5)	0.82 (0.33)	2550 (1032)
Varde, 57-61·	1940	23.5 (58.0)	1.13 (0.46)	2910 (1177)
Billund, 57-61	1940	13.2 (32.6)	0.57 (0.23)	2380 (963)
Röke, 58-61	1949	2.3 (5.7)	2.39 (0.97)	4160 (1683)
CS (Sweden) (B)				
Kungsängen, 55-61	1955	161.0 (397.9)	0.51 (0.21)	1320 (534)
Nossebro, 55-61	1944	34.0 (84.0)	1.33 (0.54)	2400 (971)
Järna, 58-61	1946	32.0 (79.1)	0.61 (0.25)	1560 (631)
Tobo, 55-61	1949	18.5 (45.7)	0.82 (0.33)	2020 (817)
Vetlanda, 55-61	1946	15.5 (38.3)	0.81 (0.33)	2470 (999)
CS (Norway) (B)				
Jevnaker, 58-61	1951	52.0 (128.5)	0.76 (0.31)	2460 (995)
Mysen, 58-61	1930	8.6 (21.3)	1.30 (0.52)	3160 (1278)

*1000 PU are equivalent to 31.1 bu wheat/acre, e.g. the yields vary to PU equivalent 17.2 to 72.7 bu wheat/acre.

These results can also be illustrated graphically as is shown in the following diagram in which the vertical axis shows the average yields and the horizontal axis the stocking rate in cattle units/hectare.

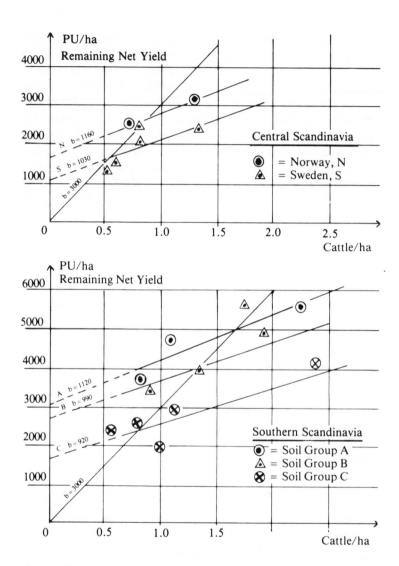

Lines of regression showing the relationship between stocking rate and yields in the various climate and soil groups. Line b = 3000 is the border line for self-sufficiency in fodder.

We see by the lines drawn through the different points that within a given framework the yield of a farm increases approximately parallel with its higher stocking. This linear function is not necessarily to be expected because the increasing use of manure means increasing yields but at a decreasing rate of increase. In this case, however, a higher stocking rate was accompanied by a shift toward a higher proportion of clover grass and beet crops that give a slightly higher yield than cereals when expressed in PU. The straight lines in the diagram are also called regression lines. The regression coefficient is the figure that expresses the yield increase in PU when the stocking increases by 1 cattle unit/ha. (see table).

Regression and correlation coefficients and yields in various climatic zones and soil groups

Group	Regression coefficient	Correlation coefficient	Influence of manure on yields, %
SS A	1120	0.77 ***	37
SS B	1440	0.58 **	48
SS B (without Ostendal)	990	0.54 *	33
SS C	920	0.71 ***	31
CS (Sweden) B	1030	0.59 ***	34
CS (Norway)B	1160	0.72 *	39
All groups	1070	0.73 ***	36

This table also contains the correlation coefficient. The highest (positive) value of this index can be $+1.0$. It expresses how closely related PU and cattle/ha are. If it is far below $+1.0$ but greater than zero, then the two, though related, are showing a greater divergency than usual. The closer it lies to $+1.0$ the more reliably one can state by how many PU the yield will increase if the cattle density is increased. (The asterisks express the statistical significance of the results: * = 95%; ** = 99%; *** = 99.9% level).

The regression coefficients in the above table lie between 920 and 1120. (The figure of 1440 is the result of a high proportion of root crops in this farm; a second figure excluding the farm in question is therefore given for the group.) These figures, averaging 1070, state that the manure produced by one cow increases the yield by 1070 PU per year. As the cow eats 3000 PU, her manure brings about a yield increase of about 36% of what she eats. Better soils, such as SS A respond better to manure than poorer soils. The yield increase brought about by one animal lies between 30 and 40% of its fodder consumption, the lower amount applying to the poorer sandy soils and the higher to the fertile, loamy soils. The average yield increase brought about by one cow can be taken as 35%.

The line b = 3000 in the diagram indicates self-sufficiency in fodder production. If the farm is placed above and to the left of b = 3000, it is or can be self-sufficient. If this is not the case, it has to buy in fodder (or manure).

A further piece of information can be derived from the diagram by studying the structure of the yields. At CU/ha = 0, that is, on a stockless farm, the lines of regression do not cut the ordinates at zero but for SS A at 3100 PU, for SS B at 2700 PU, for SS C at 1700 PU, CS B (Sweden) at 1100 PU, and CS B (Norway) at 1600 PU.

This expresses the fact that yields can also be expected without any manure or fertilizer. These yields consist of two components that can be approximately estimated: the potential of the soil itself and the carry-over effects of previous cropping and crop management, which may be called the "cultivation fertility." Experiments have been going on in various experimental stations for years, including on the loamy soil of Lyngby in Denmark, to determine how much grows on ground where the crops are removed and no manure at all is given back. From 1910 to 1933 in Lyngby, Denmark, the soil capacity amounted to about 2000 PU/ha per year. We see that without any fertilizer the yield does not drop to zero but operates at a certain level, such as the above figure for Lyngby or the lower figure of 900 PU/ha on the sandy soil of Askow (Denmark). The loamy soil of Lyngby corresponds to Pettersson's grade SS B. We thus obtain the following estimated components of the yields for this group:

| soil capacity | 2000 PU |
| cultivation fertility | 700 PU |

The sum of 2700 PU arises from the diagram where the line of regression of SS B farms cuts the ordinate at PU 2700. The table on page 73 shows however that the actual yields of these farms are considerably higher. This is then the effect of the current manuring, possibly also other measures such as tillage, and perhaps other short-term carry-over effects.

The performance factor described as "cultivation fertility" contains the sum of previous cultivation measures. Without continued manuring these effects would dwindle away in about 10-15 years. The "cultivation fertility" has to be cared for. More is said about this in the section on old field experiments.

The next two tables contain some further evaluations of the figures.

Regression coefficient and yield increments due to manuring of different crops (inner regression coefficient on all climatic zones and soil groups)

	regression coefficient	yield performance of the manures in %
Cereals	860	29
Grasses	1010	34
Root Crops (only SS)	730	24

This table shows that the effects of manures vary according to the crops. The variations between the farm groups are also much greater in this respect than with the total yield. Grasses and pastures respond best, even more than cereals. In bio-dynamic farming it is a well-known phenomenon that legumes, in this case clovers, react positively to the method. This is also expressed here. With root crops the regression coefficient is smaller; possibly there is not quite enough nitrogen available in these northern latitudes. The difference between the separate climate and soil areas has already been discussed in connection with the table on page 78. In the following table the figures are broken down into the individual plant groups.

Average yields in PU/ha, reduced to a stocking rate of 1 CU/ha (= 0.4 CU/acre)

	all crops	cereals	clover grass	root crops
SS A	4200	4000	5000	6800
SS B	3800	2900	3700	6800
SS C	2600	2200	2300	4200
CS B (Sweden)	2200	2100	2300	--
CS B (Norway)	2800	2400	2800	5100

These investigations show that the farm manure produced by one animal increases the plant production by 30-35% of the amount eaten by the animal. Starting in the fifties, the less productive farms have, to date, increased their capacity to 3000 PU/ ha per year. The farms with high yields have maintained these yields.

These correlations, found in Northern Europe, cannot automatically be applied to the climate and soil conditions in other parts of the temperate zone. The correlations arose on mixed farms. Manuring is not the only yield-determining factor since crop rotation, the choice of crops, local climate, etc., also play a part. Every farm has its own limit up to which it can be self-supporting as regards fodder. An investigation of this kind can highlight the possibilities where there are reserves for a higher fodder production that in turn would improve the yields. If higher yields are desired once the self-sufficiency level has been reached, fodder and fertilizers will have to be bought in. This is justified as the farm also markets plant and animal products that reappear on the market as organic fertilizers (horn, hoof, hair, bone meal, leather waste, etc.) or as the by-products of food production (cakes of oil seeds, by-products of milling, etc.). In the light of what was said in the first chapter, it will be important that these substances be recycled regionally, that is, within a country, instead of cheap animal foods being bought from the Third World. One can use oats and barley as feed on one's own farm. It cannot be presumed that fodder crops everywhere produce the highest yields expressed in PU. In other climatic conditions the combination of fodder and cereals, or cereals, fodder and root crops, might be more favorable.

Farms that utilize the whole of their crop production via their livestock are found in Europe in the grassland belt of the Alpine foothills and along the coasts. In North America they occur in the dairy and corn belts. There permanent pastures play only a minor role. Many farms utilize the whole of their crop rotation of corn, oats, (formerly also wheat), clover grass via their dairy or beef herds. This is not so in Central Europe where cereal and fodder crops and, more recently on a decreasing scale, root crops are found in one farm with the emphasis more on the one or the other farm enterprise depending on the climate and soil conditions.

To complete the picture, we shall now mention a few more points concerning the types of bio-dynamic farm that have been developed in Northern Europe, specifically Denmark, Finland, Norway and Sweden.

1. Intensive growing of fresh vegetables

These market gardens are usually situated within a fifty mile range of a larger market. Fresh vegetables and soft fruits are supplied between May and October. Greenhouse growing and cultivation in the open are usually combined. 10-30 different crops are grown. Stable manure can usually be obtained from a farm in the neighborhood, though the market garden may also produce some of its own. This is combined with horn meal, meat meal, bone meal and similar concentrates. Since the total acreages are small, manuring can be intensive and production is high. After a few years the quality of the products is very good and the demand so great that it can be only partly satisfied. The growing of fresh vegetables in winter greenhouses has so far not proved viable biologically because of the lack of light and economically because of the cost of heating.

These market gardens are economically stable. Their weak point is that the growing season is short and labor intensive, with working hours for the gardener—usually they are family enterprises—of 12 to 15 hours a day. To compensate for this there are 2 to 3 months of total rest in the winter.

2. Farms with fodder crops and livestock combined with winter vegetables and soft fruit

These farms, too, are relatively small on the whole, usually about 25 ha (62 acres) or less. The main part of the land is given over to fodder crops with pastures, grasses, feeding grains and fodder beets for milk production. The manure for these and for the crops sold is produced on the farm. 10-30% of the area is used for winter vegetables (carrots, beetroot, various cabbages), potatoes, strawberries or black currants. Only one or two of these are usually grown.

These farms are situated further away from market outlets, usually in agricultural areas. There is on the whole no opportunity to buy in stable manure. Therefore farms have their own livestock but they do buy in animal residues for the cash crops. These are either included in the normal fodder crop rotation or, more frequently, they are grown on a part of the farm that has appropriate soil. They then have their own crop rotation. Some of the farms using the latter method are in a transitional stage, growing their fodder crops by ordinary methods and their vegetables and fruit according to bio-dynamic principles.

Since marketing conditions for the vegetables and fruit are good, this type of farm organization is economically quite stable. It is labor-intensive, and conflicts between the needs of the field crops and those of the vegetable sections are unavoidable during unfavorable weather, even when the planning is otherwise excellent. The situation is eased by introducing mechanization in the cultivation and harvesting of the vegetables and customer picking of the fruit. A disadvantage of this farm type is that it leaves no free time in winter as the livestock requires more attention then. It is also best run as a family farm.

3. Farms with fodder crops, livestock, and cereals or potatoes

Farms of this kind become economically viable when they are bigger than 25 ha (62 acres). Their main products for sale are wheat on loamy soils and potatoes and rye on sandy soils. 20% of the total area must be available for wheat and at least 10% for potatoes and rye on sandy soils. The remaining area produces fodder for the livestock, which is mostly dairy cows, though some beef is produced, particularly if the farm has large areas of marginal pasture. Quite often dairy and beef production are combined. Breeding and fattening pigs, which would suit this size of farm quite well, is infrequently found in bio-dynamic farms in the North. The same goes for larger scale poultry-keeping.

The manure requirements of this type of farm are sometimes difficult to meet. Concentrated animal residues can be bought in for the potatoes but it is unprofitable to do so for cereals. Thus one has to depend on a harmonious crop rotation and well-growing clover and alfalfa fields. These are grown prior to wheat and sometimes also to potatoes.

Weed control can pose problems. Perennial weeds are a particular nuisance on the loamy soils of Central Scandinavia. If the crop rotation does not include any root crops and if the short growing season leaves no time for working the stubble after harvest, Canada thistle spreads on loamy soils and couch grass on sandy soils. In Southern Scandinavia, stubble can be properly dealt with, but instead farmers have to contend with annual weeds in spring grains. Catch cropping is hardly possible in Scandinavia.

As regards labor, these farms are in the best position of all the types mentioned so far. Modern machinery can be properly made use of and the work load is evenly spread over the year. In Central and Northern Scandinavia, winter forestry work provides a further balance.

From an economic point of view some sharp calculating is necessary. Intensification as it is appropriate for the farm types already mentioned is rarely possible here, for instance, by enlarging potato growing or the number of livestock, etc. Frequently acquiring additional land seems to be the best answer to the need for rationalization.

Household farms

A special form of bio-dynamic farm has come into being in conjunction with institutions of various kinds such as curative homes or educational establishments. It might almost be called an enlarged kitchen garden intended to supply a family with vegetables, berries and fruit, except that the "family" has become rather extended. It is difficult to include these farms in any of the previous categories. The smallest aim at self-sufficiency in fresh vegetables; the somewhat larger ones produce a full supply of winter vegetables, berries and fruit; the largest include potatoes and milk.

There is no point in making economic comparisons between these farms and those that market their produce and have to be economically viable. Also their labor situations are different. Within the bio-dynamic movement on the whole they play a considerable role.

Some types of farm do not yet exist in the North. The harder climate of Central and Northern Scandinavia makes it impossible to grow some crops. In Denmark and Southern Sweden, however, the potential would be there for some bio-dynamic forms of production that have yet to be developed. These include fresh vegetables during the cold season, fruit, seed production for agriculture and horticulture, tree nurseries, flowers, and egg production.

Farm report

We are including this rather detailed report of a farm in Southern Germany working on marginal soil and under difficult conditions for three reasons. Firstly, it describes improvements in production achieved by bio-dynamic methods with minimal buying-in. Secondly, it shows how the farm as a whole with all its branches must be developed in proper sequence out of the natural conditions. Thirdly, it contains a great many practical suggestions. The farm concerned is the Talhof near Heidenheim, Baden-Württemberg. (F. Sattler 1968, 1969, 1971, and personal communication 23 May 1975.)

1. Description

The farm has 50 ha (124 acres) of arable land. The clay-rich rendzinas and kolluvial soils on jurassic (malm) limestone show a topsoil 5-18 cm (2-7") deep. The fields lie between 495 and 570 m (1624-1870 feet) above sea level. They stretch out to a length of 5 km (3 miles) along a wooded valley, which means that shade lingers well into the morning on the eastern side and falls early in the afternoon on the western side. Average annual temperature is 5.5°C (42°F). Night frosts (down to -5°C [23°F] in July) can occur in any month of the year. Average annual rainfall is 750 mm (29.5") varying between 400 and 1250 mm (15.7-49.2"). The slope of the fields on both sides of the narrow valley bottom varies between 5 and 20 percent.

About 60% of the farm's total area is cultivated. The permanent grasses are half pasture and half meadow. Of the cultivated area, 46% are taken up by cereals, 46% for field fodder and 8% for root crops. The cattle kept are Simmental: on average 25 dairy cows, 1 bull and all the heifers born on the farm. Per year 6 bulls are raised to be sold as breeding animals. 120 hens from farm-hatched chicks are kept. The stocking rate is barely 1 CU/ha (0.4 CU/acre): for each cattle unit about 75 ar (1.8 acres) are required to produce the feed; part of this is no better than marginal grassland.

2. History

From 1921-29 the farm was intensively cultivated according to the standards of those days with:

55 kg/ha (49.0 lbs/acre) N	as sulfate of ammonia
32 kg/ha (28.5 lbs/acre) N	as nitrate fertilizer

84

75 kg/ha (65.8 lbs/acre) P_2O_5 as basic slag
90 kg/ha (80.1 lbs/acre) K_2O as kainite or potash fertilizer

In addition, sewage and liquid ammonia fertilizer from the gasworks were used. Apart from small amounts of N, the grassland was treated with 40 kg/ha (35.6 lbs/acre) P_2O_5 and 50 kg/ha (44.5 lbs/acre) K_2O.

Conversion to bio-dynamic methods in 1930 was preceded by accumulating large amounts of compost during the previous year. Also, as there were two liquid manure pits, a plentiful supply of a stinging nettle and liquid cow manure mixture were prepared ready for intensive application and also to water the composts that contained a large proportion of soil. An arable forage crop was grown and there was ample use of the bio-dynamic preparations.

Average yields on the Talhof farm in tons/ha (bu, t/acre)

	1922-29 (mineral)		1930-37 (bio-dynamic)	
Winter wheat	1.89 t/ha	(27 bu/acre)	2.39 t/ha	(34 bu/acre)
Barley	2.07 "	(34 ")	1.63 "	(26 ")
Oats	2.13 "	(46 ")	2.12 "	(46 ")
Potatoes	15.4 "	(6.2 t/acre)	16.3 "	(6.6 t/acre)
Beet	47.3 "	(19.1 ")	50.3 "	(20.3 ")
Hay (2 mowings)	5.78 "	(2.3 ")	7.96 "	(3.2 ")

After conversion, the buying in of concentrates was reduced to one third of the previous volume. Even so, in 1934, that is, after 5 years, the average yield of the 17-cow herd was 4288 kg (9722 lbs) milk with 153 kg (337 lbs) fat content. In 1938 a new manager was found to take over from the previous man who had fallen ill. Yields began to fall and by 1947, after the war that is, they had reached 2570 kg (5654 lbs) milk with 96 kg (211 lbs) fat content.

Then another manager took over and once more began to intensify the use of bio-dynamic methods. The present manager and author of the original report took over in 1952. In the following table the development of the crop rotation is reported.

3. Crop rotation

The main change is the transition from lucerne, which did not develop well in wet, cool years, to the clover-grass-herb mixture, which is ploughed up after two years since the carry-over effect is better than after three. The extreme temperature fluctuations in the area in March and April are often hard on the winter grains. This encourages weed growth. The transition to spring cereal crops had beneficial effects.

In the clover-grass-herb mixture approximately the following amounts of seed (lbs/acre) are used. Where grazing is near the farm buildings, plants such as tall oat grass are omitted as they suffer from being walked over by animals.

1.8 lbs/acre each of red clover, white clover and lucerne; 11.7 lbs of bird's-foot trefoil (*Lotus corniculatus*); 7.1 lbs of sainfoin (*Onobrychis sativa*); 0.9 lbs of kidney vetch (*Anthyllis vulneraria*); 2.6 lbs of timothy (*Phleum pratense*); 1.8 lbs each of tall oat grass (*Arrhenatherum elatius*), cocksfoot (*Dactylis glomerata*), meadow fescue (*Festuca pratensis*), Italian ryegrass (*Lolium multiflorum*), perennial rye grass (*Lolium perenne*), and Kentucky bluegrass (*Poa pratensis*); 0.9 lbs of red fescue (*Festuca rubra*); 1.3 lbs of yellow oat grass (*Trisetum flavescens*); 0.4 lbs of scented vernal grass (*Anthoxanthum odoratum*); 0.4 lbs of caraway (*Carum carvi*); and 7.1 lbs of burnet saxifrage (*Pimpinella*).

This mixture is sown diagonally across the rows of the cover crop. It reaches a height of about 0.5 m (20") by harvest time. Either the grain is mown over the top and the mixture and straw are subsequently cut against the direction of the combine harvester, or the whole field is windrowed and then threshed. The resulting hay is used as bedding for young livestock and part of it is eaten.

Development of the crop rotation on the Talhof farm (X = composted manure, Y = liquid manure, L = legume)

1948-1952		1953-1964		1965-1974	
1 lucerne (alfalfa)	L	1 clover-grass-herb mixture	L	1 clover-grass-herb mixture	L
2 lucerne (alfalfa)	L	2 clover-grass-herb mixture	LXY	2 clover-grass-herb mixture	LXY
3 lucerne (alfalfa)	L	3 fodder beet		3 potatoes	
4 maize	Y	4 wheat		4 barley	
5 potatoes	X	5 oats		5 Persian clover	LXY
6 winter rye		6 clover-grass-herb mixture	L	6 wheat	
7 oats-barley-pea mixture		7 clover-grass-herb mixture	LXY	7 rye/oats	
8 fodder beet	XY	8 potatoes		8 clover-grass-herb mixture	L
9 oats		9 spring barley followed by rape catchcrop	XY	9 clover-grass-herb mixture	LXY
10 red clover	L	10 spring rye		10 wheat	LX
11 winter wheat	Y	11 Persian clover (*Trifolium resupinatum*)	LXY	11 peas	
12 potatoes	X	12 spring wheat		12 oats	Y
13 spring barley		13 oats-barley-rye mixture undersown with clover-grass	X	13 vetch-peas arable forage	L
14 winter barley undersown with lucerne	X				

In the same year there is then a second cutting for green chop or silage. A smaller application of liquid manure is made. In its first year as a main crop the first cutting is taken late and made into hay; in the following year it is taken early and used for making silage. After the second mowing, composted manure is applied at a rate of 20-24 t/ha (8-9.7 t/acre) and also 20-30 m³/ha (2140-3200 gal/acre) of liquid manure. This is also the time when it is possible to increase the ploughing depth and to successfully control weeds by following this with the disc harrow, harrow and shallow autumn ploughing. (In warmer areas this autumn working of the soil could lead to heavy nitrification and nitrate leaching during the winter. From mid-October at the Talhof farm it is too cold for any appreciable nitrification which is drastically reduced when soil temperatures drop below 8°F). Persian clover (*Trifolium resupinatum*), which has rapidly been gaining ground in bio-dynamic farms for a number of years, deserves a quick mention here. 15 kg/ha (13.4 lbs/acre) of Persian clover (Portuguese seed), together with 5-10 kg/ha (4.4-8.4 lbs/acre) of annual ryegrass and about 120 kg/ha (107 lbs/acre) of oats are sown early in the year. After emergence the clover can withstand temperatures down to -10°C (14°F). The first cutting before the heading of the oats provides a good low moisture silage when predried to about 40% water content. Subsequently, there can be several cuttings of the clover, which produces high yields.

4. Tillage

With the farm's difficult soils the advantages of careful, shallow ploughing are obvious. At the time when clover-grass is ploughed under a gradual deepening of the topsoil can be achieved. Until 1962, when the last horses were sold, the soil was worked from spring to autumn with horses only. Now they have been replaced by a Fendt tractor (4-wheeled self-powered tool bar and transport vehicle) that does its job well so long as you have the patience to wait till the soil is dry enough. As the growing season is short and spring sowing has to be done before the emergence of weeds, mechanical weed control is restricted mostly to the time in summer after harvesting. Working a field four or five times with the disc harrow at this time effectively reduces couch grass (*Agropyron repens*), provided the following crop does well.

5. Fertilizing and bio-dynamic measures

The horn-manure preparation (Preparation 500, see Chapter Four) is applied in autumn and spring on the whole area and on meadows and pastures if possible after every cutting or grazing. Potatoes and beet are also given this treatment once or twice in May/June when these fields are worked. Preparation 501 is used in the summer as often as possible. Manure is stacked daily in a manure pit and sprinkled thinly every day with good soil. It is also treated with bio-dynamic preparations. After 6-8 weeks it is moved to piles in the fields and again treated with the preparations. Half-fermented, not matured manure is used except for the pastures. Organic commercial materials have so far only been used in small amounts on pastures as some of these are too steep to be driven over by tractor and trailer. Bone, horn, basalt and seaweed (algae) mixed with guano are broadcast by hand. The meadows are fertilized with ripened composted manure, compost and liquid manure if possible in the summer, or otherwise in winter on frozen ground without snow or on crusted snow. Up to 1963 rotation grazing was used with short daily grazing periods. From 1964 onwards close fodding was adopted with electric fencing that is moved twice daily. The same number of animals as before now stay on the pasture day and night. But they are given some hay from the previous year at milking time to supplement the low fibre content of the pasture grass.

As far as available storage space allows, the liquid manure is also applied during the growth season, also on the pastures. The liquid manure is stored in an open pit with a thin floating cover of straw. The bio-dynamic preparations are added in a ripened compost that is put into the pit before filling begins. In spite of this liquid manure application, the cows relish the pasture grass. While the same amount of liquid manure is applied as before, the yields are now better.

The following table gives some of the results of soil tests made in 1972 after 42 years of bio-dynamic soil management. These are average figures from the fields, excluding the marginal slopes, etc.

Talhof farm soil test, 1972						
	organic matter, %	total nitrogen, %	pH	P_2O_5 ppm	K_2O ppm	Mg* ppm
crop land	5.7	0.36	7.0	340	260	50
permanent grasses	7.6	0.51	6.6	130	250	70
pastures	6.7	0.50	6.6	330	510	70

*The magnesium content appears to be low; this is due to the testing method used. There is no deficiency. The figures for organic matter, N, P, and K are high. A shallow soil layer is the limiting factor for growth.

Seed is dressed in seed baths: Wheat with preparation 507; rye with 502; legumes, rapeseed and clover grass with 503; beet and carrots with a compound preparation stirred in milk. Seed potatoes, which are spread in a thin layer under a roof about 6 weeks before planting, are sprayed 2-3 times with preparation 507 and /or 503. One variety has been used on the farm without interruption or bringing in new planting material since 1955, and another had been in use for 6 years at the time of the report. (About the preparations see Chapter Four).

Treatment with ash for pest control as recommended in the Agricultural Course was carried out on Colorado beetles and subsequently none could be found on the potatoes. The ash treatment was also carried out against the seeds of thistles; after it had been applied for three years in succession followed by a clover-grass crop the thistles were vanquished. This method uses the ash of animal pests or of the seeds of weeds to restrict the reproduction and propagation of the pests and weeds. It is described in the Agricultural Course. There are numerous positive reports about this method, but a number of questions still await clarification.

6. The dairy herd

The herd originated from two cow families dating back to the thirties. The two mothers were distinguished by good and steady performance, good constitutions, and the absence of any serious defects of inspection points. Good bulls are brought in. All female offspring, without putting too much emphasis on points, are reared if possible until first calving. The criteria looked for include: regular oestrus cycle, good conception rate, easy parturition, good milking, good feed intake. Even in winter the animals are allowed some exercise in the open, if only for a quarter of an hour. Since this was started, injuries and broken horns hardly ever occur when the herd goes out to pasture in spring. In spring, while still on full winter rations, the herd spends 2 hours in the afternoon out at pasture. Gradually the silage and fodder beet are reduced and the grazing time is increased. In the autumn, hay is first given at milking time. Before full winter feeding starts, green fodder is still fed for a while.

The winter ration consists of about 18 kg (40 lbs) low moisture clover-grass silage; 5 kg (11 lbs) corn silage; 2 kg (4.4 lbs) home grown concentrate consisting of oats, barley and peas; 10-14 kg (22-31 lbs) hay, aerated in the barn. The 4-year average of the hay analysis is: crude protein 11.6%; digestible protein 7.1%; nitrogen-free extractives: 377; phosphorus 0.24%; fibre 25.2%. The mineral supplement consists of Weleda lime supplement; bran (to evenly distribute the small quantities); bolus alba; algae meal; some sea salt; herbs. Salt has been added to the silage. On the pasture salt is available ad lib.

Herbs are chopped and ground in a hammer mill. Apart from being part of the mineral mixture, they are added before milling to the grains in the concentrate. They are farm grown and are as follows:
  wormwood (*Artemisia absinthum*)
  ladies love (*Artemisia abrotanum*)
  rue (*Ruta graveolens* hort.)
  caraway (*Carum carvi*)
  fennel (*Foeniculum vulgare*)
  coriander (*Coriandrum sativum*)

balm (*Melissa officinalis*)
sweet marjoram (*Origanum majorana*)
sage (*Salvia officinalis*)
camomile (*Matricaria chamomilla*)
peppermint (*Mentha piperita*)
common basil (*Ocimum basilicum*)
dill (*Anethum graveolens*)
thyme (*Thymus vulgaris*)
common goats rue (*Galega officinalis*)
garden chervil (*Anthriscus cerefolium*)
hyssop wort (*Hyssopus officinalis*)
lovage (*Levisticum officinale*)

with stinging nettle (*Urtica dioica*) making up about 50% of the total amount.

Feeding is not for top yields but also for health. If needed, in summer during grazing the animals get additional hay or green chop, but no concentrates. The following figures show the development since 1953, when the registered herd began:

Talhof farm registered herd

	number of cows	milk kg	per cow lbs	butterfat kg	per cow lbs
1953	19.4	3110	6842	123	271
last 20 years	21.6	4324	9513	169	372
last 4 years	24	4888	10754	205	451
1973	24.7	5275	11605	222	488
1974	25.3	5064	11141	235	517
average 1974 of the local registered herds (several counties)		4393	9665	179	399

(The figures are on average lower than for the larger animals of the Friesian herds in the US dairy belt.)

Sick animals, difficult calving, retention of placenta, calf-rearing illnesses and udder inflammations are rare. Calves are reared on farm-produced whole milk, crushed oats, hay or grazing, and in winter carrots. First calving takes place at 27-32 months. The bulls raised for breeding, up to 680 kg (1500 lbs), had daily weight gains of 1296 g (2.85 lbs) (5 year average).

For poultry there is a hen house divided into two and a large run also divided into two. The chickens are raised at the farm by brood hens. A great deal of tea made from stinging nettles, chives, camomile and peppermint is used in rearing. The hens produce on the average 235 eggs per year.

The yields on this marginal land in the year before the report was written were:

spring wheat	3.8 - 5.0 t/ha	(53.7 - 70.6 bu/acre)	
spring rye	2.0 - 3.0 "	(30.9 - 46.0 ")	
hulled barley	2.8 - 3.8 "		
hulled oats	2.4 - 4.0 "		
potatoes	18.0 - 25.0 "	(7.3 - 10.1 t/acre)	
meadow and field	· "		
fodder (hay)	8.0 "	(3.2 t/acre)	

In 1952 the gross income was 500 German Marks/ha and in 1967 1,800 DM/ha. The labor income rose in the same period from 3,500 to 15,000 DM. In 1968 there were 5 permanent farm workers and in the summer sometimes one or two Waldorf school pupils. The following are the figures for 1973 (US $ in parentheses, assuming an exchange rate of 1 $ = 2.33 DM):

gross income	DM/ha	2,648.00	(459.80 $ acre)
expenditure	" "	1,224.00	(212.50 " ")
income	" "	1,424.00	(247.30 " ")
income per labor	DM	20,278.00	(3,521.00 $)
net income	DM/ha	670.00	(116.30 $ acre)
expenditure for com-			
mercial manuring			
materials, including			
preparations and			
some straw		14.00	(2.40 " ")

This detailed report describes some of the arsenal of bio-dynamic measures and their practical applications.

5 Lessons Learnt From The Old Field Experiments. The Value Of Animal Manure

Owing to the farsightedness of some researchers, field experiments were started by a number of experimental stations in the early days of chemical fertilizers. In part the questions put then are now considered to be out of date. Yet it is just these experiments that can give us information on the long-term changes in soil fertility. The present flood of new materials and, though often only seemingly, new methods, is not always accompanied by sufficient awareness of the fact that soil fertility undergoes long-term changes. Every form of cultivation damages, maintains or improves soil fertility. The direction in which change is likely to occur cannot be predicted within too short a time. E. von Boguslavski (1965) holds that it takes a 10-year field experiment to measure the soil fertility of a habitat. A farmer caught up in the day-to-day running of a farm is only too inclined to assume that a difficult field has always been that way.

A situation such as that described above in the Report of the British Ministry of Agriculture develops over a number of years. Examples even exist that show that with exceedingly carefully balanced mineral fertilizing yields can be maintained for 50-89 years (Millar 1955).

Scheffer (1946) took soil samples from the plots of the continuous rye growing experiment in Halle, Germany, and tested them for two years in pot experiments with mineral fertilizers. In the following table the fertilizers that had been used on the soils for 50 years are shown, for instance, as NPK 0-56-90 (NPK 0-49.8 - 80.1), i.e., 0 kg nitrogen (0 lbs/acre), 56 kg P_2O_5 (49.8 lbs/acre), 90 kg K_2O (80.1 lbs/acre) per hectare. The significant point about these results is that they were influenced neither by lack of nutrients, nor of water, nor of air in the soil. The plants grew in pots, were well fertilized and stood in the most favorable conditions. Yet there is a factor in the soil substance that causes a yield loss compared with stable manure of about 10% in soils that have long been unfertilized or have been treated with mineral nitrogen fertilizers.

50 years of fertil- izing in the field		relative yields in the pot experiment
12.0 t/ha stable manure	(4.8 t/acre)	107.2%
NPK 0-56-90	(0-50-80)	103.7%
NPK 40-56-90	(36-50-80)	97.9%
NPK 40- 0- 0	(36- 0- 0)	97.7%
control		96.0%

The Morrow Plots in Urbana, Illinois (USA), provide an interesting example of the fertile prairie soils (Circular 777, 1969). Three rotations had existed since 1876: 1. continuous corn (maize); 2. corn, oats; 3. corn, oats, clover.

From 1904 until 1954 one half of the plots received a manure-limestone-phosphorus treatment. From 1904 to 1909 manure was applied at a rate of 2 tons to the acre. Since then the amount added has equaled the dry weight of the crops removed. Five applications supplied a total of 12.85 tons of limestone. Also 6.6 tons of rock phosphate and 1.65 tons per acre of bone meal were applied from 1904-1925. In 1955 the plots were further divided and a quarter of each plot received additional treatment, lime and chemical fertilizers for corn at the rate of N: $P_2O_5 : K_2O = 200 : 40 : 30$ lbs/acre (220 : 44 : 33 kg/ha). The following table shows the corn yields for 1907-55 and the corn yields up to 1967 so far as we know them. ·

The yields are higher from 1955 onwards. This is the result of growing modern varieties. The earlier manure, limestone, phosphate fertilization has maintained the soil to such an extent that even a high application of mineral fertilizers does not show any effect. But this holds only when the rotation includes clover. As is shown by the figures for the period 1907-55, crop rotation plus fertilizer leads to the best results. The two can partly substitute each other, though not entirely. As regards yields, the effect of non-fertilizing or of onesided rotations can later on be

Corn yields on the Morrow Plots in bu/acre (t/ha)

Treatment of soil		Crop rotations		
		corn	corn-oats	corn-oats-clover
1907-1954				
control		26.0 (1.7)	35.0 (2.3)	57.0 (3.8)
manure, lime- stone, phos- phorus		54.0 (3.6)	71.0 (4.8)	80.0 (5.4)
1907-1967	**1955-1967**			
control	-	46.2 (3.1)	49.7 (3.3)	72.7 (4.9)
control	NPK	104.0 (7.0)	114.4 (7.7)	126.4 (8.5)
manure, lime- stone, phos- phorus	-	78.1 (5.2)	125.0 (8.4)	133.9 (8.9)
manure, lime- stone, phos- phorus	+ NPK	117.2 (7.9)	129.0 (8.7)	132.7 (8.9)

partly, though not entirely, replaced by high applications of chemical fertilizer. There do not seem to be any results from quality tests. Manure, limestone, bone meal and rock phosphate are first and foremost soil-building materials rather than fertilizers. The loss of natural fertility in the soil, which is expressed in these results, can be partly measured by investigating the nitrogen losses observed between 1904 and 1953.

Nitrogen losses on the Morrow Plots, 1904-53, kg/ha (lbs/acre)

Crop rotation (see above)	corn	corn-oats	corn-oats-clover
control	1980 (1800)	1100 (1000)	1200 (1100)
manure, limestone, phosphate	999 (900)	0 (0)	330 (300)

These figures show above all the favorable effects of manure on the total nitrogen content of the soil and thus on its natural fertility. For further information see also W. V. Bartholomew (1951).

Similar observations are also available from other long-term experiments, e.g., the Sanborn Field in Columbia, Missouri (USA), which has existed since 1888. Only the plots treated with animal manure still had the original amount of nitrogen in the humus after 50 years. Plots treated with mineral fertilizers had similar yields but had lost nitrogen. In Rothamsted, England, farmyard manure was applied for 20 years and then stopped. The carry-over effects on the yield were still noticeable after 50 years, even though much reduced toward the end. In 1898 the so-called Cylinder Experiments were started in New Jersey, USA. Sixty metal cylinders measuring 1.20 m (4 ft) in length and 60 cm (2 ft) in diameter were placed on open ground and filled with subsoil and 20 cm (8") of topsoil. Various fertilizers such as cow manure, mineral fertilizers, green manure and limestone were applied in various combinations. After 40 years, 56 of the cylinders showed a negative nitrogen balance. Despite the application of nitrogen fertilizers, the soil also had to supplement nitrogen to the crops. This was not the case in the four cylinders that had been treated with animal manure. Sixteen years after the cessation of the manure applications the carry-over effects were still discernible in the soil's capacity to store plant nutrients (cation exchange capacity) (Millar 1955).

Bronner and Janik (1974) studied changes in the soil's humus content in the sugar beet regions of Upper Austria. Though the applications of organic matter were equally large, only stable manure was able to maintain or raise the soil's humus content. When green manure, straw and beet tops were used the humus content decreased.

These observations show that built into the solid substance of the soil there is an "old strength" that is particularly well maintained or improved by animal manure. For awhile, the loss of "old strength" can be covered up by applying chemical fertilizers. But this is not without hazards, particularly when we

97

consider that the gradual decline of the natural soil fertility cannot be measured in the early stages either in practice or by scientific experiments. Farmers are familiar with the good productivity of the fields near the farmyard, which ordinarily receive more stable manure than those further away.

The separation of livestock from the soil that produces their food and that receives their dung in return is one of the most disastrous developments adopted by modern, specialized agriculture, particularly in temperate humid zones. These are the regions in the developed countries where the highest yields are being achieved today. (The problem is different in humid-warm and in arid-warm regions.) Scientists almost without exception and also state agricultural administrations on the North American continent are far from realizing the negative effects of mass livestock rearing on the productivity of the soil. Or else they are not prepared to admit that generations to come will have a heavy price to pay for the temporary economic advantages arising from mass animal rearing. The soils in these regions have often been in agricultural use too briefly to show immediately the damage that is done. Changes in soil fertility take decades. It must be said in all seriousness that the advocates of mass livestock rearing are taking on a heavy responsibility. The history of agriculture in the "old world" speaks clearly enough. The author knows quite well that the opinion expressed here can apparently be refuted by dozens of experimental results. The operative word is "apparently."

The following two sections will discuss the farm organism in its relation to its rural environment.

6 Is Soil Erosion Still A Problem Today?

The cultural task of the farmer and forester can be viewed from two angles. On the one hand, if we leave the cultivated landscapes of Central Europe behind us and go up into the mountains or perhaps to the far North, or if we experience the wildernesses that still exist in America or on other continents in all their untouched ruggedness, we may find that such landscapes are apart from man and belong to other dimensions. We feel at home in landscapes that have been shaped and tended by the hand of man whose work brings the realms of nature closer to us.

Yet, on the other hand, man has for thousands of years been destroying fertile land by cultivating it, keeping animals on it and cutting down the forests. In every region cultivated by man there are more parts he has destroyed than improved. Destruction of the soil and of natural communities mark out the path of civilization in many places. Soil erosion, i.e., the removal of topsoil by water or wind, is one of the chief factors in the destruction of soil.

Most recently the North American continent has become an example of the rapid process of erosion. After somewhat more than 200 years of settlement, the best soils of the continent had been claimed by about the turn of the last century. Hugh Hammond Bennett, whose name will forever remain linked with that of the Soil Conservation Service, produced the following figures for the United States in the thirties (areas in millions of acres):

practically destroyed or heavily damaged		282
moderate or beginning damage		775
cropland practically destroyed	50	
heavily damaged	50	
half the topsoil lost	100	
beginning erosion	100	
slightly damaged (woods, bogs, marshes)		702
deserts, mountains, etc.		145
total		1904

Now we know that America is a continent with much higher rain intensity than Europe. This has, of course, played its part. Therefore, the areas particularly affected are the southeastern and southern cotton and tobacco growing parts with their intensive rains, but also all the other States where crops are

99

grown. On the other hand, settlers from the upland regions of Germany have maintained fertility for centuries by strip-cropping on the slopes of Pennsylvania. The arid northwest, where large areas of wheat are grown, is affected by the dust storms that have occurred in approximately ten-year cycles since the middle of the last century. This was the situation when the Soil Conservation Service was established in the middle thirties. Its program, built on voluntary participation by the farmers, has to a considerable extent contained erosion. Farmers wishing to participate receive a report. This surveys in aerial maps the classes of land for all or for limited uses. It then describes soils and contains suggestions as to crops, rotations, fertilizing, etc., that are suitable for the respective fields. A great deal of other information, such as waterways, outlets, etc., is given. This report is found on more or less every farm and its suggestions are taken seriously. In the dry west, strip-cropping, stubble mulching, and so on are curbing wind erosion.

Let us examine a few more examples. The Mediterranean region, the cradle of Western civilization, is the best-known historical example of erosion. In Greece four fifths of the land is mountainous, reaching heights of 1500 m (4500 ft), and yet 72% of the land is less than 35 km (22 miles) from the coast. The steepness increases the force of run-off water. The climate, rainy in winter and dry in summer, does not favor the growth of soil protection crops such as clover and grassland. Cereals, maize, rice, grapes, olive trees, tobacco, vegetables and cotton are grown. Thus, the soil lacks year-round cover and is exposed to rainfall between crop rows. There is not much left today of the unbroken forest cover of the late neolithic period. Of 13 million hectares (32 million acres) a little more than one quarter is cultivated; 2.4 million ha (5.9 million acres) are wooded, though only a third of this is closed forest. It is estimated that since the days of ancient Greece 67% of the country's surface has been affected by erosion (Grigorakis 1967). As of now, only a few thousand hectares have been protected.

L. Jung (1962) reports that since 1940 one million hectares (2.5 milion acres) of forest have been lost in Western Anatolia through lumbering and slash and burn methods. Grazing of the forests is now preventing regrowth. The increase in population has meant that since 1927 the number of sheep has risen by 175% and the number of goats by 126%. From 1453 to 1912 the laws of the Ottoman sultan, Mohammed II, forbade overgrazing and the intensive use of land subject to erosion. Within 21 years, 40% of a large reservoir near Ankar was filled with silt. After World War II, on large areas specially prepared for the purpose, the ancient ploughs, which left a lumpy surface, were replaced by disc ploughs, which work fast and leave a fine surface. Heavy erosion was the result. These examples show that in the Mediterranean region soil erosion is still in progress.

In 1965 G. Richter's report for the German Federal Ministry of Agriculture was published. Since agriculture began about 2000 years ago, and always when forested areas were cleared, erosion, sometimes heavy, sometimes less so, has occurred periodically. This is proved by the young sediments in river valleys, some of which have now been dated. Many fields are today still subject to more or less rapid sheet erosion, i.e., the annual washing away of a surface layer. Formation of gullies is also observed. In areas north of Brunswick, in the Lüneburg Heath region and the northwest German coastal areas, wind erosion in spring destroys newly planted fields. With regard to the present situation in Central Europe the report says:

As opposed to the destruction of arable land, the downgrading of soils and devaluation of arable land are here much more in the fore- ground. Farmers are often of the opinion that the main damage caused, for instance, by heavy rain lies in the fact that plants and fertilizers are washed away or buried in mud. They think they can simply mend the damage by smoothing over the erosion ditches, adding more fertilizer and, if the worst comes to the worst, replant- ing. For the rest they consider that sloping land has always been poorer and yielded less than level land and there is nothing more to be said. In this way the slow, creeping progress of soil degradation renders the initial stages of the soil erosion process important and dangerous in Central Europe. The danger lies in the fact that these early stages are not noticed in their full extent and that thereafter hardly anything is done to remedy them. In the USA this process of the destruction of cultivated land took place rapidly before the eyes of virtually one generation of farmers. Because in Europe it goes on slowly through several generations it is often made light of or ignored. The opinion easily arises that a particular piece of land has always been stony and infertile. If suitable countermeasures are not undertaken in the USA, the end of the degradation process marks the end of using the land at all. In Central Europe, on the other hand, the degradation process usually continues to the limit of arable cropping. Then plot by plot the land is given over to grassland or afforestation. The examples described show how slowly and imperceptibly the change of use moves up and down the hillsides, starting at the steepest places.

Available space does not allow for a full survey of the worldwide problem as it exists in the developing countries with their rapidly growing populations. Erosion is spreading rapidly both in the slash and burn areas of the humid tropics and the grazing land of the dry regions. Land reforms are compounding the problems. Water and wind erosion is still acute today. It destroys or diminishes urgently needed cultivated areas; in the whole world more land is destroyed than can be compensated for in terms of yield by increased fertilizing and other improved cultivation methods. Erosion also does other damage. Low-lying land is buried and reservoirs, rivers and deltas are filled with sand and silt.

Bio-dynamic farming contributes in its sphere to the slowing and control of erosion. Of course, organic and bio-dynamic farmers cannot claim to have done or to be able to do the bulk of the work needed to reduce erosion. But the way in which their farms are worked and run does contribute to soil conservation while their personal attitude also motivates them to do what is necessary in this direction.

Water erosion comes about when, during rain storms or longer periods of heavy rain, surface run-off occurs on slopes. The even distribution of permanent plant covers such as woodland or grassland, or fodder fields decrease the run-off, slow down its speed or even completely retain the water. This is particularly the case where the run-off from long slopes is interrupted by a permanent plant cover.

Organic farms with their varied crops and tendency for medium-sized fields divide their land into sections with differing vegetation. This is less the case with highly mechanized large farms specializing in cereals. Assorted shrubs, bands and hedges reduce run-off.

Coarse textured soils with a crumbly structure and no hard pan absorb more rainwater and carry it more rapidly into deeper layers than finely textured, compact soils. On soils with a stable structure the surface roughness brought about by ploughing, provided it lasts through the winter, slows down the speed of run-off and by letting small puddles form on the surface allows more time for the water to seep into the earth. Sand particles and also stable crumbs are hardly ever eroded; run-off water only reaches sufficient speeds to carry particles if the structure has broken down under the impact of rain drops or the water from melting snow. The speed of the run-off drainage and the detachment of smaller particles from soil aggregates are the deciding factors regarding water erosion. A crop rotation that includes hay and clover fields and catch cropping increases the length of time in the year when the soil is protected against the mechanical impact and splashing of rain. The degree and duration of the ground cover influence erosion. Organic substances and limestone contribute to the stability of the structure. They reduce the detachment of soil particles and increase infiltration. Deep-rooting fodder plants break up hard pans. Organic manuring contributes to the formation and stabilization of structure as do also the proper use of crop residues and deep-rooting plants. Thus it is obvious that a farm with livestock, which uses ripened farmyard manure, crop rotations and also permanent grassland, has the "side effect" of building soils that are able to resist erosion. According to Kohnke the soil factors that reduce water erosion are (Kohnke and Bertram 1959):

Increasing stability; much active organic substance; clay content of the soil; bivalent cations (calcium); water-stable structure; good nutrient supply (produces strong plant cover);

medium soil humidity when rain starts (detachment is then reduced).

Decelerating transportation: coarse texture of soil; water-stable aggregates; much organic substance increases stability.

Reducing run-off: coarse texture and not too fine structure; high infiltration and drainage capacity; absence of hard pans; temporary surface storage of water.

Wind erosion takes place when fine sandy soils are exposed unprotected during dry periods to strong winds. In as much as cultivation and the layout of fields favors drying and exposure it also favors wind erosion. We can influence neither wind velocity, nor the texture of the soil nor dryness. Organic substances in the soil stabilize soil aggregates that are then not subject to wind erosion. In the corn belt of the United States the author has seen drifting on what were once rich prairie soils that have also lost a good part of their humus as a result of cultivation. The basic measure for controlling wind erosion is to create a mulch zone on the surface and shorten the open stretches across which particles can be blown. Mulch can be created by partially working in crop residues. In the above instance, which was observed on a loamy soil, even working the capped surface would provide enough surface roughness to control drifting. Wind protection by wind breaks has always been used in areas where erosion is a danger. It is important to interrupt open land that has been prepared for sowing by protective strips of stubble, grass or mulch running across the prevailing wind direction. Wind causes the "saltation" of fine sandy particles (0.1 - 0.5 mm in diameter), which, upon falling back to the surface, cause the coarser particles to "creep" along the surface and also bring about the formation of dust. This may then be lifted to great altitudes and travel over long distances. If these saltant particles are caught at intervals of 50 - 100 m (150 - 300 feet), the process is prevented from intensifying over longer, unprotected stretches. Dividing up an area and mulching are the means that effectively curb wind erosion.

Both water and wind erosion can be controlled by creating in the landscape diversified farm units that are in themselves further subdivided and that as far as possible maintain closed cycles of organic substances. Such biological units within a landscape are a fundamental concern of organic and bio-dynamic agriculture.

In earlier years of bio-dynamic work there were quite frequent reports on the reduction of erosion. Fields where rain had caused the formation of gullies so wide that they could not be traversed by tractor and plough no longer showed erosion after 5-7 years of bio-dynamic management, even after heavy rainstorms of 100 mm (4 inches) precipitation in a few hours. Stellwag and Schröder, for instance, reported such incidents frequently in connection with their experience in Eastern Europe before the war.

7 The Farm As A Part Of The Organism Of A Landscape

One of the eight lectures by R. Steiner on agriculture is devoted to the relationship between the plants and animals living in a landscape. Trees, bushes, plants, birds and insects, etc., form a community. It is important for the freshness and variety of a rural district that it should be interspersed with sufficient wood and grassland. In these more natural soils a varied fauna of worms, arthropods, and their larvae, etc., bring about the formation of soil and humus. Pastures should be surrounded by bushes and shrubs, which offer the farm animals small amounts of leaves and twigs of the most varied kinds. The bank of a stream stabilized by the roots of trees, the belt of reeds around a pond, a marshy lowland, all these have a part to play in the life of a landscape. If a farm is to flourish, it is essential that it should be embedded in a healthy and varied environment.

In view of the increasing environmental crisis, more notice is now being taken of these relationships. As has been recently described (Koepf 1972), it is a fact that a farm unit with its different lines of production can be likened to an organ within a greater organism. The one cannot function in a healthy way without the other. Yet development in agriculture as a whole is opposed to the application of this principle.

The fact that agriculture has become more technical and that its utmost profitableness is more and more accentuated has, however, the consequence of demixing the uses that are now concentrated and intensified in big areas. For the human theory of environmental control emanates from the idea that less variety induces more control over nature to the welfare of man. But according to the prevailing knowledge, we have the opposite result in natural ecosystems; there variety is resulting with "live and let live" in an effective biological steering and stabilization of the environment and at the same time of high biological and also economical value (W. Haber 1969).

This is a general expression of principles that find their practical application in mixed farms and gardens, in integrated pest control, and so on.

The change in environmental quality brought about by agriculture is only a part of the total problem. The utilization and development of the land must be seen in a greater framework and it concerns us all. The population is growing. Increasing prosperity is augmenting demand for goods that in turn breed refuse. Urbanization of space, of occupations and of spiritual attitudes is proceeding. Demands made on the available living space and its goods are becoming ever greater and more varied. Man's interference in nature's household is increasing in intensity and extent. Through his technology and his chemical industry, man has become a threat to the stability of the kingdoms of nature. It is for these reasons that demands are made for overall planning of the environment that must take account of the rights of biological, sociological and economic concerns. It need hardly be added that even the most perfect plans with corresponding rules and regulations depend for their effectiveness in the last resort on the understanding, insight and responsible actions of every individual. This is particularly the case in agriculture where every seemingly insignificant decision has consequences for the totality of life.

Present environmental problems are involved in manifold ways with farming and horticulture. With suitable brevity we shall attempt to summarize a few of these relationships in the following.

Survey of the capacity of the ecological system and its parts, and the impact of modern civilization.

1) Surface Area

Utilization: for agriculture and forestry; towns and villages; industry; highways and other installations for traffic; small and large scale open-cast mining (gravel pits, brown coal); recreation; refuse tips.

2) Soils

Functions: basis for plant production; sanitizing of a limited amount of wastes; important factor for the water supply of an area.

Burdens: food, fibre and timber production; effects of agricultural chemicals; absorption of oil, lead and other contamination and of industrial effluents.

Main and side effects: positive or negative influences on water supplies, quality of the atmosphere, nutrient contents and dynamics, and biological processes including the occurrence of pests and fungal and bacterial diseases; soil and erosion; accumulation of undesirable substances; influence on drainage, groundwater replenishment and water quality.

3) Water

Functions: fish production; water for drinking, general and industrial use; sanitizes a limited amount of waste waters; universal natural carrier of beneficial and harmful substances; habitat for a section of wildlife; recreational waters.

Burdens: extraction for use; regulation of water courses; open-cast mine shafts; pollution from residential and industrial wastes and sewage, e.g., detergents containing phosphates; poisonous substances; discharges from farmyards and silos; nitrate and phosphate from agricultural land; oil wastes.

Main and side effects: frequent excessive demands on supplies; lowering of the groundwater table; oxygen deprivation; poisonous substances; high content of salts; unhygienic, undrinkable water; fish kills; changes in the

107

water's plant and animal life including commercial fish; undesirable nitrate concentrations in groundwater; eutrophication of inland waters; influence on suitability for irrigation; changes in aesthetic and recreational values; excessive waste heat from power plants.

4) Air

Functions: life element of all organisms; bearer of weather conditions; worldwide carrier of beneficial and harmful substances including airborn plant nutrients; bearer of diffuse sun radiation.

Burdens: contamination by industrial wastes such as dust, oxygen compounds of sulphur, oxygen compounds of nitrogen, fluorine, lead, carbon monoxide, carbon dioxide, PAN (substances brought about by the effects of light on polluted air), smog, pesticides, odors, products of nuclear reactions.

Main and side effects: nuisance and/or damage to health of man and animal; loss of light caused by the blanket of smoke over towns and industrial areas; impairment of or damage to growth caused by dust, sulphur compounds and fluorine; extensive, often worldwide, spread of pesticides and other harmful substances including radio isotopes; supply to the soil of plant nutrients such as nitrogen and sulphur compounds, (rain contains small beneficial and harmful quantities of the last two and some trace elements, depending on industry in the area).

5) Vegetation

Functions: food production; production of industrial raw materials; influencing of macro-and microclimate; soil cover; habitat for animals; hygiene factor; recreation; aesthetic and spiritual values for human beings.

Burdens: changes of the natural plant community; felling of clumps of trees in a landscape; exploiting of forest stands; cultivation methods in agriculture, horticulture and forestry including methods for forcing high yields.

Main and side effects: modification of ground cover; regional water supplies; local climate; incidence of fungal, bacterial and animal pests; "improvement" of the landscape, its beauty and value as an object for study; the incidence, usually impoverishment, of botanical species.

6) Fauna

Functions: member of the natural environment; food production; production of industrial materials; aesthetic and scientific values.

Burdens: far-reaching changes of habitat; reduction of undisturbed natural areas; disturbance from technological equipment and means of transport; progressive pollution of the biosphere with substances that are hostile to life.

Main and side effects: loss of natural communities; extinction of some species; massive increases in pests.

This list, which is incomplete in many ways, is intended simply to give some impression of the manifold ways in which a human population with a relatively sophisticated technology influences the kingdoms of nature by which not only its physical but also its recreational and other needs have to be met. Recognition of this situation has led to a flood of books and articles of which only a few shall be mentioned here: White House Report 1965; German UNESCO Commission 1969; M. Nicholson 1970; NN Solid Waste Disposal 1969; World Health Organization 1971; National Academy of Science Publication 1400; and many others. Our concern here is merely to give an impression of the innumerable influences originating in the human population. There are countless interrelationships within the natural kingdoms and in their connections with man. A purely technical mode of thinking will be insufficient if this complicated network is to continue functioning properly; technical thinking will have to be placed at the service of an inner attitude. "In living nature nothing happens that is not connected with the whole," said Goethe in his essay "Der Versuch als Vermittler zwischen Objekt und Subjekt" (The experiment as a mediator between object and subject). And he continued: "and if the phenomena appear to be isolated, if the experiments are looked upon as isolated cases, this does not mean that they are in fact isolated; the question is simply: How do we find the connection between these phenomena, these events?" We are invited by such words to adopt a certain attitude of recognition toward living nature. In view of the present environmental crisis it is clear how important such an attitude is, above all when it is a matter of practically applying the knowledge gained. To utilize the kingdoms of nature for the benefit of mankind we need specifically ecological and biological categories of thinking. The idea of the farm as an organism is such a category.

Much that can be said about the environmental crisis is being said today. We shall now simply introduce a few examples of how a varied agriculture adapted to the local environment can reduce or even remove many of the environmental problems encountered.

First Example: What is required near urban areas?

This example spotlights the conflict of interests always encountered by planners. The planning of whole regions, which is much discussed today, is often a matter of zoning. Bleak conurbations are to be prevented from growing any further. Green belts will be inserted between towns of medium size. And in these green belts agricultural and horticultural production of a suitable kind will take place.

It is interesting to see what ideas are gaining ground in this connection. S. Herwijer (1969) quotes the view of an American planner concerned with the special marketing situation in such green belt areas. He comes to the logical conclusion that what is needed are highly specialized farms for which he gives examples: large-scale dairy farms; battery production of poultry and eggs; specialized market gardens. In other words he suggests the kind of enterprise that would be able to afford the extremely high price of land in such areas. Quite apart from the unavoidable problems of waste disposal, pest control, chemical plant protection, and sanitary engineering, we must simply ask

110

ourselves: Is this the kind of production the city dweller wants to find on his back doorstep? Do we want to complement urban life by placing these specialized production units and all their environmental problems in the immediate vicinity? Would it not be far more meaningful in these areas close to towns to establish a biologically sound form of farming and bring about a really cultivated landscape? Both for adults and for the young people growing up in the towns this would be far more important than the sight of factories for food production on the one hand and "nature" artificially preserved in nature reserves on the other.

Such ideas can lead to far-reaching consequences from which those practical people who aim only for what is "possible" might justifiably shrink. Many planners are nevertheless thinking of the consequences. Thus in his above-mentioned lecture S. Herwijer says: "Meanwhile at an international level the question has been aired as to whether continued private ownership of land should not be regarded as the weak point in maintaining agriculture in urbanized regions." It is not our intention to deal here with this emotive question in a few remarks. We quote it because at the beginning of this chapter we mentioned the conflict between onesided economic thinking and biological thinking. Protection of the environment and care of the landscape will in future demand increasing attention, whether it is a matter of green belts around towns or other areas such as marginal soils. The biological, economic and legal questions that will arise in this connection are far from being solved.

Second Example: Land use, farm organization, water supplies

In the management of water supplies and water quality we are confronted particularly clearly with the relationship between environmental quality and land use. In industrialized countries water is becoming scarce. Consumption is rising while part of the supplies are rendered useless or at least reduced in quality by pollution. In the Federal Republic of Germany, water consumption by the population and industry lies between 30 and 570 litres (8-150 gallons) per head per day. Future consumption for the population is expected to be about 350 litres (93 gallons) per head per day. In 1954, 90% of drinking water was still obtained from groundwater. But large groundwater reservoirs are not found everywhere. The situation is similar in other countries and continents. Large groundwater bodies occur mainly in the great glacial sand and gravel beds in and around the once glaciated areas, and in the gravel and sand deposits of the great river glacial sand and gravel beds in and around the once glaciated areas, and in the gravel and sand deposits of the great river valleys. Hilly country and mountain districts are often well supplied with many springs and on the whole have sufficient water for local populations, but without artificial storage basins they cannot contribute to the supply needed by the large cities. So they are depending more and more on processing water from lakes and rivers. Replenishing groundwater reserves is a precondition for future supplies as well as for the maintenance of present supplies to water courses.

The above situation stands quite independently of the fact that only a fraction of total precipitation, about 4-5%, is used for human consumption including industrial usage. This makes it clear that only a small part of the total precipitation can ever be pumped. Water is also, however, the most important "plant nutrient." To produce 1 kg (2.2 lbs) of plant dry matter, 200-800 kg (53-212 gallons) of water is needed. This water has to be available in the root zone. Sufficient rainfall is not enough. What matters is that sufficient water is stored within the root zone.

The influence of agriculture on water supplies should be at least indicated in broad outline. It becomes plain that mixed farming can claim considerable significance in this respect. The path taken by water through the landscape is shown in the following diagram.

The diagram indicates that the water supplied by rain follows two different routes: 1) A small proportion (1-3 mm per each rainfall) evaporates straight from the foliage. 2) Some of it leaves the catchment area as surface run-off. 3) And some percolates down into the soil whence it is either taken up by

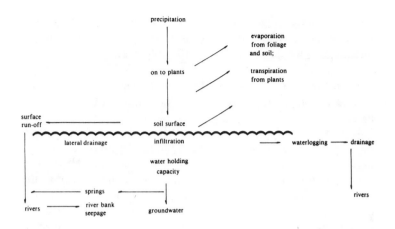

plants or feeds the springs and rivers as groundwater of subsurface run-off from slopes. We see that the third path is the one that enables plant production and replenishment of groundwater to take place. The water holding capacity referred to in the center of the diagram denotes the water that is retained by the soil and gives life to the plant cover. In what respect are infiltration, storage and drainage influenced by the way the land is used? The greatest infiltration takes place in the forest with its littered floor, which usually has a high proportion of humus in its friable topsoil. Strip cropping and contour farming on slopes diminish the surface run-off. A high humus content in the soil increases infiltration and the water holding capacity of the soil. Surface run-off increases if the soil has become compacted by onesided monotonous crop rotations, lack of humus or the use of heavy implements at the wrong time, etc. The same applies if the layout of a catchment area does not contain as much variety as possible with woodland, grasses, fields, hedges, terraces, etc. Without this variety, flooding is also much more likely after heavy rain or when snow melts. In short, all measures of farm organization and soil cultivation that allow rainwater to enter the soil will increase the amount of water available for people and plants. Externally the process is quite inconspicuous. In

Germany the average rainfall is 770 mm (30.3") per annum. If 1.3% more of this water were to enter the soil than does so already, this would be 10 mm, or 100,000 litres, per hectare. For 100 square kilometres (62 square miles), the area of a medium-sized town, this would be 1 million cubic metres (264.2 million gallons), enough to supply a population of 10,000 for a year.

Even when water is available in sufficient quantities, if need be from rivers, good water is becoming increasingly scarce, and in part it is agriculture itself that has brought about this situation. Both liquid and solid manures were once valued as the basis for every farmer's fertilizing program, but this is less the case today. With factory farming of poultry, pigs and cattle the manure has become a troublesome waste product. The cost of manure disposal from animals held in large numbers is indeed expensive. There have been reports from the USA, Germany and Switzerland of fish kills caused by the discharge of manure. G. E. Smith (1967) found 5500 kg/ha (5000 lbs/acre) of nitrate in the soil under feedlots. It is clear that quite apart from the loss of valuable manures, animal wastes cannot be dealt with satisfactorily either in lagoons or in sewage treatment plants. Every farm with 100 dairy cows would need a sewage plant equal in capacity to one for a human population of 1500. Even if this were feasible it would lead to excessive salt contents in rivers. The only reasonable way to dispose of valuable manure is to make the best possible use of it on the land and in order to do this, livestock keeping must be decentralized. Care must also be taken in the way it is applied. S. A. Witzel (1969) found that fresh manure spread on frozen ground in January lost 8.8 times as much nitrogen, 8.5 times as much phosphorus, and 2.3 times as much potash as fermented manure applied in May.

Everywhere in general agricultural practice the interest in farm manure is decreasing while that in mineral fertilizers is increasing. In Baden Württemberg, West Germany, the use of phosphates rose by 106% between 1955 and 1965, from 23.2 kg/ha (20.6 lbs/acre) to 47.4 kg/ha (42.2 lbs/acre) P_2O_5 (Brugger 1966). Phosphate fertilizers remain in the topsoil so whatever is washed away also increases the amount of phosphates in rivers and streams. As has been shown for a number of surface waters (Koepf 1970), a positive correlation (r = + 0.3 to 0.8) exists between the concentrations of total and soluble phosphate. Silt from topsoils will increase the concentration of soluble phosphate in inland lakes and encourage the formation of aquatic blooms. Farm manures and composts, on the other hand, cannot contribute to over-concentration of phosphates in water unless they are carelessly applied. In an 8-year experiment on the application of garbage compost to steep vineyards, J. Bosse (1968) measured the reduction to soil erosion. Application of 400 and later 200/ha (160 and 80 t/acre) reduced soil losses as follows:

clay	from 49,000 to 8,500 kg/ha (44,000-7,600 lbs/acre)
humus	from 1,350 to 800 kg/ha (1,200- 700 lbs/acre)
potash	from 95 to 38 kg/ha (84- 34 lbs/acre)
phosphate	from 112 to 43 kg/ha (100- 38 lbs/acre)
magnesium	from 38 to 12 kg/ha (34- 11 lbs/acre)

Dawes, Larson and Harmeson (1967) report that NPK application in Illinois (USA) increased from 19-13-29 kg/ha in 1954 to 60-20-30 kg/ha in 1964. During the same period it was found that in five rivers draining a large agricultural area of the State the nitrate and phosphate concentrations had increased one and a half times.

Koepf (1968) has pointed out that the nitrate concentration in groundwater is related to the proportion of arable land, fruit and vegetable growing in an area. Undesirable nitrate concentrations in groundwater supplies come from tilled arable land and hardly from grasses or woodlands. Indeed, arable land, vegetables, etc., are heavily manured with chemical fertilizers, as is shown by Schwille (1962, 1967). Because of the dangerous nature of nitrates, most States have legislation establishing drinking water standards (50 mg nitrate per litre), but these are frequently exceeded. Leaching of nitrates cannot, however, simply be prevented by restricting the use of fertilizers. It is far more a matter of using plants to stabilize nitrogen in the soil and diminish leaching (Koepf 1969; Klett 1968; von Wistinghausen 1971).

E. von Wistinghausen (1971) has shown that on a sandy brown soil nitrate leaching in winter can be diminished by 30% by means of undersowing and catch cropping. This is a most important observation since it shows that an essential fertility factor, nitrogen, can be retained in the soil by plants, a situation that at the same time contributes to a decrease in water pollution. It has already been pointed out (p. 55) that nitrogen added to the soil in organic instead of mineral form drastically reduced nitrate leaching.

To summarize these few examples, we can say that agriculture can play a large part in improving water quality and also in increasing water reserves. The required measures are: mixture of woodland, grassland and arable land in the catchment areas; reduction of surface run-off; crop rotations; ample use of catch cropping and green manures; careful storage and application of manures and composts. An additional important measure on bio-dynamic farms is the total avoidance of chemical fertilizers.

Third Example: Disease and pest control

"All pest problems are created by environmental conditions that favor the pest. Altering the environment to make it less favorable may control pests, or at least reduce their rates of increase. Environmental manipulation for pest control includes plant spacing, species diversity, timing, crop rotation, plant hormones, water management, fertilizers, soil preparation, and sanitation." These sentences are quoted from the White House Report "Restoring the Quality of our Environment" published by the President of the United States of America in 1965. At the time, this report did much to draw public attention to the existence of environmental pollution. It also stimulated research into methods of biological pest control. Ever and again we find, however, that there are no wonder drugs. The sentences quoted above are entirely valid. A varied environment, i.e., a healthy landscape, variety in types of farm, and balanced organic

116

fertilizing are, together with cultivation methods, the basis for the specially timed selective use of remedies in particular locations. These are the principles of an integrated pest control program. In a varied landscape and in a farm that functions as an organism there are stability factors that cannot be replaced by anything else. Natural communities are mostly self-regulatory, because every organism interacts with many other organism at the same habitat. A farm is not a natural community. Therefore self-regulation through multilateral interaction must be supported by man. The need for this approach is increasingly felt in scientific circles, as shown by a quotation from a recent paper by Ferris and Ferris (1975):

> Turnbull (1969) evaluates three strategies of control in terms of (the) concepts of community dynamics. The first strategy is that of applying controls from without, as with pesticides. This type of control, as we have seen, evolves the maximum homeostatic response and is likely to be self defeating in the long run. Moreover the danger of environmental contamination is of increasing concern. The second strategy is the concept usually evoked by the phrase, "biological control," and includes attempts to restore the natural states of homeostasis by environmental manipulation. Unfortunately, since the objective of agriculture is to supply the crop species with overwhelming and permanent advantage, this strategy too ultimately will fail.
>
> The third strategy suggested by Turnbull (1969) is to use the natural forces, and not to ignore them or to submit to them. We know what we want from communities and what forces are at work or should be. Attempts at integrated pest management which incorporate the use of pesticides as an integral part of the system, to lend a bias, and not just as an emergency supplement, are an attempt in this direction. For the main pest controls, however, the system relies on natural phenomena and habitat modification.

Chapter Three

Plant Life, Soils, Fertilizing

The two introductory chapters of this book have dealt with the general foundations of bio-dynamic work. Now we shall see how they can be applied to the care of soil life, fertilizing and other environmental conditions of growth. Among the growth factors are warmth, light and air. These are no less important than water and nutrients for the yield and quality of products, so let us turn to them first.

1 Cosmic and Atmospheric Growth Factors

The macro-climate determines the amount of warmth and light plants will receive and also how much water will be at their disposal. Factors such as precipitation and how it is distributed over the year, average temperatures and temperature extremes, length of days, average cloud cover, wind speeds, and so on, provide the framework within which we have to adapt land use. Within this framework, however, warmth and light, humidity and air movement among stands of plants, depend in return on the plant communities and the nature of the topsoil. How far the roots of plants are able to utilize the warmth, air and humidity of a particular habitat is anyway regulated to a great extent by the soil.

Light and warmth, the cosmic growth factors, air and water, the atmospheric growth factors, and their daily and seasonal rhythms are influenced by the upper layer of the soil and its vegetation. These can be helped to act most favorably by shelterbelts and windbreaks, mulching, cultivation, suitable crops and spacing of plants (Geiger 1950). To improve the micro-climate and the warmth, air and water household of the soil is a major concern of those who want to grow high quality crops.

The surface of the soil and its vegetative cover are the point of interchange for the sun's radiation on plant and soil. The spaces above and below the soil surface are differentiated clearly from one another. Diurnal and seasonal rhythms of light and warmth and thus also of air movements and evaporation have their origin in the surface layer, of either the soil or the foliage. Of the approximately 2 calories that arrive in the earth's atmosphere per minute and square centimeter, about 58% reach the earth's surface. The amount of radiation received by plants and soil changes almost at every step. In northern latitudes less energy can penetrate because of the oblique angle of the sun's rays. Also in the north the days are long in summer and short in winter. High humidity content of the air, clouds and also the carbon dioxide in the atmosphere absorb radiation. Thus it is more intense at high altitudes, or in dry regions and during dry periods. South-facing slopes receive more and north-facing slopes less direct sun radiation. East and west-facing slopes are in between but those that face west are warmer on average because on the east-facing slopes morning warmth is used up for drying before the sun can start having its full effect.

During daytime, plant cover and soil surface are warmed by the influence of sunlight and diffuse radiation. Dark, humus-rich soil and also damp soil, since this is also darker in color, absorbs and stores more warmth than light colored soils. Light colored sand increases the amount of solar radiation received by the plants. Light trellis walls behind fruit and vines intensify fruit formation and the taste and color of fruit because they reflect the light. Where there is no protective plant cover, the air near the ground, young plants that have not yet spread to cover the ground, and also the micro-flora and fauna of the topsoil are exposed to higher daytime and lower nighttime temperatures than is the case under opposite conditions. During daytime the air directly above the earth's surface or in the upper third of the foliage is warmest. Dry, sandy soil increases this effect because its surface, being a poor conductor of heat, grows hotter than damp, loamy soil.

On clear nights the warmth radiates back from soil and leaf surfaces. Especially in spring when rows do not yet cover the surface and soil and air are still cool, night frosts can then occur. Cloudy skies and a high water vapor content of the air shroud the earth in a protective mantle of warmth rather like a greenhouse. The canopy of a forest or the crowns of individual trees, closed plant cover in field and garden, in a lesser degree also the proximity of hedges, buildings, fences or slopes that mask off a part of the sky, reduce midday insolation and nighttime outward radiation. In these cases the temperatures near and under the soil surface are more balanced. Under dense plant cover the temperature is almost even. The damp and crumbly condition of living soil under a leafy canopy is the result.

Within the soil too, diurnal and seasonal oscillations of temperature are greatest near the surface. Since one litre of water needs four or five times more heat than 1 kg of soil or rock to become 1°C warmer, damp soils take longer to warm up than dry soils, but they then also store the heat longer. Dry sand also conducts heat more slowly than loam or damp loam or even damp sand. Thus dry sandy soils, for instance desert soils, become extremely hot on the surface during the day. At night they cool down equally rapidly and become cold especially because there is little conduction from deeper layers. In loamy and damp soils the daily temperature variations are less extreme. At night they have more warmth stored and ready to flow to the surface. On the other hand, high humidity such as that of marshy ground does cause temperatures to drop more because of evaporation.

The conditions that favor late frosts are clear skies; dry air; light, dry sandy soils or a dry, light layer from hoeing or harrowing; dry fallen leaves or dry mulch.

The daily temperature variations near the surface of the soil are transmitted to the air within the growth zone. In the daytime this is warmest near the ground or just above the leaf cover but it is also the most turbulent because of irregular air currents flowing up. Evaporation is at its highest because in daytime the air is at its warmest, but its relative humidity is low. At night when the surface of the soil is cool the air near the ground is coolest and heaviest. As it settles in layers the nightly air stillness falls, i.e., on flat land local winds drop, while on slopes the steady downwind starts. This is a current of air that flows down slopes and valleys. During the day warmth streams from the surface of the soil downwards, while at night it flows from the depths back to the surface. This daily wave reaches down to 25 to 35 cm (10-14'') within the soil. During the summer half of the year the warmth from the sun is gradually conducted deeper and deeper into the earth, though at 2 to 3 meters (app. 2-3 yards) there is hardly any annual fluctuation. The rise in temperature reaches its greatest depth in the earth in late autumn. From winter onwards until it begins to grow warm in spring, the warmth from the past summer streams back from the depths to the earth's surface. The flow of warmth from the inner layers of the earth is so low that it does not show in the temperature fluctuations. The warmth rhythms in air and soil are governed by the sun. In addition to precipitation and average temperatures it is above all late frosts in spring that limit cultivation and determine the time for spring planting and tilling. The length of the growth period depends upon this. The following are some of the causes of late frosts:

1. Low general temperatures.
2. Clear night: no cloud cover and dry air, i.e., absence of the protective atmospheric mantle; no wind, so the layers of cold air near the ground remain undisturbed.

3. Only slight warmth from below the surface if the soil is cold and/or dry and loose.
4. High evaporation such as often occurs in damp low-lying land.
5. Cold air flowing into hollows or collecting in pools behind banks, hedges, woodlands, buildings.
6. No protection from neighboring trees, hedges, slopes or buildings.

From this follow the normal protective measures:

1. Deflection of the cold air, either by placing obstacles to catch it or by removing obstacles to prevent the formation of pools of cold air.
2. Choice of medium slope for crops (fruit, vines) that are sensitive to frost.
3. Later mulching in spring.
4. Postponing hoeing and harrowing for a few days.
5. Artificial protection such as covering, plastic cloches in gardens; protective irrigation; fires among valuable crops; ventilators.
6. Prophylactic spraying with BD Preparation 507 the evening before frost is expected.

Protection against wind in landscape and garden

The flat lands swept by the westerly gales from the Atlantic and the winds from the North Sea, i.e., Northwest France, Holland, Northwest Germany, Schleswig Holstein, Denmark, Southern and Central England, have since time immemorial had their delightful hedges as a protection against wind. In some districts it would be almost impossible to grow any crops without wind protection. On grazing land, trees in addition provide shade for animals and the few twigs and branches nibbled are a valuable additional food. In the prairie areas of some countries, which have been under cultivation for only some decades or a few centuries, systems of windbreaks and shelterbelts still need to be established. J. Santon (1967) lists the following effects of wind protection:

1. Shelterbelts reduce wind velocity;
2. Shelterbelts reduce evaporation;
3. Shelterbelts increase the soil temperature;
4. Shelterbelts increase the air temperature.

In some regions, particularly dry steppes, the greater volume of snow that falls behind hedges contributes to the water supply and also helps to reduce the depth to which frost penetrates the soil. Also, since the speed of wind is reduced, more rain falls behind hedges. Dew formation can be increased. From investigations in Denmark, J. Santon presented the following improvements on yield increases achieved through protection against wind.

Yield increase through protection against wind (%)

	to the East	to the West
	of the shelter belt	
four cereals, grains	17.1	11.7
turnips and potatoes	15.0	10.3
clover grass	24.1	23.3

Between 14 and 50 m (46 to 164 feet) from a shelterbelt, 627 kg (1382 lbs) of apples were harvested. Between 194 and 280 m (630 and 919 feet) from the same shelterbelt 483 kg (1065 lbs) of apples were picked. In the dark prairie soils (chernozem) of Russia, wind protection led to yield increases of 20-30% in normal years, over 200% in dry years and about 19% in wet years. As a general rule wind protection in Central Europe is reckoned to increase yields by about 10%. This includes the regions further from the coast that are not in such need of protection as are coastal areas. Many of the highland regions, however, are dependent on shelterbelts. It is reckoned that these will have to occupy 2-6% of agricultural land.

Different from shelterbelts though as a side effect also serving as shelters are field plantations, single trees, groups of trees, bushes, hedges, rows of trees along the banks of streams and rivers. The area they occupy should be at least 1-2% of the land (not counting actual forest). Together with coppices and small patches of woodland they contribute greatly to the beauty and liveliness of landscape. Hedges are not in the first place hiding places for mice, caterpillars and slugs, as is so often said, but habitats for birds, weasels, hedghogs and also beneficial insects. From the point of view of wind shelter alone, a properly laid out shelterbelt is no different than a stretch of woodland of the same height. But woodland also serves other purposes in a landscape. To a certain extent it develops its own climate and mitigates extremes of temperature in its neighborhood. Woods take in the water from mist that partly falls to the ground in drips and partly evaporates. Forests and woodlands are surrounded by belts of water vapor and they retain water in the soil, thus improving the local water household. They also filter dust out of the air.

As has been said above, windy areas and those tending to be dry are the first that need protection from wind. But plantings are also useful where wind erosion is a problem, where snowdrifting occurs in winter and where mechanical wind damage is done to plants. The plantation is angled vertical to the prevailing wind direction. In hilly country, wind conditions have to be carefully studied in order to determine the proper siting and spacing of plantations. Quite often it is a matter of protecting sensitive crops from cold air and care must be taken not to create a cold air trap instead. Air must be able to flow away, if possible to the nearest body of water. Gaps in shelterbelts or wrongly sited openings can become places of increased wind speeds (jet effect).

Protective plantations bring about short stretches of damming in the direction from which the wind comes and decreased wind velocity on the other side of the barrier. The effect is most marked in the space measuring six times the height of the barrier. By a distance of 15 times the height of the barrier, wind velocity again reaches 80% of the original speed.

Shelterbelts can take the form of plantations with widths of 10-15 m (10-15 yards) or rows (1-3) of trees. The purpose in both cases is the same. There should be no large gaps but 30-50% of the area must be permeable by wind. Impervious obstacles cause damaging eddies and turbulence when what is required is a slower, laminar flow. The shelterbelt should consist of trees and bushes that offer the wind decreasing resistance toward the top. High trees are thinned to an extent where their crowns do not intermingle. New plantations consist of tall-growing trees, low-growing trees, and shrubs and bushes in the lower and middle areas. According to Santon, 50-70% should be nurse or filler trees planted to help establish the permanent shelterbelt trees. Coniferous trees are useful for this since they cope better with grass, though high shelterbelt trees are usually deciduous. It should also be decided whether timbering is intended or whether bee pastures, wild fruits and berries are to be included. The suitability of plants and their role as shelters for wildlife must also be considered. Some tall trees with shallow root systems influence a wider area and impoverish the soil, e.g., poplar and birch. There is little point in including in this general discussion suggestions for the choice of trees since this can only be decided locally. It is best if a plantation resembles the natural woodland and coppice communities of the locality, particularly the woodland edges. The saplings themselves should be local, or certainly not from milder parts of the country.

Orchards and gardens are often surrounded by rows of trees or hedges as a protection against wind; with small gardens the shade from the trees should fall outside the garden or on the compost heaps. Here too it is a matter of creating mildly damp spaces with even temperatures while reducing winds that dry everything out. In a garden the place of a planted hedge can be taken by a fence, or reed matting or some other dead screen. But here too about 40% of the screen should be permeable by wind. In a vegetable garden the space can be further subdivided by planting taller plants at intervals, e.g., rows of currants, gooseberries, raspberries, runner beans, maize, peas, artichokes, sunflowers, etc. Shrubs and flowers at the ends of beds provide another subdivision. Too much humidity is just as undesirable as too little wind protection. In windless spaces the soil is warmed more thoroughly during the daytime. But air turbulence at night prevents the air next to the earth from cooling off too much. Where there is moderate sunshine, or damp soil, or at times when the temperature is falling, soils that are protected from wind are usually warmer. So long as the surface is damp, protection against wind also protects against evaporation, i.e., it saves water. This effect is hardly noticeable if the soil surface is dry.

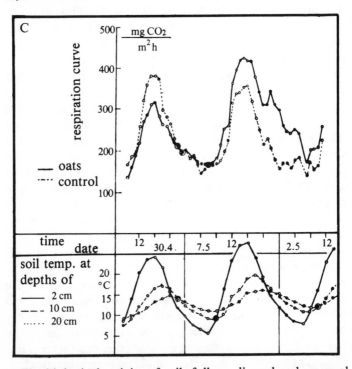

The biological activity of soils follows diurnal and seasonal temperature and humidity rhythms (Koepf 1966 b).
a) and b) = "respiration curve" of the soil of a field (grams of CO_2 per m² hour.)

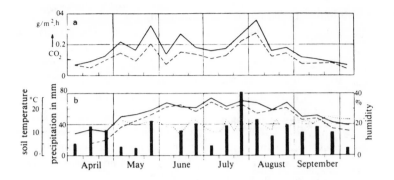

a) solid line: with stable manurc (ten-day mean)
   broken '' : without stable manure (ten-day mean)
b) solid line: air temperature in °C (ten-day mean)
   broken '' : soil temperature in°C (ten-day mean)
   dotted '' : soil humidity in % (ten-day mean)
   columns: precipitation in mm (ten-day mean)
c) process of soil respiration during two sunny days.

Mulching

Mulching serves to protect the soil and improve the climate directly above and below the surface. In regions threatened by wind erosion, the western part of the Great Plains of the United States, for instance, stubble mulching is the most effective form of protection. Stubble and other harvest residues left on the surface or only partially worked in also provide a certain amount of protection against water erosion.

Gardeners often apply stable manure to newly turned soil in the autumn, which is certainly not the best way to use this valuable fertilizer. It is better to work it in when it is mature. Horse manure with plenty of straw, today most easily obtainable from riding stables, is favored by gardeners for autumn application to strawberries. By spring the straw has been washed clean and then serves to protect the berries from splashing.

Mulching in the narrow sense is giving the soil a cover that is not necessarily intended to serve as a manure. The effects are rather:

1. Particularly in warm and semi-arid regions and during hot, dry summers evaporation from the soil surface is reduced. This is particularly important with light soils that do not store much water.

2. With the surface deprived of light, weed seeds are prevented from germinating and young weeds cannot make it to the surface. This method of weed control, however, works with thick layers of mulch.

3. Mulch protects the soil from the impact of raindrops. Disintegration of aggregates is particularly intense when rain falls directly on soil, particularly dry soil.

4. During rain, mulch protects low-growing plants like strawberries, bush beans, lettuce, and also herbs that are to be used fresh or dried from being splashed with mud. Only a thin layer of mulch is needed for this.

5. A covering on the soil in winter protects the ground against frost penetration, particularly when there is a lack of snow. In cold climates, the lifting of the surface soil by frost, which damages the root systems of plants, is reduced (Edmond 1964).

6. Mulch applied in spring can obviate a good deal of hoeing both for weed control and loosening of the soil.

7. The soil microorganisms as well as worms remain active in a surface layer that is protected from drying out and from temperature extremes.

H. Grotzke (1966) stresses that mulching procedures cannot be standardized but have to be applied in accordance with the local climate and the nature of the plants in question. In general, a mulch will keep the soil somewhat damper and cooler. Therefore in northern parts it should be remembered that plants like cucumbers, melons, tomatoes and peppers, which come from warmer areas, prefer warmer soils. In warmer areas, on the other hand, plants like lettuce, different varieties of cabbage, and carrots will react favorably to a protective layer of mulch. On the whole, mulching improves aeration of the soil. There are a few materials, however, that prevent it. In wet years, particularly on heavy soils, it is advisable either to remove the mulch completely or at least from the close vicinity of plants. The same may be the case if slugs start to play havoc during damp weather. The click beetle (*Agriotes* sp.) sometimes lays its eggs, from which wire worms emerge in due course, under a thin layer of hay or straw mulch. It avoids thick layers of mulch.

The following are a few general remarks about the more important mulch materials:

Old boards suffocate persistent weeds and grass. They are easily laid but of course have to be removed for working the soil. Weed-free compost is a manuring, protecting and enlivening material for covering the soil round trees, fruit bushes, roses, shrubs and herbs in winter. Under young grass and other green plants left to wilt, aeration tends to be good, and soil life develops favorably. This mulch can be used for lettuce, beans, peas, varieties of cabbage and other vegetables. The same goes for hay, though this does not lie so snugly on the ground. It must of course be free of weed seeds. Chopped straw is an equally useful material, particularly if it has started to rot during winter. Thin layers of baled straw can also be used. Leaves either partly rotted or put through a shredder to prevent them from being too easily blown away are suitable for mulch. In the autumn they are removed and used for compost. Stinging nettles in any amounts can be used favorably for mulching. Paper, which of course has to be weighed down, is also used. Peat is highly esteemed but not exactly cheap. Black plastic sheeting, which raises soil temperature, is being increasingly used. It favors the soil life; the soil moisture that condenses on the underside during the night falls back and keeps the earth damp. It is useful for strawberries, peppers, tomatoes, etc. Saw dust is not recommended as a mulch, except perhaps for cultivated blueberries. A stone mulch is sometimes used for trees, but this can lead to mouse damage of the trunks. Finally, it should be mentioned that the loose layer resulting from regular hoeing acts in a similar way to a layer of mulch with regard to aeration, evenness of soil temperature, and reduction of evaporation. The following table will serve as a summary.

The influence of weather conditions and of natural and artificial ground cover on soil climate and the air near the ground (after Koepf, 1966a, adapted).

weather conditions	change	influence	direction of influence
clear sky, dry air	absorption in the atmosphere	radiation toward and away from soil of light and warmth in daytime & of warmth at night	higher during daytime and nighttime
		atmospheric reduction of outward radiation at nighttime	lower
		soil & air temperature	higher in daytime, lower at night
cloudy sky, high air humidity	absorption in the atmosphere	radiation toward and away from soil	lower during day and nighttime
		atmospheric reduction of outward radiation at nighttime	higher
		soil & air temperature	less difference between day and night

130

ground cover	changes	influence	direction of influence
dry straw, leaves, etc.	conduction	soil temperature	lower during daytime, higher at night
	conduction	air temperature	higher during daytime, lower at night
	movement of water vapor	evaporation from soil	lower
loose, crumbly surface, dry	conduction	soil temperature	lower during daytime, higher at night
		air temperature	higher during daytime, lower at night
	movement of water vapor	evaporation	lower
plastic sheeting	heat absorption	soil temperature	higher during daytime and nighttime
	movement of water vapor	evaporation	lower
white dust, light soil color, dry soil	reflection	soil temperature	lower during daytime
		light	more intense
dark dust, dark soil color	reflection	soil temperature	higher at night

plant cover	absorption	radiation	lower near ground
		soil temperature	less contrast between day and night
		evaporation from soil	lower
plastic cloches	radiation	soil temperature	higher day and night
	movement of water vapor	evapo-transpiration	lower
light trellis wall	reflection	radiation of light	higher during daytime
hedge	wind	wind speed	lower
		evaporation	lower
		rainfall	often higher

2 Soils and Soil Building

If healthy food is to be produced in a healthy countryside, then soil conservation and a healthy soil life must be one of the main concerns of those who work the land. When bio-dynamic work began in 1924, agriculture in general was chiefly concerned with supplying plants with soluble nutrients. Since then soil research has come up with much valuable information on soil structure, soil microorganisms, the role of humus, etc. Yet, in agricultural practice today, methods of fertilizing, appreciation of humus, crop rotation and tilling often conflict greatly with what is now known about soil biology. This knowledge is, however, valuable for soil building in bio-dynamic farming and it is advisable to gain an idea of the complex relationship between soils and plants.

Soil characteristics and growth

1. Depth: The root space stores water and nutrients. The depth to which roots can grow is limited by rocks, hard pans, plough pan, clay pan, poor aeration due to a high clay content, poor structure or waterlogging, dead subsoil, anaerobic rotting of organic substances caused by poor aeration.

2. Infiltration: Surface run-off decreases and infiltration increases the replenishment of soil and groundwater reserves. Infiltration is slow on capped, heavy soils and rapid on sandy, crumbly soils.

3. Water-holding capacity: Unlike drainage, the water-holding capacity is lowest in sandy soils and increases through loam, silt, clay and finally peat where it is greatest. In the same sequence, however, the amount of water which the soil holds more strongly than the suction of the roots increases. The "usable storage capacity," i.e., the difference between water capacity and water content at wilting point (= water content at pF 4.2)* is thus 5-10% of the soil volume in sandy soils, 20-25% in loamy soils, but only 10% in clay soils (because of the high wilting point). The share of usable water can be improved by a porous soil structure. This is achieved by good humus management, root penetration of the subsoil (by deep-rooting plants such as clover, clover grass), activity of earthworms, stable aggregates, careful tillage, gradual deepening of topsoil, good lime content. To put life in the subsoil with the help of deep-rooting plants is one of the most important measures for achieving high and reliable yields. The higher the usable storage capacity, the less are yields dependent on precipitation during the growth season.

4. Drainage: Porosity of soil also prevents waterlogging. Hard pans limit deep root penetration. The soils above them easily become waterlogged while during droughts their water store is soon exhausted. Only deep roots can guarantee an even supply of water to the plant.

5. Aeration of the soil: Roots and soil microorganisms need a

---

*The pressure that must be applied to remove, through flow, water from the soil is measured in "pF units"; pF 4.2 corresponds with the water content that plants are no longer capable of using and they begin to wilt.

constant supply of the life element, oxygen. Loose, crumbly structure and the absence of waterlogging ensure good aeration.

6. Soil temperature: Soils that are porous and not too wet warm quickly. Spring planting in good time lengthens the growing period.

7. Nutrient and lime supply: The store of plant nutrients in soil depends on:
   a) its silica, clay and calcium content;
   b) quality and amount of humus present;
   c) fertilizing with compost, stable manure, crop residues, commercial organic materials, ground rocks, slowly soluble minerals, e.g., phosphates, etc.

From this store, the available nutrients are "mobilized" by the soil microorganisms with the help of the soil moisture. If manuring is done properly, the soil microorganisms will ensure a harmonious supply of nutrients to plants, deficiencies and excessive amounts are avoided and loss through leaching stays small. Good base saturation and desirable pH are maintained with the help of compost, farm manure, good aeration of soil and occasional applications of lime.

8. Soil microorganisms: Many effective organic substances are produced by roots, soil bacteria, soil fungi, etc., while they are alive and also left as residues when they die. These substances are also contained in compost. Among them are enzymes, vitamins, antibiotics and a number of nutrients. Their significance for yield, quality and pest control has as yet hardly been recognized. The more active, numerous and varied the bacteria, fungi, worms and other animal and plant organisms in the soil can be, the better can the plant and animal pests in the soil be contained within their normal rotation and organic fertilizing.

Increasing mineral fertilizers can for a time apparently conceal weaknesses in soil life and soil structure. But relapses eventually occur in the form of unreliability of yield, decreased effectiveness of fertilizer applications, reduced quality in products, and outbreaks of pests or diseases in field or stable. That bio-dynamic farms with the amount of plant nutrients they use, which is considerably lower than usual, can achieve good and in some cases excellent yields, is due to their all round development of all aspects of fertility. This includes choice of crop, organic fertilizing and tillage. Furthermore, it is necessary to have a thorough knowledge of the soil, which can be judged in three categories:

1. Warmth, air and water supply in the root space.
2. Supply of nutrients to plants.
3. Protection of crops against bacterial, fungal and animal pests and diseases.

So many excellent books about soils are available today that we shall here only go into a few points that immediately touch on the theme of this book.

Rocks and minerals in soil formation

These are evaluated according to their three main constituents:

1. Quartz, also called silica or silicic acid, SiO_2. Sand usually consists of quartz.
   Promotes aeration and warming of the soil.
2. Clay, a so-called silicate consisting of quartz, alumina (aluminum and iron oxides), and the alkaline elements of calcium, magnesium, potassium and sodium.
   Promotes supply of nutrients and water to plants.
3. Lime, calcium carbonate, $CaCO_3$.
   Regulates, together with the fore-mentioned alkaline elements in the silicates, most biological and chemical processes, and stabilizes humus substances and soil structure.

The following tables encompass some data on soil-forming rocks and minerals.

Sedimentary rocks		
loose	solidified	remarks
loam		mixture of sand and clay, often with lime

135

loose	solidified	remarks
clay	shale	makes for heavy, clay-rich, often wet and cold soils
alluvial sediments		e.g., in valleys, from soils in the catchment area, in deltas, along coasts, etc.
sand	sandstone	silica, becomes increasingly rich in nutrients, depending on whether the binding agent consists of silica, iron, clay or lime
pebbles	conglomerate	from other weathered rocks
angular debris	breccia	
glacial till boulder clay		ice age deposits of limestone and silicates weathered from the surface; lime may or may not be leached from the top layer
glacio-fluvial sand and gravel		primarily quartz, poor in nutrients
loess		wind deposits of fine sand, silicates and lime
lime tuff	limestone marl dolomite	mainly animal and plant deposits; limestone with more than 30% clay calcium magnesium carbonate
bog soil		mainly organic matter (15-30%), eutrophic fen, rich in lime and nutrients, or acid low moor
peat		raised moss, sphagnum peat moss, acid, poor in nutrients

Ninety-five per cent of the earth's crust consists of primary, that is, igneous and metamorphic rocks. Seventy per cent of the earth's surface, however, is covered with sedimentary or secondary rocks that have come into being as a result of the weathering of the primary and also older sedimentary rocks.

Igneous and metamorphic rocks (figures after Schroeder 1969)				
1. rocks				
course-grained	granite		diorite	
fine-grained	quartz-por-	melaphyre	diabase	basalt
porphyritic	phyry			
2. their mineral composition:				
a) quartz	20-30%		< 1%	
b) clayforming silicates:				
orthoclase (potassium) feldspar	30-50%		< 1%	
plagioclase (calcium-sodium) feldspar	30-40%		50-60%	
augites, hornblendes, olivine	1%		45-60%	
mica	5-8%		< 1%	
c) apatite, containing phosphorus	traces			
3. their most important elements:				
silicon	30-35%		20-25%	
calcium, magnesium	2-8%		10-15%	
potassium	4-6%		< 1%	

4. characteristics:

color	light	dark
specific weight	'light'	'heavy'
character	'acid'	'alkaline'
	rich in silicic acid	rich in calcium and magnesium

5. Metamorphic rocks with varying composition:
gneiss, mica schist, glauconite, phyllite, schist

Textural classification of soils

Soil textures differ depending upon whether the rocks are more silicious or whether they contain more clay and lime. They have been classified according to the size of mineral grains: clay 0.002 mm; silt 0.002-0.02 mm; sand 0.02-2 mm. Larger particles are called grit, pebbles, stones. Distinction is made between:

sand	coarse textured soils
sandy loam	
loamy sand	
silt loam	medium textured soils
loam	
clay loam	fine textured soils
clay	

The more sandy a soil is, the greater is its silica content; the more clayey it is, the greater is the content of silicate. Water seeps into sandy soil rapidly, and aeration and warming are favored. Clay and also silt make for finer pores in the soil and thus greater retention of water, in extreme cases lack of aeration, water-logging and difficulty in working.

Quartz contains no plant nutrients. These are supplied by clay and clay-forming silicates. The soil's supply of potassium occurs mainly in micas, potassium feldspars (orthoclase), and plagioclase. Calcium, magnesium and some potassium also come from calcium feldspars, augites, hornblendes, dark mica, chalk and dolomite. Other elements also come from the silicates, e.g., iron, which accounts for 1-4% of the soil, manganese, zinc, copper, molybdenum. Natural phosphorus is contained in small amounts in rock as apatite.

Without humus and soil microorganisms, however, the wealth of nutrients in soil minerals would remain inaccessible to plants. Clay and loam without humus are compact and hardly penetrable by roots. Sand without humus is almost dry only a few hours or at best days after a rainfall. Only organic substances and soil microorganisms make the dead minerals suitable for growth. Humus makes heavy soils more loose and gives light soils some physical characteristics of "heavier" soils, e.g., improves their water holding capacity.

Soil life and humus

An organism constantly exchanges its substances, yet nevertheless retains its shape. Soils contain characteristic amounts and qualities of organic substance that are continuously turned over. It is just this continuing turnover that is the source of all renewal in soil fertility, a fact not always properly appreciated.

What do we mean when we speak of organic substance, or humus?

The word denotes three things:

1. *Stable humus.* This gives the soil its specific dark brown coloring. It arises from the activity of microorganisms in the effective humus. When we speak of humus content or humus quality we mean this stable organic substance that has been created during the course of many years or even centuries. It is, though, not entirely stable, as is shown by the partial breakdown of the permanent humus, to be seen in many places, which has been brought about by agricultural malpractices.

2. *Effective humus*. Roots, leaves, twigs, needles, stubble, straw, dead grass, beet tops, potato tops, weeds, compost, animal manures, green manure and dying soil organisms are "mineralized" and become water, carbon dioxide, and plant nutrients such as ammonia, phosphate, calcium, etc. Only a small proportion remains as a stable humus.

3. *Edaphon*. To this belong microorganisms including such "specialists" as nitrogen fixing bacteria, nitrate forming bacteria, sulphur oxidizing bacteria, etc.; furthermore there are actinomycetes, fungi, algae, earthworms and some animal organisms such as protozoa, lower invertebrates, annelids, mites, springtails and many other soil insects. The live (not dry) weight of these organisms varies between 5,000 and 25,000 kg/ha (4450-22000 lbs/acre).

There are abundant varieties of humus on the earth. Their proportion in the soil and also their value for growth (humus quality) depend on:

1. *The silica, clay and lime content of the soil*. Lime and clay-rich soils produce grasses, herbs, clovers and also leaf litter; these are rich in proteins and minerals. Their residues produce nutrient and nitrogen-rich stable humus. Quartz-rich soils generate a more fibrous plant cover that is poorer in nitrogen and minerals but rich in carbon, for instance, coniferous woods, heathland, etc. The resulting humus is poorer and acid.

2. *The climate*. In temperate and more so in warm zones, humid climates favor rapid growth that supplies raw materials for humus in large quantities. Cool climates, and above all periods of frost or drought, delay decomposition and transformation into stable humus. This gives rise to an accumulation of only partly transformed humus materials. Waterlogging and lack of aeration in the soil also inhibit the life processes and lead to an accumulation of unfavourable forms of humus. In warm climates, decomposition is accelerated. Because of the low production of vegetable matter in drier areas, the soil has a low content of organic

140

matter. Balanced medium temperature and humidity favor the formation of valuable humus varieties in sufficiently large quantities.

3. *Cropping and soil management.* Wherever plough or spade loosen virgin prairie or forest soil, or even grassland or hay-fields that have been kept for several years, thus allowing more air to penetrate the soil, humus decomposition is accelerated. Nutrients in the humus and minerals are mobilized and this benefits the cultivated plants which may be subsequently sown. However, if the soil is not to become exhausted, the speed of decomposition must be matched by replacement through manures and humus-producing plants. Proper field cultivation means intensifying substance turn-over through decomposition and recycling.

How much humus should the soil contain?

In the topsoil of most arable land the amount of humus usually varies between 1.5 and 4.5%. As a group these soils are also referred to as "mineral soils" as opposed to moor or peat. The latter cover only a few percent of the earth's surface. 1.5-4.5% of humus is about 45-135 t of organic matter per hectare (18-54 t/acre). Grassland and forest have a higher proportion, bogsoil (eutrophic fen and acid low moor) has 15-30% and peat or raised moss over 30%. It is the aim of the bio-dynamic method to raise the humus content of the soil above a certain threshold level. Beyond this level unharmonious plant nutrition, deficiencies and lack of structure formation hardly occur at all; plants remain healthy. This only applies, however, so long as manuring is continually carried out properly. Below this minimum level it is hardly possible to manage without mineral fertilizers. Growth disturbances caused by the weather, poor soil structure, etc., appear frequently. Sandy soils should contain over 2%, loamy soils 2.2-3.5% and clayey soils 3-4.5% stable humus. The quality also plays a part, however. Fibrous humus, which often occurs in higher percentages, is found in soil rich in silica, in cool, humid climates, at higher altitudes, and in northern latitudes. Mature, nitrogen-rich humus is an even, dark brown color and there are no noticeable plant structures. It forms a unit with the soil's mineral substances to which it is life-giving. This is the protein-rich humus of good arable and grassland. The difference between carbon and nitrogen-rich varieties of humus is often expressed in figures as the carbon : nitrogen ratio. In the topsoil of good arable land the C : N ratio is 10-14 : 1, in poorer soils it is 15 : 1, and in heath and coniferous forest soils it is 15-30 : 1. The following table contains a number of indications about the role of humus in plant growth.

The influence of humus on soil fertility		
1. Organic soil substances	plant residues animal excretions	
	soil life	
	effective humus is decomposed and nourishes the microbic population of the soil	stable humus is permanent and combines with the soil minerals
2. Influence on nutrient supply to plants	microorganisms fix nitrogen out of the air; effective humus gives off nitrogen, phosphorus, sulfur and other mineral substances during mineralization; effective humus mobilizes nutrients including trace elements from the soil minerals	stable humus contains supply of nutrients that are gradually released (old fertility). It stores available nutrients and prevents leaching

3. Soil structure	microbic population of the soil helps to create and stabilize soil structure	stable humus makes light soils more plastic and increases their water-holding capacity; makes heavy soils more porous, improves aeration and their warming-up; it forms the crumbly structure of the topsoil
4. Health and quality	effective humus supplies plant hormones, antibiotics, and stimulates biological activity that inhibits the spread of pathogenic microorganisms	stable humus contains numerous active substances, stimulates biological activity; it includes a rich variety of microorganisms that inhibit the spread of pathogenic organisms

To supplement this table we shall quote a few sentences from Krasilnikov's book *Soil Microorganisms and Higher Plants* (Washington, D.C. 1961) that deal with the varied significance of soil life. He holds the view that plants growing in good humus soil are far less dependent on mineral nutrients than is usually assumed. This view, that in humus-rich soils organic substances play a much greater part in supplying nutrients to plants than the adherents of the mineral theory believe, is supported by recent observations made by others.

The biologically active substances of the soil not only enhance the growth and increase the yield of plants but also confer on the plant better nutritional qualities. The fact that plants can grow in pure mineral nutrient media in the absence of microorganisms cannot serve as a proof of the uselessness of the latter in the nutrition of plants. Plants can indeed be grown in mineral media and yield seeds without the participation of microbes. Is it possible, however, to secure the vitality of such plants in subsequent generations? We have presented our observations on the growth of green algae and duckweeds. Grown under sterile conditions, in mineral media without the addition of composts or metabolic products of bacteria, these plants lose their vitality and eventually die out. Plants grown in the same media but supplemented with composts or metabolic products of bacteria (non-sterile conditions), have been kept in our laboratory for more than twenty years without any visible decrease of vitality. The assumptions of some authors concerning the fact that soil contains only small amounts of organic compounds of phosphorus and nitrogen, which cannot, therefore, be considered as nutrients of any great importance, are also groundless. The knowledge we do possess allows us to assume that the processes of synthesis of various organic compounds proceed incessantly in the soil. Owing to such uninterrupted synthesis (be it as it may in small amounts) the total production of these compounds may be sufficient to meet the needs of plants. Bacterial life in the soil is known to be very short: it is counted in hours. Even during their life, the dying cells are subjected to autolysis. The selective action of the vegetative cover may be directed toward the selection not only of useful, but also of harmful microflora. Where agrotechnical rules are not observed, with an incorrect choice of crop rotation, the fields are contaminated by phytopathogenic bacteria and fungi and other harmful microbes and weeds. Especially after prolonged repeated cultivation of plants on the same field in monocultures, one observes this effect. The accumulation of an undesirable microflora under monocultures is most often caused by the insufficient growth of microbial antagonists in the rhizosphere. Many outbursts of epiphytotic diseases, as for example fusarial infections of cereals, cotton, saplings of woody plants, as well as cotton wilt, are in our opinion caused by the above phenomenon. This was proved by microbiological studies made of cotton fields which where afflicted by the fungi *Verticillium dahliae* and *Fusarium vasinfectum*. These fungi are removed as soon as mycolytic bacteria begin to grow in the soil. These bacteria, as was shown above, grow under lucerne and certain grass mixtures. In agricultural practice since time immemorial crop rotation has been used as one of the methods of increasing crop production. It was empirically found that with certain crop rotations, not only crop production increased, but disease decreased.

Recycling organic matter

To recycle organic matter is, as has already been shown, the mainstay of good farm organization and management. To do it successfully often needs years of observation and careful consideration. Nowadays a considerable body of experience is documented and available in tables and text books. Such figures can be a valuable aid in assessing the state of one's own farm. The planning of the recycling of organic matter starts from the following points:

1. Cropping system, i.e., the proportion of organic matter exhausting and organic matter restoring crops and their influence on nutrient supply and soil structure. To maintain a balance a certain amount has to be recycled.

2. The required amount is met by
   a) cropping (crop rotation, intercropping and undersowing cereals, arable forage crops, leys)
   b) farm manures
   c) bought-in materials.

When planning for the farm's needs with regard to organic matter, it must not be forgotten that to replenish humus it is not only the amount of organic matter that is important but also its subsequent fate in the soil. In particular it must be taken into account that the loosening and aeration of the soil resulting from tillage accelerates the decomposition of effective humus. On the other hand, periods when the soil is not ploughed favor the building-up of humus. The same applies to mixed crops rather than stands of one crop. Lucerne or clover grass fields of several years standing are most effective crops for improving the state of humus. Materials that break down readily or evenly distributed fertilizer, e.g., young green manure plants or liquid manure,

145

undergo rapid and almost complete decomposition. Older, matured parts of plants or green manure that has been growing over the winter are more slowly decomposed and contribute, especially when worked on by soil-inhabiting animals, to the restoration of humus. Animal manure, particularly manure well rotted in heaps and also composted manure, are significant for their longer lasting effects on soil fertility, though the immediate nutrient effect of fresh or liquid manure is higher. This often makes it tempting to give preference to the latter. It is often said that with fresh manure or straw manure more organic matter is added to the soil than is the case with heap manure or composted manure. This is numerically correct, but the humus building effect need not necessarily be positive. Also some organic matter is lost when the manure is stored in piles and so is some of the nitrogen. But little attention is paid to the fact that these losses are merely transferred to the fields where even the decomposition of soil humus is likely to be stimulated as a result. The net consequence over a number of years of applying fresh manure is no increase, or even a decreasing humus content in the soil, whereas applying composted manure results in a slow but steady increase of soil organic matter up to a climax state. In general it can be relied upon that mature stable manure has less direct nutrient effect but is a greater stimulus to soil building than is the case with, for instance, liquid manure or fresh stable manure. If soil fertility is poor, yields will of course depend on relatively fresh or half fermented manure. This only goes to prove that one must improve the soil with humus restoring crops and as much well rotted manure as one can possibly produce.

In the intensively farmed fields of the loess regions, humus content is often only 1.5% or less. This is too low. Good cropland in temperate humid climates should contain 2.5-4% humus in the topsoil, depending somewhat on the habitat. Copious fodder crops on arable land in highland regions lead to humus contents of over 4%. 2% humus in the topsoil is equivalent to about 60 t/ha (24 t/acre); 4% humus corresponds to about 120 t/ha (48.5 t/acre) of humus dry matter, not counting the humus content of the subsoil. The rate of decomposition depends primarily on the crop rotation system. According to Scheffer, 15, 25 or 35% root crops in a rotation use up 2, 3 or 4 t/ha (0.8, 1.2, or 1.6 t/acre) of humus dry matter per annum. In impoverished fields such amounts are not even available as decomposable matter, so growing such crops means making do with less than the minimum requirement of humus.

Root crops themselves leave crop residues of only about 0.8-1.0 t/ha (0.3-0.4 t/acre) with about 20 kg/ha (18 lbs/acre) of nitrogen. Cereals leave 1.5-1.7 t/ha (0.6-0.7 t/acre) with about 15 kg/ha (13 lbs/acre) of nitrogen. Clover grass left for a year or more leaves residues of 4.6-5.0 t/ha (1.6-2.6 t/acre) including 60-200 kg/ha (53-178 lbs/acre) of nitrogen. Added to this are a further 100-230 kg/ha (89-205 lbs/acre) of nitrogen, that remain on the farm if it uses its own fodder. Selling lucerne hay or lucerne meal is one of the quickest ways to deplete a farm's store of nitrogen, calcium, potassium, phosphorus and other minerals. Five tons of straw contain about 3.5 tons of organic matter, though this can only become effective if used chopped in conjunction with clover undersowing. On loams that contain sufficient lime, black medic (*Medicago lupulina*) is excellent for this purpose, otherwise other legumes are used.

The annual application of 10 t/ha (4 t/acre) of stable manure, corresponding to 30 t/ha (12 t/acre) in three years, adds 2 t/ha (0.8 t/acre) of humus dry matter. Ten tons of stable manure is roughly what is produced by one cattle unit (CU) in a year. The amount of 10 t/ha (4 t/acre) needed on the farm each year could thus be supplied by a proportion of 1 CU/ha (1 CU per 2.5 acres) if livestock were held in stables all the year round with medium amounts of bedding. More figures will be given when the different fertilizers are discussed.

Bought-in organic materials, except actual manure or composted urban wastes, are less effective for the replenishment of organic matter in the soil unless used to improve the development of humus restoring plants, particularly fodder crops.

Soil Types

A ditch dug with vertical sides down from the surface to

existing little changed or unchanged rock reveals a profile with layers called horizons that are designated by capital letters A, B, C, G, S, T, etc. The A horizon at the top is the loosest and contains the most humus and plant roots. This is followed by horizons that are bleached and impoverished through leaching, or enriched by deposits from higher layers, or mottled from waterlogging, etc. In short, a profile is a history book of the way the soil came into being. It shows the formation of humus, the enlivening of mineral substance, root development, aeration, natural drainage, i.e., a great deal more than the mere division into textural soil groups. This modern system of classification according to the characteristics mentioned and the manner in which they came into being gives us the soil types. Some of the more frequent types are rendzina, brown-earth, para-brown-earth, podsol, chernozem, pseudogley, gley. For more information see a modern textbook on soils. In the USA, the seventh approximation to soil classification has been suggested (Soil Survey Staff 1960). Anyone interested in soil will find in these modern systems of classification according to soil types not only an interesting subject for study but also a practical help. This classification takes into account the fact that rocks, climate, vegetation and in cultivated regions man himself, all play their part in the formation and development of soils.

3 Fertilizing

It would not be appropriate to restrict the debate between organic and non-organic methods in agriculture solely to contrasting farm manure with chemical fertilizers. As has already been seen, fertilizing is a part of the total farm organization, i.e., livestock, crops, crop rotations, natural habitat and so on. The subject of fertilizing takes up a large part of the space in R. Steiner's lectures on agriculture, though not much is said about mineral fertilizing. In the interests of clarity about what he means and what others mean, we shall quote the more important passages on this subject in full:

Revitalizing the earth: "Hence you will find it easier to

permeate ordinary inorganic mineral earth with fruitful humus substance, or with any waste product in process of decomposition—you will find it easier to do this efficiently if you erect mounds of earth, and permeate these with the said substance. For then the earthly material itself will tend to become inwardly alive—akin to the plant nature." In a controlled way this process takes place in a compost heap.

The life of plant roots and soil life: "It is simply untrue that life ceases with the contours—with the outer periphery of the plant. The actual life is continued, especially from the roots of the plant, into the surrounding soil. For many plants there is absolutely no hard and fast line between the life within the plant and the life of the surrounding soil in which it is living. We must be thoroughly permeated with this idea, above all if we would understand the nature of manured earth, or of earth treated in some similar way. To manure the earth is to make it alive, so that the plant may not be brought into a dead earth and find it difficult, out of its own vitality, to achieve all that is necessary up to the fruiting process. The plant will more easily achieve what is necessary for the fruiting process if it is immersed from the outset in an element of life. Fundamentally, all plant-growth has this slightly parasitic quality. It grows like a parasite out of the living earth, and it must be so."

Mineral fertilizers: "There is one fact that can already give you a strong indication of what is needed. If you use mineral, purely mineral substances as manure, you will never get at the real earthy element; you will penetrate at most to the watery element of the earth. With mineral manures you can influence the watery content of the earth, but you do not penetrate sufficiently to bring to life the earth element itself. Plants, therefore, that stand under the influence of mineral manures will have a kind of growth that betrays the fact that it is supported only by a quickened watery substance, not by a quickened earthy substance." In a parallel passage he says: "A second thing is this: As a result of materialistic tendencies, once more it has been thought well in modern times to treat the manure in various ways with inorganic substances—compounds or elements. Here too, however, people are learning from experience. It has no permanent value. We must in fact be clear on this: So long as we try to ennoble or improve the manure by mineralizing methods, we shall only succeed in quickening the liquid element—the water. Now for a firm and sound plant structure it is necessary not only to quicken and organize the water—for from water that merely trickles through the earth, no further vitalization proceeds.''

During one of the question and answer sessions it was asked what kind of potassium could be used in the transition period of a farm. The answer was: "Sulphate of potash magnesia.''

With regard to the usual farm fertilizers, Steiner made the following point: "Existing customs can surely be retained, at any rate to begin with. The point is simply to add what I have indicated. As to other usages of which I have not spoken, you surely need not begin by representing everything as bad—trying to reform everything.'' The question then followed: "What if one uses inorganic manures?'' "Mineral manuring is a thing that must cease altogether in time, for the effect of every kind of mineral manure, after a time, is that the products grown on the fields thus treated lose their nutritive value. It is an absolutely general law. Precisely the methods I have given, if properly followed, will make it unnecessary to manure oftener than every three years. Possibly you may only have to manure every four or six years. You will be able to dispense with artificial manuring altogether. You will do without it if only for the reason that you will find it much cheaper to apply these methods. Artificial manure is a thing you will no longer need; it will go out of use. Nowadays, opinions are based on far too short periods of time. In a recent discussion on bee-keeping, a modern bee-keeper was especially keen on the commercial breeding of queens. Queens are sold in all directions nowadays, instead of merely being bred within the single hives. I had to reply: No doubt you are right; but you will see with painful certainty—if not in thirty or forty, then certainly in forty to fifty years' time—that bee-keeping will thereby have been ruined. These things must be considered. Everything is being mechanized and mineralized nowadays, but the fact is, the mineral world should only work in the way it does in nature itself. You should not permeate the living earth with something absolutely lifeless like the mineral, without including it in something else. It may not be possible tomorrow, but the day after tomorrow it will certainly be possible, quite as a matter of course."

Since 1924, when these indications were given, opinions on the value of soluble chemical fertilizers have changed somewhat. Practice has remained virtually unchanged, however. Only the amounts have increased. The degree of doubt as to the rightness of present practice was expressed once again in a doctoral thesis (Jahn-Held 1971) that appeared recently. It mentions about 10 nitrogen or nitrogen phosphorus formulae with "slow solubility." The attempt is made to imitate by technical means the flow of available nitrogen compounds that originate in the life processes of the soil. This is now being extended to include "potassium compounds with slow solubility rates." In accordance with what has already been quoted from Krasilnikov (p. 143) it can be assumed that the varied effects of the biological complex and of mineral weathering are inadequately imitated.

Two points made in the passages quoted from Steiner should be emphasized:

1. Plants should grow in soil that is alive and permeated with organic substances that are gradually decomposing. This brings us to the following remarks about nitrogen and farm manure.
2. Mineral substances (which here obviously means bought-in mineral materials) should work as they would in nature.

Handling nitrogen on the farm

This is one of the essential points in any farming method. Nitrogen (chemical symbol: N) in gaseous form accounts for 79% of the air. It is not a plant nutrient in this form. To be a nutrient it has to be available as a compound. The two fundamental substances for any life on earth, protein and humus, are organic nitrogen compounds. The nitrogen in humus is also called organic nitrogen. Plants are able to absorb only minute soluble amounts of this. So far as is known at present, the main nitrogen supply for plants is based on ammonia (NH$_4$) and nitrate (NO$_3$) that arise during the mineralization of effective humus and to a certain extent also of stable humus. Urea is a dead product excreted by living organisms. When synthetic urea is used as a fertilizer, the effect is no different from that of other mineral nitrogen fertilizers. Indeed, because of its high nitrogen content (46%) urea is a fertilizer that brings rapid results.

Of all plant nutrients, nitrogen has the most significant effects on yield, nutritive quality, taste, health value, keeping properties, suitability for processing, e.g., baking qualities of grains, disease and pest resistance. The natural fertility of good soils that has been built up over the centuries is based to a great extent on the nitrogen content of the stable humus. In soils, plants and manures, nitrogen is a highly mobile element. It constantly changes from one compound to another, from humus nitrogen to ammonia, from ammonia to nitrate. It is also fixed from the dead nitrogen in the air and can also escape again in different ways into the atmosphere. This mobility is also the reason why nitrogen escapes the usual soil analysis methods used for available potassium, phosphorus or trace elements.

In a simplified form the following diagram shows the passage of nitrogen through the farm.

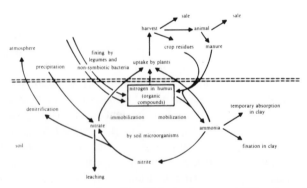

The nitrogen cycle (after Von Wistinghausen 1971, altered).

Ammonia arises through the microbial decomposition of humus. One part is gradually absorbed by plants. Another is transformed into nitrate by bacteria. A third is transformed back into humus together with strawy harvest residues. Some may be stored or temporarily absorbed by clay and humus; some is also permanently fixed in clay minerals.

Nitrate is taken up by plants, or leached, or in wet soils with poor aeration undergoes denitrification, which is also a bacterial process, and escapes into the atmosphere as gaseous nitrogen.

Organic nitrogen is brought into the cycle by farm manure, crop residues and nitrogen fixing bacteria. Only liquid manure or floating manure contain worthwhile amounts of ammonia; rainwater contains traces of nitrate.

Organic nitrogen is exported from the farm by the sale of products containing protein, leaching, denitrification and other similar gaseous losses.

The nitrogen household of the soil is a life process of which the mineral forms, ammonia and nitrate, are only passing phases. What still remains to be asked is how the supply of nitrogen that is necessary for improving both yield and quality can be secured. Only a general framework can be given. Experience and practice will show by the yield, the appearance of the plants and the way they grow whether the supply of nitrogen is right.

The topsoil of an arable field contains between 0.05 and 0.3% organic nitrogen. This is about 1500-9000 kg/ha (1330-8000 lbs/acre). The annual rate at which ammonia and nitrate are formed out of this is 1-2.5%, the higher amount occuring where there is more organic nitrogen in the first place. This means there is in the soil a supply of ammonia and nitrate of between 15 and 230 kg/ha (13-205 lbs/acre). In addition 10-15 kg/ha (9-13 lbs/acre) of nitrate are supplied by rainwater, and with an average application per year of 8-10 t/ha (3.2-4.0 t/acre) of stable manure come a further 40-50 kg/ha (36-44 lbs/acre). (An application of 24-30 t/ha (9.7-12.1 t/acre) every three years contains 120-150 kg/ha (107-134 lbs/acre) of nitrogen, making available 40-50 kg/ha (36-44 lbs/acre) in the first year. The

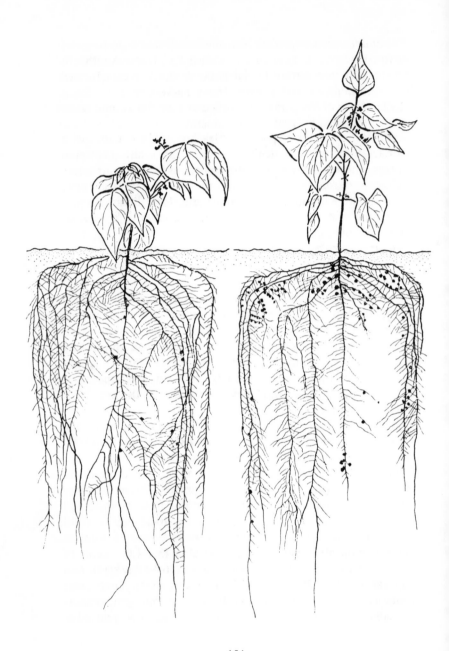

remainder becomes available in the two following years or is absorbed in the soil nitrogen.) This is the framework within which crops have to cover their requirements, which lie between 60 and 130 kg/ha/year (54 and 113 lbs/acre/year). We thus see that in addition to stable manure, increasing the nitrogen capital of the soil is a prerequisite for an adequate supply to the plants.

What figures can be used as guidelines for the annual replenishment of the soil's reserves? Clover-lucerne crops leave 60-200 kg/ha (53-157 lbs/acre) of nitrogen; pulses (large-seeded legumes) leave 30-40 kg/ha (27-36 lbs/acre). Farm manure (assuming there is a 20% loss) supplies 50-60 kg of N per year per cattle unit (110-132 lbs N per year). Non-symbiotic nitrogen-fixing bacteria supply 10-15 kg/ha (9-13 lbs/acre); rain brings 10-15 kg/ha (9-13 lbs/acre) from the air; crop residues from crops other than legumes contain 15-20 kg/ha (13-18 lbs/acre). These additions accumulate in the framework of the crop rotation during the course of several years and from the reserve in the soil.

Diagram:

French beans (*Phaseolus vulgaris*), "Sabo" strain, grown in special vessels that allow observation of root growth.
Left: Garden soil, once weekly 0.4% NPK in watering water.
0.4 0/00 NPK
Right: The same soil with compost (J. Bockemuhl).
These and the following illustrations of plants in special vessels allowing for root observation are from work at the Research Laboratory at the Goetheanum, Dornach, Switzerland, some of which is as yet unpublished.
The number of nitrogen fixing nodules shows how soil rich in compost promotes the plant's own nitrogen fixing activity. With N fertilizer the plant grows profusely but with little tendency to uprightness. The plant in compost, which is the same age, shows a harmonious balance between the vertical and horizontal. The number of buds was similar in both plants.

Nitrogen losses in individual fields, apart from the nitrogen content of the harvested crop, are brought about by leaching and evaporation of gaseous nitrogen (denitrification). Nitrogen is leached in the form of nitrate during periods when the soil has no plant cover. 20-60 kg/ha (18-53 lbs/acre) are often lost during autumn and winter. That is 20-80% of what is needed for a good harvest. The most important measure to prevent nitrogen leaching is undersowing and catch cropping in autumn and winter. Stable manure ploughed in early in autumn can, if the weather is warm and dry and thus favors nitrification, lead to nitrate losses during the winter of up to 50 kg/ha (45 lbs/acre) of N according to the author's own findings. It is advantageous to spread manure in stubble or straw together with undersown clover, thus providing protection against nitrogen losses. Liquid manures undergo a rapid turnover because of their thin and even distribution. Considerable losses are the result if nitrogen is not stabilized by growing plants, including green manure and catch crops, or copious harvest residues. The investigation in the American corn belt referred to on p. 55 shows that organic management alone can reduce nitrate leaching to tolerable levels.

Probably the most thorough investigation into the influence of agriculture on the nitrate pollution of water has in recent years been undertaken in Illinois, USA. Nitrate pollution rose approximately parallel with the use of mineral fertilizers, and modern isotope techniques revealed the probability that in that area about 55% of the leached nitrate originated in mineral fertilizers. Nevertheless, taking all the circumstances into account, the experts considered it hardly possible to reduce drastically the nitrate pollution of water by introducing legal limitations on the use of fertilizers (Illinois Pollution Control Board 1972, Kohl 1971). The reason for this is simply that so long as no change is made in the present permanent cropping of corn or the "crop rotation" of alternating corn and soybeans, the existing situation would only change to a small proportion anyway. There are no fodder crops to reduce the amount of open ground that is bare of any vegetation for most of the year

156

and thus a prey to leaching of fertilizer and valuable soil nitrogen.

Nitrogen fertilizer in general agriculture is often given in amounts far in excess of need. Apart from leaching, this leads to gaseous losses of nitrogen in the atmosphere. In pot experiments, thus eliminating leaching as a factor, between 0 and 50% of nitrogen in fertilizers was not recovered either in the soil or the harvested crop. In field experiments, where leaching also played a part, up to 50-75% of nitrogen was lost. Data exist that indicate that 0-85% of nitrogen in fertilizer evaporates from the soil. The amount depends on the weather, the organic soil substances present, etc. Considering that it takes energy equivalent to 5 tons of coal to produce 1 ton of fertilizer, it is obvious that we have here another example of the wastage incurred in the profligate way we run our affairs. The greater the amount of fertilizer applied, the higher are the losses through denitrification. (Allison 1955, 1966; Fried and Broeshard 1967). Westermann and Hauck (1972), who experimented with Sudan grass and sorghum that they fertilized with labelled nitrogen in urea and oxamine, concluded that 20% was lost as gaseous nitrogen. They suspect that the loss is even greater in actual agriculture. Fliege and Capelle (1974) applied to spring barley and oats 80 and 74 kg/ha respectively (72 and 67 lbs/acre) of labelled nitrogen fertilizer. In these field plot trials the gaseous losses of nitrogen amounted to 51 and 31% respectively.

Taking the farm as a whole, it has been shown that losses through harvesting are considerably higher than the amounts actually exported from the farm. A cow that gives 5000 kg (11,000 lbs) of milk per year eats during the same period, which includes the time when she is dry, about 75 kg (165 lbs) of nitrogen in digestible protein or 115 kg (253 lbs) of crude protein nitrogen. 25 kg (55 lbs) are exported from the farm in the milk. So theoretically in the solid and liquid excrement about 90 kg (198 lbs) of N per year are excreted and should be returned to the land with as little loss as possible. A wheat harvest of 4 t/ha (56 bu/acre) contains about 120 kg/ha (107 lbs/acre) of N. About 80 kg/ha (88 lbs/acre) are exported from the farm in the grain.

A rapeseed harvest exports 60-70 kg/ha (53-62 lbs/acre), though the cakes can of course be taken back as feed.

Straw is too poor in nitrogen to be directly advantageous for improving effective humus (C:N ratio = 50-100:1). But over a period of time all the substances it contains do have their effect if as much of it as possible is expertly returned to the soil. Wheat straw contains about 3 kg (6.6 lbs)/ton of nitrogen; 2 kg (4.4 lbs) of phosphate; 15 kg (33 lbs) of potassium; 3 kg (6.6 lbs) of lime; and 2 kg (4.4 lbs) of magnesium (Nieschlag 1963). The content of these substances in barley, oat and rye straw is even higher, which means that these straws, even if they cannot be directly used as fertilizers because nutrients have to be added, nevertheless provide a considerable capital of organic matter and plant nutrients that circulate on the farm.

These considerations are of value only if the farm returns absolutely all its waste with as little loss as possible to the soil. If manures and fertilizers are handled wrongly and farm waste treated carelessly, irreplaceable losses of nitrogen and other nutrients occur. Proper handling of nitrogen is based on:

1. biological nitrogen fixation on the farm by growing clover and clover-grass mixtures and large seeded legumes;
2. return to the soil of absolutely all manure and waste by methods most suited to avoiding losses;
3. reduction of nitrogen leaching by undersowing and catch cropping.

For private gardens it is always possible to buy in enough organic fertilizers, either manure or commercial organic fertilizer. Market gardens depend on buying in and carefully composting materials containing nitrogen. It is advisable to alternate nitrogen consuming garden crops with crops that replenish the soil.

Against the above discussion of the nitrogen supply of farms it could be maintained that in general agricultural today, yields cannot be maintained without the use of mineral nitrogen fertilizers. A study of the traditional agricultural regions does indeed reveal that the organic content of the soil is rather low, being 1.5-2.2% humus corresponding to 2800-3500 kg/ha

(2500-3100 lbs/acre) of nitrogen in the top layer of the soil. Experience has shown that larger farms with 0.5-0.6 CU/ha (0.2-0.25 CU/acre), despite the copious use of mineral fertilizers, achieve lower yields, particularly with more demanding row crops, than do small farms with a larger proportion of fodder crops and 1.2-1.5 CU/ha (0.5-0.6 CU/acre). On light soils, poor sandy soils, etc., the build-up of a farm's own supply of nitrogen depends on avoiding all losses in the careful use of manure. In general such farms have to buy in organic fertilizers for a number of years.

A further point may be made here. It must be remembered that the nitrogen balance sheet of soil and plant is fundamentally altered by the use of mineral fertilizer nitrogen. Bio-dynamic farms routinely produce good average grain harvests. In contrast, the need for nitrogen where mineral fertilizers are used is disproportionately high. According to the evaluation of the relevant literature undertaken by S.R. Aldrich (1972), 37% more nitrogen is needed in the corn belt of the USA in a corn-soybean rotation than is taken out of the soil by the crop. This figure refers to "economically optimum yields." And the straw there is left on the fields. This optimum requirement corresponds more or less to the present average application of 130 kg/ha (118 lbs/acre) of nitrogen (these figures refer to Illinois in 1969), but not counting the nitrogen that is added via lucerne, clover and animal manure.

In the USA from 1945-1970 corn yields (America's most important grain crop) have increased from 2.3 to 5.4 t/ha (34 to 81 bu/acre). The energy requirement for nitrogen rose from 58,800 to 940,800 kcal/acre, and that for all mineral fertilizers together from 74,600 to 1,055,900 kcal/acre. In 1970, 32% of the total energy input for corn production was used for nitrogen and 36% for mineral fertilizer. The amount applied equals about the nitrogen content of the manure from 1 cow. Spreading 10 t/acre of manure takes about 398,475 kcal/acre. To produce and spread the same amount of fertilizers (112 lbs of nitrogen, 31 lbs of phosphorus, and 60 lbs of potassium) takes, including spreading, 1,451,425 kcal/acre. An economically significant amount of energy could thus be saved if farm manure were properly utilized. For supplying plants with nutrients, decentralized as opposed to mass livestock farming saves calories (figures according to Pimental and colleagues 1973).

Finally, one further example will show the relationship between nitrogen content and varying fertility in four soils of the North American corn belt. All four are high yield chernozem soils. The samples are taken from a farm that was an amalgamation of several farms, so the previous treatment of the fields was not uniform (Koepf 1964). The soil grading is based on several years' experience of the farm and its yields.

Soil fertility and nitrogen content of 4 soils, graded according to decreasing fertility				
	1	2	3	4
clay and silt content %	66.5	69.0	64.8	60.2
humus content 0-15 cm (0-8 inches) %	4.09	4.06	3.23	2.58
nitrogen content, topsoil %	0.180	0.200	0.175	0.154
nitrogen content 20-25 cm (8-10 inches) %	0.195	0.191	0.182	0.161
total nitrogen content to a depth of 48-55 cm (19-22 inches) kg/ha (lbs/acre)	8550 (7600)	7920 (7000)	6300 (5600)	5760 (5100)

Fertility is directly related to total nitrogen content. Field 4, which is the weakest, is partly eroded. Fields 1, 3 and 4 have lost part of their nitrogen content, as is shown in the comparison between topsoil and the layer below the ploughed layer. Only field 2 shows a normal decrease of nitrogen with increasing depth, which is as it should be. For a number of years field 1 was used for fodder crops and grazing. Field 2, having been used in a corn-corn-oats-clover rotation for forty years as part of a farm with livestock, is in perfect balance.

Compost

In European countries, where the land has been tilled in places for over 2000 years, it is generally recognized that it is necessary to add organic manure to the soil in addition to crop residues. In North America, where the land has merely been in use for a few hundred years or even in some places only for a few years, this is not held to be necessary. Scientists and farmers until recently were not inclined to believe that the proper storage and application of farm manure was worth the expense. It was cheaper, they used to hold, to use nutrients from a sack or tank. Of course, with the exception of the eastern States, the use of large amounts of mineral nitrogen fertilizer has only occured for the last 20 years or so. Many "experts" still regard the removal of animal manure as a matter of hygienic waste disposal. Because of the absurdly high concentration of livestock in factory farming the disposal of excrement is certainly becoming increasingly costly. Mass livestock husbandry in Europe—which is indeed being increasingly copied—is a far worse mistake than in the USA. Loss of soil fertility and increasing pollution will become noticeable far more drastically and rapidly in European countries. "Manure disposal" is now also an object of research in Europe because it is becoming such a problematical aspect of animal husbandry.

In all warm, temperate and cool climatic zones with sufficient precipitation (semi-arid to humid), proper humus replacement supported by the use of farm manure forms the backbone of any horticultural and agricultural production that sets store by quality, freedom from residues and lasting soil fertility. In warm, humid regions this is achieved by permanently keeping a green cover on the soil and shading it, since shade slows down the decomposition of humus. The main bulk of the matter returned to the soil is plant substance. In temperate zones, where intensive cropping is widespread, the aim is reached by a combination of crop rotation, catch cropping and animal manures. In the cooler, more northern zones the organic content of the soil is often sufficient but there may be a shortage of nitrogen. It is then a matter of supplying sufficient animal manure and other animal wastes. The method of organic matter replacement varies according to climatic zones.

This is not the place to discuss organic farming in hot, arid zones, so it will suffice here just to mention a historical point. In former times composts were also used in these regions. Older forms of agriculture had at their disposal liquid manures made from animal and plant wastes of all kinds, which probably achieved rapid effects. A further characteristic trait has already been mentioned, namely, the recommendation to let compost ferment for 3-4 years before use. This is out of the question in our time, but it is to the point in warm regions with little rainfall or in Mediterranean climates where humidity occurs mainly in winter, since here organic substances that are insufficiently rotted and therefore not sufficiently stable become entirely decomposed within a short time. In other words, expressed in modern terminology, those who used the land long ago knew how to improve the exchange capacity of their soils. (See reference on p. 57 to Ibn Al-awam's book).

Types of compost

Garden composts consist of fresh and wilted weeds, grass clippings, dead leaves, spent mulch, vegetable trimmings and other kitchen wastes. The proportion of soil they contain is usually quite high, up to one third or more.

Composted manure consists of solid and some liquid excrements of domestic animals with varying amounts of straw, old hay and other litter, sometimes even sawdust or wood shavings.

Garbage compost consists of domestic and some industrial waste, either with or without sewage.

In addition there are potting soils, leaf mould, and standardized soils usually used in market gardens. The latter, usually on a peat base, can also be produced in accordance with bio-dynamic principles. In this case one uses rock phosphate, sulphate of potassium-magnesium and the quantity as otherwise of organic nitrogen, mostly hoof, horn, etc.

However different the materials used, the consistency of the finished product or the various uses may be, the basic rules for the production and use of compost always remain the same. Yet, even though they are simple, mistakes are frequently made. The end result aimed for is an even, dark brown, humus rich, hygienic, living substance with a pleasant smell. Success with composting depends on:

a) the right mixture to start with;
b) controlled fermentation;
c) treatment with bio-dynamic preparations.

Point c) will be discussed in the next chapter while the principles regarding a) and b) will be dealt with here.

Compost mixture

The older parts of plants, including dead leaves and straw, have a low water content and consist mainly of woody substances, i.e., cellulose and lignin. They are rich in organic carbon. The younger parts of plants contain more water and, like legume straw, more nitrogen compounds. The right amount of nitrogen speeds up fermentation and produces a richer compost. Insufficient nitrogen delays fermentation and leads to an acid end product. So what matters is the ratio of carbon-rich (C) to nitrogen-rich (N) matter usually termed the C:N ratio. In the initial mixture it should be 25:1 and in the end product 15-20:1. Too much nitrogen in the initial mixture leads to losses that can usually be recognized by the smell of ammonia. Straw has a C:N ratio of 50-100:1; legume straw 15:1; oak and beech leaves 40-60:1; animal excrement 14-16:1; fresh stable manure depending on the amount of bedding 20-25:1; fermented manure and composted manure 15-20:1. In practice, of course, one does not actually measure the C:N ratio but instead one observes how the mixture works. In the following, raw materials that can be used are listed in order of increasing nitrogen content, i.e., the C:N ratio grows closer toward the end of the list:

waste paper; sawdust; straw; spent mulch; dry leaves; old hay; wilted green plants; fresh weeds; vegetable trimmings; young grass; legume hay; kitchen waste (excluding packaging).

Nitrogen-rich materials that are used to create a favorable C:N ratio:

fresh or dried animal dung; guano; blood, horn and bone meal; wool, hair and leather waste (rots slowly); fish waste; oil seed cakes or meal, etc.

As a rule, garden composts contain enough soil. It is favorable to have at least 5-10% soil in any kind of compost, but there should not be more than 30% except in the case of compost made of sods or similar materials. Topsoil is used, or subsoil that has been prepared and frozen through and kept damp. If the compost is intended for use on sandy soils, any added loam is useful. The soil should be fine and spread evenly and thinly in the heap so that all vegetable matter is in contact with a thin dusting of soil. Basalt meal can be used instead of soil, especially where the compost is to be used on light soils.

Sawdust, leaves, coniferous needles and green plants, i.e., waste that produces acid in fermentation, work better if lime is added to the mixture.

Stable manure or dried manure without too much straw is by far the most valuable addition to garden and garbage compost. The amount of chicken manure has to be limited, however, to a level where no putrefying, foul smelling patches occur in the heap. Fresh cow manure with bedding can be added in any amount. Dried manure can amount to 20-50 kg (50-110 lbs) per cubic meter or yard.

Controlled fermentation in compost

This occurs through the whole cross section of the heap if air permeates slowly and there is a medium water content. The right conditions are brought about by particle size, water content and shape and size of heap. Large lumps of earth, or layers of young grass or leaves or unwieldy sticks should be avoided. The latter should be mixed with finer material with particles ranging from the size of sand to 1-2 cm (about ½ an inch) in diameter. Evenly mixed material is better than layers. Even for medium-sized gardens it is worthwhile obtaining a mechanical shredder, which need not be too costly. Water content should be between 50 and 60%. Everything is nice and damp: leaves, stalks, etc., are fully steeped in water but nothing can actually be squeezed out. Proper watering must be done during the building of the heap, not afterwards. Dry litter must be repeatedly soaked a few days before the building of the heap. When the materials are coarse and contain less water the heaps are larger than when the material is wet and fine-grained.

Turning and the mixing this entails promotes the fermentation process. Properly built heaps in garden or farmyard should need only one turning or not at all. For large-scale composting turning is done mechanically. Space and rapid turnover are factors of importance only in large-scale commercial composting while they are not too urgent in private gardens or farms. If the outer layer of the heap (about 20 cm or 8 inches) is too hot and dry, it is sufficient to loosen and dampen this layer. More water for the heap as a whole can only be added during turning. If a foul-smelling blackish-blue centre core forms, the only thing to do is loosen the whole heap by turning. Material that is too loose can be dampened and the heap made larger. Sometimes twigs, boards or pipes are used to make air tunnels along the inside of the heap. Watering through holes slanting inwards from the sides is also frequently mentioned.

A thin cover of straw or old grass helps protect the heap from too much drying by the sun or wetting by rain.

It is an advantage to set aside a permanent site in the garden for the compost heap, since a population of organisms that help fermentation will become established. The spot should be well drained and if possible slightly sloping. Waterlogging from running or stagnant water should be avoided. Three-quarters shade is best and, if planting has to be done, elder, hornbeam and hazel shrubs are especially suitable. Nothing is planted on the heaps themselves. Heat is only generated properly if the proportion of soil in the heap is not too high. The temperature rises to 30-50°C (86-122°F) within a few days and then gradually reduces again until it is more or less that of the surrounding air.

If the heap is not damp enough it may overheat and possibly dry out too much. Garden and garbage compost needs 6 to 18 months for fermentation, depending chiefly on the amount of organic nitrogen in the mixture.

Manure composts usually ferment during a period of 6 weeks to 6 months, actually mainly depending on when they can be best applied to the land. Composts should have finished phase I of heating and rapid decomposition and entered phase II of humus formation and ripening. On commercial composting sites and when digesters are used, the attempt is made—unsuccessfully— to stabilize the material in 5-7 days. Due to the heat, pathogenic germs can be killed during that time but stabilizing under these conditions can only be achieved by drying. When the material gets damp it starts to smell again.

Evenly brown, fine compost that smells pleasant can be used in almost unlimited amounts, either sieved or unsieved. After properly controlled aerobic fermentation it is perfectly hygienic. The seeds of some weeds can survive a short period of fermentation but most lose their ability to germinate. This is the case when the heap generates the proper amount of heat and is kept properly damp. Further indications on composting will be given in Chapter Eight.

Stable manure and liquid manure

The evaluation and application of animal manure in agriculture has been characterized by two trends over the last twenty years.

First, there is the concentration of large numbers of poultry, pigs and cattle in one place, so that the masses of dung that accumulate have to be "disposed" of. While it is known in the long and medium term approximately what it means to an individual farm to reduce or give up its livestock altogether and how intensive livestock farming affects the bank account of the private entrepreneur, no quantitative assessment exists of the short and long-term damage such practices do to a nation's economy. Meanwhile ways and means are sought by altering buildings and machinery to simplify dealing with the enormous amounts of bedding and manure that accumulate. All sorts of new inventions appear on the market and yet it will take years before the advantages and disadvantages both from the point of view of biology and of agricultural engineering will finally have been sorted out. So far mechanical barn cleaning, loading, spreading and pumping have indeed greatly simplified the handling of large masses of manure.

Second, in bio-dynamic farming emphasis is placed on the importance of animal manure for building up permanent soil fertility. The widespread view, that so long as organic matter is recycled it does not much matter what sort, is met with reservations. Organic and bio-dynamic farms do not do without livestock. The very basis given by nature makes it necessary to keep animals, for if land is not left actually fallow, some grassland is everywhere present. Anyway, the organic growing of crops is hardly possible in the temperate zone and also not meaningful without including the production and consumption of fodder on the farm itself.

In a cowhouse, 10-12 t/CU of manure are produced a year, depending on the amount of bedding. In a covered yard the amount is 14-17 tons per year. Ten tons of farmyard manure contain 40-45 kg (90-100 lbs) of nitrogen, 30-35 kg (66-77 lbs) of phosphate, 50-60 kg (110-130 lbs) of potash; and in addition calcium, magnesium and the trace elements manganese, copper, boron, zinc, cobalt, molybdenum. The amount of organic dry matter that can be called effective and stable humus is 2-2.5 tons. When bedding is used, the amount of manure from sheep, horses and pigs is similar, or in some cases a little lower. The following table (after Nieschlag 1963) shows the average proportion of components in animal excrement.

One reason for quoting these figures here is to show the considerable amounts of nutrients contained in liquid excrements that are often neglected: about 30 kg (70 lbs) of nitrogen per CU/annum and twice as much potassium. Loafing yard manure (with bedding) contains both the solid and liquid excrements. With sufficient bedding, good fermented manure is created in covered yards or loafing yards. This contains about

Average composition of animal excrement

		water %	nitrogen %	phosphorus %	potassium %	calcium %
solid ex-	cattle	80	0.3	0.2	0.1	0.1
crements	pig	80	0.6	0.5	0.4	0.05
	sheep	80	0.6	0.3	0.2	0.4
urine	cattle	90	1.04	0.10	1.5	
	pig	90	1.00	0.05	1.00	
	sheep	95	1.50	0.10	1.80	

75% water, 0.75% nitrogen, 0.35% phosphate, 0.75% potassium, 0.6% lime, and 0.2% magnesium. The amounts of nutrients in farmyard manure are somewhat lower: 75% water, 0.4-0.5% nitrogen, 0.3% phosphate, 0.5-0.6% potassium. Depending on the type of stabling and the amount of water used for barn cleaning, 3-5 cubic meters or yards of liquid manure are produced per CU per year.

Reference is here made to the systems of housing and manure handling as they are used by a majority of farms in continental Europe. The aim is to preserve as much as possible of the manuring value of the animal excrements and to avoid the pollution of ditches, rivers and streams.

The term farmyard manure refers to the solid excrements plus bedding as it is produced in cowhouses with standings and tying (occasionally also in cubicles or free stalls). The dung channel behind the standings includes a gutter for the urine and some washing water; these are collected in a storage tank. In America the system of separate handling of solid and liquid manures hardly occurs.

The term loafing yard manure refers to covered yards in which enough bedding is spread to absorb both solid and liquid excrements without any effluent running off the manure. The droppings without bedding scraped from yards or free-stall houses are mostly collected in pits and handled as slurry.

Occasionally they are piled up in an uncontrolled fashion and remain exposed to rain or drying by sun and wind. Considerable nutrient losses and pollution hazards occur in this case.

100 hens produce 6 tons of dung per year, containing approximately 90 kg (200 lbs) of nitrogen, 80 kg (180 lbs) of phosphate, and 50 kg (140 lbs) of potassium (Tietjen 1965). Dried chicken dung contains about 12% water, 3.6% nitrogen, 3.5% phosphate, and 1.6% potassium. Chicken dung from deep litter contains 32% water, 1.7% nitrogen, 2.1% phosphate, 1.3% potassium (Advisory Leaflet 320,1969). As these figures show, chickens on a farm produce valuable manure for gardens and fields. If free ranging is required, not more than a few hundred can be kept, however. Runs can be kept relatively green by subdividing and using the portions alternately. For floor houses for raising broilers an effective step against disease is to submit the litter to a short period of composting before re-opening the house with a fresh batch. As much water as necessary is added to the manure, which is then mechanically turned; a reasonable amount of good, mature bio-dynamic compost is added (Hancock and Escher 1965).

The wellbeing of the animals and the quality of the manure are both improved considerably by using appropriate amounts of bedding materials. The question of whether to go over to slurry usually only arises if one is in the process of altering buildings or building new barns. Even if the high cost of investing in slurry techniques is no deterrent, one should take into account that they bring with them far-reaching changes both for the livestock and for the soil life. Technical developments have made the handling of solid manure economically competitive from the labor point of view, so the biological advantages of solid manure can really be made the most of.

The majority of bio-dynamic farms use cowhouses with sufficient bedding material, mechanical barn cleaners and separate storage tanks for liquid manure. The manure is collected outside the cowhouses in flat or conical piles for intermediary storage and treated with bio-dynamic preparations. The compound preparations, described on p. 206, work well in starting the fermentation. When these compound preparations are spread in the barn they help control odor. At the earliest possible opportunity the manure is then transported to compost piles in the fields. The size of the heaps is the same as with other compost. Then they are again treated with the preparations. Soil up to about 5% of the total mass added to the piles greatly improves the manure. On light soils it is important to add loam or basalt meal. The heaps should be covered with potato tops, earth or other litter. The effluent that arises during storage near the cowhouses should be collected. Rock phosphate or bone meal can be added to the manure. 50 kg (110 lbs) per CU/annum is usually recommended. This technique for cowhouses is simple and proven and is even sometimes used where there are free stalls with bedding in the passage.

Loafing yard manure requires the right amount of bedding to prevent the manure from becoming either too dry or too wet. It is treated bio-dynamically in the stable. 6-12 kg (15-30 lbs) per CU of bedding daily are recommended. When well fermented, the manure is either taken straight to the fields for heavy feeders or stored in compost piles on the fields. N. Remer (1968) describes more developed composting processes.

In Scandinavian countries the following additions are often made to manure. Toward the end of fermentation, about 20 kg (45 lbs) of meatmeal and 10 kg (22 lbs) of bonemeal are added per ton of manure. It is not possible to use more than 10 kg (22 lbs) of sulphate of potash magnesia as higher amounts have a detrimental effect on fermentation. For plant composts, up to 40 kg (90 lbs) lime per ton of plant matter are added.

To use liquid manure successfully, enough storage space must be provided in advance. Existing storage is usually only 3 m³ (100 ft³) per CU. This should be increased to 5-6 m³ (170-200 ft³). The more recent above-ground storage tanks have proved to be quite usuable and of themselves lead to the two-chamber system, which is most desirable. One tank is used for collection and the other for fermentation, the contents being pumped from pit to tank. The addition of chopped stinging nettles and small amounts of mature composted manure has proved useful in the production of good liquid manure. The bio-dynamic preparations are suspended from a float in porous bags and thus held in the liquid. This treatment is also carried out on mountain grassland farms where there is a shortage of straw and all excrements are used as slurry. Vacuum tanks have proved useful for haulage.

The only method that produces no liquid manure is the loafing yard with bedding. In nearly all other cases pits are needed to collect and distribute the liquid manure, which contains a good deal of nutrients as is shown above. How concentrated it is depends on how much water is added. When cattle are out at pasture a proportion of course remains on the meadow. The nitrogen in the urine is in the form of urea, which rapidly decomposes into ammonia and carbon dioxide. As the smell shows, ammonia easily escapes into the atmosphere. Air-tight tanks have been used to avoid this loss, though considerable losses will still occur when the liquid is distributed. In bio-dynamic farms another system works well. Open containers with walls that will withstand frost are less costly. A thin floating cover of chopped straw or stinging nettles on which some rock phosphate or basalt meal is scattered (about 1% of the total content of the pit) does a great deal to prevent the loss. The speed of decomposition of the cover increases as time goes on. Sawdust or wood shavings, which are quite good for fixing nitrogen, can also be used. In the top layer adjoining the air the excess of carbon compounds gives rise to bacteria that stabilize the rising ammonia. The process can also be improved if, at the start of filling, well-rotted compost is added in an amount of about 5% of the pit contents. An effective fertilizer is thus produced of which the one-sided effects can be balanced by adding rock phosphate and basalt meal. But liquid manure is not only valuable because of the nutrients it contains. Because it is a product of animal metabolism it is important for the microorganisms in the soil. Either fresh or fermented it provides a number of physiological effects. The most permanent advantages from liquid manure are achieved if it is applied to the soil with grain stubble or catch crops.

171

It is usually said in favor of slurry, which includes solid and liquid excrements (sometimes also referred to as liquid manure), that it prevents or reduces the loss of nutrients in animal dung. When solid manure is stacked, some of the carbon is lost, which is desirable. What is not desirable is that, if the heap is not properly handled, some of the nitrogen is also lost. By proper handling, nitrogen losses should be kept to below 20%. The result of piling or composting is ripened manure that brings about lasting improvements in the soil. In contrast, slurry supplies the soil with nutrients that work rapidly. Noticeable yield improvements are achieved, particularly with root crops.

Not counting added water, about 18-20 tons, cubic meters or yards) of mixed solid and liquid manure per CU are produced if the cattle are stabled all the year round. Proportionally less is produced if they are put out to pasture. One cubic meter or yard of liquid manure from cattle contains about 4-5 kg (9-11 lbs) of nitrogen, 1.8 kg (4 lbs) phosphate, and 7 kg (15 lbs) of potassium. So the need for sufficient storage space, expensive though it is, is high, even if emptying takes place several times a year. The farmer should not find himself in a situation where he has to put the manure on the fields at the wrong time just because the storage tank is full. Straw, which is an important factor in healthy livestock husbandry, should be used as little as possible in slurry to avoid a number of technical difficulties in the gutter, with pumping and with application to the fields. Yet problems with handling are frequent. Over fertilizing occurs often because the nutrients are so easily accessible. A number of problems can arise from the supposed advantage of not using any straw at all. As with all things, moderate applications are what is important for the soil to develop. It is advisable to apply slurry to as many fields as possible once a year. Also liquid manure can be used together with crop residues, straw and green manure for sheet composting. Technical details can be found in many advisory service leaflets.

A good deal of positive experience has been gathered in bio-dynamic farming in connection with the use of slurry on permanent grasses. Regular use of bio-dynamic preparations coupled with sufficient storage duration for the manure have led to a varied plant population of grassland including red clover as part of the sward. Even extremely shallow pebbly and gravelly soils have remained green in times of drought. In other cases, hard pans in the subsoil became so porous after a few years that even after heavy rain no water collected on the surface as hitherto. The fermentation of slurry can be almost entirely controlled by modern stirring equipment.

Experiments are in progress in Switzerland and Germany to aerate both liquid manure and slurry in order to stabilize some of the nitrogen. No final results are available as yet. But it looks as though it would be possible through bio-dynamic treatment to avoid losses of nitrogen and yet produce a material that has much fewer detrimental side effects on plants due to too much available nitrogen in concentrated slurry.

It is best to treat the excrements bio-dynamically while they are still quite fresh. The so-called compound preparations have also proved useful for controlling the fermentation of floating manure. (See p. 206 about how to make them). If one bucketful per 20 animals is sprinkled on the gutter twice a week, the result is an odorless fermentation with a soft instead of a hard floating cover and a fertilizing action that is not too strong. Also in cowhouses with stands one can spread a small shovelful of this material on the manure before the barn is cleaned. A shovelful spread every week in the pit for the liquid manure improves the fermentation. Usually this treatment is applied in addition to the treatment with preparations as described above. As can be seen from this discussion, there exist various ways of treating liquid manuring materials according to bio-dynamic principles. (The compound preparations stirred into water can also be used for seed treatment. In addition they are useful for compost making when sprinkled over the materials while the heap is being built.)

Observations concerning bio-dynamically treated floating manure reveal facts that are the polar opposite of experience gained in many ordinary farms with new cowhouses built to collect floating manure and slurry. It has been known to happen that even in farms with 2-2.5 CU/ha (1 CU/acre) additional nitrogen fertilizer has been spread, even though 200-250 kg/ha (180-220 lbs/acre) of nitrogen have been available in the animal excrements. This leads to a deterioration in the quality of the fodder crops since their sugar content decreases and the quality of the protein is poor. If the floating manure is not stirred properly and treated bio-dynamically it often builds up a dangerous level of disease germs and parasites (Schaumann 1970).

Some farms that are equipped for handling liquid manure can choose to produce composted manure. If the standings are as usual 5 ft long with a grid that covers the gutter, the animals do not have to stand and lie on the gutter. A small, approximately two-inch step up to the dung passage makes it possible to lay planks across the grid, supply the cattle with bedding and thus prepare for solid manure. If the stall gates permit, mucking out can be done with a front-loader. This choice means that solid manure can be made for certain crops and also, for instance, sick animals can be provided with bedding. The biological advantages of matured manure can also be preserved for the farm by providing a loafing barn with plenty of bedding for young animals and then composting the manure.

Garbage composts and sewage sludge

The 180-300 kg (400-600 lbs) of garbage produced every year by every member of the population poses considerable sanitary and disposal problems. So far coordination of effort has not been achieved between communal authorities interested in getting rid of the garbage and possible users of garbage compost. With the exception of Holland, no country has made much progress in establishing composting of garbage as a possible alternative to burning or land fills. This is unfortunate, for though garbage only amounts to one-third of the available stable manure, there are nonetheless many sites where it could be used to advantage if composted. Burning is expensive since steps have to be taken to prevent air pollution and there is also a residue that still has to be disposed of. Usable sites for sanitary landfill are growing scarce while groundwater pollution from garbage tips is becoming a threat in many places (Noring et al 1968; Farkadsi et al 1969).

Various qualities of treated garbage are, however, available. In the form of mulch compost that has been stored only briefly it has proved useful for erosion control in steep vineyards (J. Bosse 1968). E. Spohn (1971) has been indicating for some time that matured garbage composts are most versatile in application, being suitable for land reclamation, horticulture, fruit, hops, etc. Good garbage compost can also be used as fodder to remedy iron deficiency in piglets. For large amounts to be used in agriculture as a whole it would be essential that only good quality composts be produced and that the cost and labor of haulage be shared by producer and farmer. Sanitary disposal and recycling of organic matter to the soil bring advantages not only to the farmer but to society at large and progress will only be achieved if the costs are shared.

Garbage composts contain 15-40% organic matter, 0.4-0.6% nitrogen, 0.2-0.3% phosphate, 0.2% potassium, and 2.5-5% lime. The latter is important for acid soils. In addition this compost contains trace elements that raise the amounts in poor soils. It has not proved useful to add mineral nitrogen fertilizers during fermentation (Popkema 1967). The use of garbage compost is in harmony with the principle of returning organic matter to the soil. Crumbly, evenly sieved composts help to improve soil structure.

Sewage sludge in amounts proportional to the garbage produced by the population can also be sanitized by composting. Between 10 and 40 tons per hectare (4-16 t/acre) of compost can be used both on farmland and in the garden. For land reclamation a layer 0.5-1.5 inches thick is spread on the ground, i.e., 100-200 t/ha (40-80 t/acre). Matured composts are preferable. These are produced either by milling and storing for several months in heaps that have to be turned, or by a modern aerating process (Blaubeuren technique, Voith-Müllex, Fairfield, etc.). The latter also require longer periods for the compost to mature, e.g., the treatment in the digester has to be followed by composting.

On bio-dynamic farms, human sewage is only used in amounts

175

that are actually produced on the farm. Meanwhile, rapidly increasing populations and the general availability of mains drainage have led to a situation where large amounts of sewage are collected that somehow have to be disposed of by the treatment plants. In untreated sewage and also in sewage that has been treated mechanically and biologically a number of potential dangers can lurk, such as anthrax bacilli, salmonella, tuberculosis bacilli, enteric fever, intestinal parasites such as roundworms and tapeworms, viruses such as polio and infectious hepatitis. In various ways these infections can find their way to people and domestic animals and sometimes they are strong enough to set off the disease. It is thus understandable that many experts on sanitation are demanding that untreated sewage should not be allowed to leave the sewage works, though at present their demands are somewhat utopian since untreated waste from some sources is still finding its way into rivers. In our opinion, a practical way of solving the problem would be to burn the sludge. A perfectly hygienic product can be made, however, by composting sewage sludge together with other waste, such as municipal garbage. To do this, a temperature of 60°C(140°F) must be maintained in the compost heap for six days whereby careful turning must ensure that the outside parts are also subjected to this heat. (Garbage composts with sludge generally reach higher temperatures than garden composts, etc.). With the exception of anthrax bacilli, most germs that cause diseases can be suppressed by composting for sufficiently long periods.

Because of its nutrient content, sewage sludge is often described as a valuable organic fertilizer. Sludge with about 70% water (which is low), contains 14-18% organic matter, 0.6-0.7% nitrogen, 0.55% phosphate, 1.7-3% lime. The amount of potassium is small. It must, however, be taken into account that every organic substance has other effects in addition to its nutrient qualities. Dried sludge or sludge that has been composted for a long period does not lead to the production of foul-smelling vegetables as used to be the case when produce was grown on sewage fields or when fresh sewage containing about 90% water was used. Nevertheless, on principle the cycle must not be too short. The influence human feces have on the live processes differ in some respects from those mediated by animal manure.

A difficulty arises here. The marketing of commercial composts containing sewage is often aimed at the private gardener although he is the very person who could most easily grow his fresh vegetables on animal manure and garden compost that is free of any feces. He need not depend on sewage fertilizer in order to conduct his gardening organically.

Nowadays garbage and sewage sludge are complex mixtures of materials. If, for instance, roadside gutters that collect rainwater are linked to the sewage mains, the sewage will contain extra lead from road dust. As a result of industrialization and the increased burning of fossil fuels, the environment is becoming speadily contaminated with increasing amounts of substances, through emissions from burning various fuels, effluents, solid wastes and pesticides. Some of these substances occur anyway in the biosphere; contamination is nevertheless being threatened on a world-wide scale by the build-up of undesirable concentrations. Copper, zinc, manganese, cobalt, cadmium, lead, mercury, arsenic and molybdenum are examples. Many of these are vital as trace elements but depending on their mobility in soil and plants can become dangerous even in marginally increased amounts. For instance, if cadmium and zinc are added to soil in soluble compounds, then applications of as little as 3 and 20 ppm respectively will impair yields. This is less so in the case of copper or lead, the latter being not particularly mobile within the plant. Developments must therefore be carefully monitored. Polycyclic hydrocarbons such as 3, 4-benzpyrene, which occur in nature, are also found in stable manures and composts. Some of these have been declared to be carcinogenic. Borneff and colleagues (1973) have recently shown that in the amounts in which they occur in farmyard manures and composts they hold no cause for concern. A new process, pyrolysis, is now being tested for dealing with garbage. The cellulose and lignin components that form the lion's share of the organic substances in municipal garbage are liquified into mineral oils (Appell and Fu 1972). The complex nature of garbage creates problems, however; homogenous wastes are more suitable.

Buying-in fertilizers

The aim of farm management is to create a farm organism. Depending on habitat as well as market and transport facilities, and also depending on his own abilities and interests, the farmer will buy in more or less fertilizers and feed. This is in order so long as he is working toward the goal of a self-contained farm organism. What has been said hitherto has surely made it clear that by this a biological principle and not a rigid dogma is meant. In how far buying-in can be reconciled with the principle of the greatest possible self-containment depends not only on the type of materials coming from outside but also on the amount in proportion to the farm's own manure and feed. The amount and type of goods bought in are taken into account in the matter of the quality guarantee (See Chapter Ten). At this point we shall restrict our discussion to the buying-in of fertilizers, though of course concentrates, seeds and livestock are also items that are usually bought.

Discussions on bought-in fertilizers can become rather bogged down in emotional ideas about the nature of what is "organic" or "natural" or "bio-dynamic." If we take into account the manner in which plants are "fed" by the life processes in the soil, however, we should be able to reach a sound enough judgment. The remark by R. Steiner quoted earlier is also to the point, namely, that "the mineral world should only work in the way it does in nature itself."
Crops are supplied with nutrients from a number of sources:

1. *the minerals in the soil:* above all potassium, calcium, magnesium, some phosphorus, silica, iron, manganese, and the other trace elements;
2. *stable and effective humus* including crop residues, green manures, etc.; above all nitrogen, sulphur, some phosphorus and also all the minerals under 1;
3. *composts and manure:* above all nitrogen, sulphur, some

phosphorus, and all the minerals mentioned under 1, which were present originally in the fodder or the plant waste;

4. *commercial organic fertilizers:* these are used chiefly for their nitrogen and phosphorus content, or, as in the case of seaweed fertilizers, for their varied compostion;

5. *minerals with low solubility:* some of these are ground basalt, greensand, lime, dolomite, rock phosphate; basic slag is also related to this group; these contain varying amounts of the minerals under 1;

6. *chemical fertilizers;* these are available on the market for all minerals in the form of more or less high-percent, soluble fertilizers containing one or more nutrients.

The substances mentioned work in the ground in the most varied ways. If they contribute to the growth and propagation of soil microorganisms, this in turn has a regulating effect on plant growth. Minerals share in this function if deficiencies occur, which can afflict plant life and soil microorganisms equally. It is most favorable if plant nutrients can be made available in organic compounds, in composts, manures or organic waste products. With regard to their composition they bear within them the ordered pattern they have received from the living organism from which they stem. They are natural compounds arising out of life processes and serving life processes.

Fertilizers that are rapidly turned over or decomposed in the soil often have unphysiological (unnatural) effects. But if decomposition is too slow the necessary nutrients might not be released fast enough for the growing season in question. So the effectiveness of manuring materials depends on when they are applied, how they are worked into the soil, how much is used, how finely ground they are, how the farmyard manure has been distributed, etc.

The necessary transformations of substance also depend on the warmth, aeration and humidity of the soil. Too much humid heat can increase the rate of decomposition in the soil too much while if it is wet, cold, compacted or too dry the manure will work poorly or too late.

The following remarks are relevant to points 1-6:

179

The minerals present in the soil are released by the dissolving power of water. This is fairly low but is increased by the substances secreted by the soil microorganisms and plant roots, notably carbon dioxide. Thus to a great extent biological activity determines the ability of the water in the soil to make available the minerals in the soil.

As indicated in points 1 and 2, in organic farming nitrogen is obtained entirely from organic compounds. Soil microorganisms produce soluble nitrogen compounds that can be used by the plants. The life processes that mobilize the nitrogen also regulate the rhythm in which other nutrients are made available and their concentration. Thus it depends on the soil life how one particular nutrient is supplemented by other nutrients also present in the humus. If nitrogen is provided by the humus, a harmonious supply of nutrients can be counted on. As a matter of comparison, we can point to the harmonious nourishment given by whole-grain bread, which offers all the nutrients in every slice and not just an excess of some.

The commercial organic fertilizers available on the market are used mainly for root crops, etc., that need large amounts of nutrients. The following table gives a few examples. These fertilizers are frequently offered in the form of blends. The purchaser should then make enquiries as to the components and the nutrient content. Oilseed cakes are used to some extent. Dried chicken manure is now also available in large amounts. Feed additives such as antibiotics or copper make hitherto useful materials more and more questionable. Bio-dynamic producers in Europe increasingly have to have their bought-in materials tested.

With regard to these commercial organic fertilizers it is not always sufficiently noted that the amount used is just as important as the kind. It is hardly possible to overfertilize with mature compost or composted manure. But overfertilizing often occurs with fresh and liquid manure and the same goes for supplementary fertilizers such as blood meal or pig bristles, etc., which work fast. *The strict principle must be remembered that the exclusive use of organic fertilizers does not in itself make a farm*

Composition of some organic fertilizers				
	organic matter	nitrogen %	phosphate %	potassium %
hair, bristles	80	12-13		
horn (meal and shavings, dust)	80	12-14		
blood meal	75-85	12-13		
feather meal	75	12		
meat scrap	70	64% protein,	6% fat	
bone meal, steamed	30	4	20	
de-glued	5-10	1	30	
guano		6	12	
chicken dung, dried	40-50	2-3	3	2

"organic" or "bio-dynamic." The amounts of organic fertilizer used must also be correct.

This viewpoint also applies to mineral substances with low availability. The nutrients from ground rocks, e.g., basalt meal, are made available to plants by the same processes as those from the minerals present in the soil. With these, overfertilizing hardly occurs. The amounts used for horticulture are 2.5-5 t/ha (1-2 t/acre). For agriculture less is needed. Most farmers can afford to use basalt meal as a supplement for compost. Its use is quite noticeable on light soils where it promotes clay formation and helps the supply of nutrients to plants. The effectiveness of such mineral substances is increased if they are applied in manure or compost. The same applies to the use of rock phosphate in stable manure. Every farmer knows that it is much more difficult to assess the proper amount of lime needed. Damage from too much lime is well-known and difficult to correct. Applied in the proper amounts it stimulates the biological processes in the soil. In this way and by improving soil structure it also helps improve yields. The effects of lime, however, have to be balanced by intensified humus replacement. Bio-dynamic farms need considerably less lime than others. It is usually found that the amount recommended by the experimental station can be reduced by 30-50%. The reason is that matured farm manure and optimal soil aeration correct acidity; furthermore, mineral fertilizers that accelerate the leaching of lime, thus making it necessary to add more, are not used.

With regard to all the nutrient sources mentioned above under points 1 to 5, the supply of nutrients to plants is regulated by the soil microorganisms. Nitrogen comes entirely from organic compounds. This is not the case with mineral fertilizers. These alter the concentrations and ratios of nutrients by by-passing the soil microorganisms. They thus also change the correlation between nutrient ions in soil and plants. This is no great difference seen from the point of view of current ideas about plant nutrition. But for thinking that is concerned with truly biological categories the difference is considerable.

The above gives some of the principles to be considered when attempting to assess the buying-in of fertilizers.

If the supply of nutrients to crops can be achieved in accordance with points 1-3, the result is a self-contained farm organism. There are a number of examples in which this has virtually been achieved.

If the organic commercial fertilizers and slow-working mineral substances mentioned under points 4 and 5 are used, then not only the type of fertilizer but also the amount has to be taken into consideration, in addition to the manner in which it is applied, e.g., direct onto the soil or via the compost heap. The more concentrated and easily accessible the nitrogen content is, the greater is the need to take into account the amounts used, in order to conform to bio-dynamic standards. Where the limit lies cannot be stated generally but depends on the individual farm and its cultivation, livestock, disease level, soil conditions, etc. For marketing and production in accordance with the standards of the Demeter trade mark, not only is advice given on crops but also maximum amounts of fertilizer that may be bought in are laid down and must be adhered to.

Finally, we must add that general agricultural developments with regard to mineral fertilizing have moved further and further away from the viewpoints stated here. Sulphate of potash magnesia or basic slag, which have plenty of balast and used to be used much more, may be more favorably assessed than modern, highly-concentrated single and compound fertilizers offered today for transport and labor reasons.

4 Silica, Clay, Limestone

In R. Steiner's lectures on agriculture, these terms denote not only three different substances but rather three complex groups of influences from the environment on plant growth. It seems appropriate, therefore, to add a few remarks on this theme.

Silica or quartz-rich, that is, sandy, soils are poor in nutrients but usually offer favorable conditions for the supply of warmth and air to the root zone. In damp, cool climates, quartz-rich soils usually contain humus that is poorer in nitrogen and tends to be acid. This, however, is not the only point with regard to the significance of silica for growth.

In a highly diluted solution as $Si(OH)_4$, this substance is present in the water in soil and plants and is important as a formative element in plants and their ability to resist fungal and animal pests (Jones and Handreck 1967). The water in soil contains silica in concentrations of 30 to 40 ppm. This amount remains relatively uninfluenced by the chemical processes, particularly the pH range determined by calcium. Iron and aluminum compounds can decrease its concentration in the soil solution. Silica is transported by water into the plant in amounts that roughly correspond to the amount of water taken in and the concentration in the soil. It may be noted, however, that there are definite "silica plants." Among cultivated crops these are the grasses and cereals. Nitrogen and phosphorus fertilizers seem to reduce the intake of silica. Japanese rice farmers like to use silica-rich slag since this helps build up resistance to fungal pests. In a natural state, grasses contain 10 to 20 times as much silica as dicotyledons, particularly legumes. The latter are "calcium plants." They prefer calcium-rich habitats or, like lupins, accumulate calcium in their ash.

In plants, silica is a formative element. It is found in the more external organs, in the beard of cereal plants more than in the spelt, in the spelt more than the leaf, in the leaf blade more than the leaf sheath, in the leaf sheath more than the leaf stalk. In the leaf blade it occurs in surfaces and cell walls. Thickened portions of cell walls in leaf veins are impregnated with silica.

Silica makes it more difficult for fungi to enter the leaf surface and also damages the masticatory equipment of a number of pests. In rice it increases resistance to leaf spot (*Helmintho-sporium*) and stem rot. In cereals it increases resistance to mildew and in wheat to frit fly attack. Too great a supply of nitrogen and water causes weakness and too little cellulose in stalks. In the straw of oats that had lodged, 0.19% silica was found, whereas in oats that had remained standing the amount was 0.71%. The use of waterglass (sodium silicate) backs up the effects of silica against mildew. For the rest, the silica in the shoot must be obtained from the soil. Probably because of its effect on iron compounds in the soil, effective humus favors the intake of silicic acid, whereas too much nitrogen and water seem to reduce it. We see here yet another delicate balance in the environment in which plants must grow. In plants some silica is stored as opal, which returns to the ground either directly or via animals and in places is present in the ground in considerable quantities. In these places it is a source of soluble silica. It seems therefore that there might be a silica cycle between plants and soil.

The bio-dynamic use of silica will be dealt with in the next chapter. The effects briefly described here take place in the main without any contribution by man but they can obviously be negatively influenced by overfertilizing with nitrogen.

While silica is a formative element that contributes above all to the ability of plants to resist disease, limestone and related substances in contrast are mainly effective in the biochemistry of soils and the metabolism of plants.

The effective constituent in limestone is the element calcium.

Limestone consists mainly of calcium carbonate. As is shown in the table on page 137, this calcium also occurs in the silicates in rocks and soil. On loamy soils, those, for instance, that have developed from basalt, plants can find a sufficient supply of calcium although there is no calcium carbonate in the soil. The same applies to originally chalky soils from which the calcium carbonate has been leached from the top layer. Calcium, or lime as it is often rather inexactly called, determines the reaction of the soil or the pH. Yet the pH tells us nothing about the calcium or lime reserves in the soil. This has to be assessed from the limestone content of the soil, the calcium content of the parent rock or by special analysis. It is known that the desirable soil reaction for heavy soils lies between pH 6.5 and 7.2; light soils should be more acid with a pH figure of 5.5-6.0. Soils under permanent grass have lower figures than neighboring cultivated fields by about 0.5-1.0 pH.

A strongly acid reaction leads to the destruction of silicates and the leaching of nutrients. In extreme cases, quartz or the clay mineral kaolin remain since they are resistant to weathering. Counts have shown that where reaction of the soil is acid there are more fungi in the soil and where the reaction is mildly acid or neutral there are more bacteria.

A high calcium content slows down leaching and the release of nutrients from rocks. Neutral or moderately acid reaction maintains the balance between leaching and fixation.

Lack of lime leads to the formation of acid types of humus. This is in part soluble in water and can be leached. The brown color of water from peat stems from this. Lime favors the formation of stable and richer forms of humus. There must be enough calcium and clay in the soil for stable humus to be built. The same applies to the stabilizing of humus in compost. If the lime content is sufficiently high, a favorable crumbly structure will occur in moderately heavy soils.

Phosphorus as a nutrient is locked up if the soil is strongly acid or if there is an excess of calcium. It is most accessible if the reaction is only moderately acid.

The solubility of iron decreases as the pH figure increases; the same applies to boron, manganese and zinc; the opposite applies for the trace element molybdenum.

These few indications show how important it is to maintain the proper calcium content. As mentioned above, this is achieved not only by means of applying lime but through organic management.

Transformation of substances in the plant takes place in the semi-liquid plasma. So that the proper reactions can take place, this must have the right viscosity and pH figure. Potassium and sodium increase viscosity, while calcium decreases it. So limestone and related substances help the plasma to retain its appropriate consistency. There is a known correlation between bitter pit in apples, a kind of tissue disintegration, and local calcium shortages in the tissue.

Clay is a fine-grained silicate that arises, as the soil develops, as a relatively permanent product of weathering out of feldspars, mica and other silicates present in the parent rocks. When dry it is hard but when moderately damp it is plastic. In the soil it is an important colloidal and binding substance and, together with stable humus, glues the coarser particles such as sand to form aggregates. Stable humus combines with clay to form organo-mineral complexes and is then stabilized, i.e., more resistant to decomposition. The most important function of clay is that it absorbs plant nutrients loosely on the surface so that the plants can take them in. At the same time they are also protected from leaching through being bound in this way. Stable humus can store nutrients in a similar fashion.

Silica is everywhere present in soils and plants. In the plant it is also a formative and structural element that contributes to disease resistance.

Clay supplies the plant with nutrients from the soil, particularly potassium, calcium, and magnesium, but also others; it stabilizes humus.

Lime regulates most of the chemical and biochemical processes in soils and plants. Lack of calcium in the soil leads to the destruction of clay, the leaching of nutrients and the for-

186

mation of acid humus; excessive amounts lock up a number of nutrients.

The importance on a regional scale of limestone for soil formation, growth and the health of livestock was studied by William A. Albrecht (1956). He compared the soils of the forested areas of the eastern USA, where precipitation is higher, with the lime-containing loams of the steppe areas, which underwent less leaching. In their natural state the latter have a good supply of nitrogen-rich humus; the luscious grass supports buffalo. The parts under cultivation yield high corn and wheat crops. It has been observed that with the gradual depletion of the natural supply of nitrogen-rich humus, gluten-rich hard wheat has slowly escaped westwards. The quality of the fodder grown in these areas can be recognized by its protein and mineral content and in particular by its effects on animal health and reproduction. This quality can hardly be achieved on the poorer soils that were formerly found under trees.

5 Soil Testing And How To Evaluate It

Soil tests can be a valuable tool for agriculture if the right conclusions are drawn from the results. Knowing what can and what cannot be deduced from the figures that come from the laboratory can be a help to advisers and farmers in their decision making. But decisions must be based on a total view of the field or farm in question, and in order to use the figures properly the farmer must also know:

1. the parent rock of the soil type.
2. the textural group, the soil profile and in particular the humus content; root development and soil life; problems of soil structure such as tendencies of capping and compaction of the topsoil; plough pans and other hard pans; earthworm density; breakdown of farm manure, etc.
3. the history of the field or farm, in order to be able to determine the basis of difficulties that might be the consequence of
   a) existing soil properties;

b) management in general;

c) a single mistake or omission.

This general information supplies a reliable basis for interpreting and evaluating both biological and chemical analyses.

The biological state of the soil: By means of laboratory tests the farmer ought to determine the humus content and possibly also the total nitrogen content of the soil from a good, a medium and a poor field on his farm. Unless one wants to follow up a special question this need then not be repeated for a good many years. Bacteriological tests can also be made. As with chemical methods, the important thing is the interpretation of the figures that should be undertaken on the farm itself and not miles away in a laboratory.

The chromatographic method developed by E. E. Pfeiffer supplies good pictures allowing the estimation of amount and quality of humus in soil, the stability of humus, the maturity of compost, etc. Seasonal influences on the soil can in some instances also be investigated by this method. It is also possible to estimate whether a soil contains enough humus in relation to its clay content. (Pfeiffer 1960, Koepf 1964). See illustrations opposite p. 196.

Chemical soil analysis is undertaken by private or public experimental stations. Some advisers also use simple field tests that yield approximate results. Tests usually include the pH and information on the available phosphate and potassium. These tests are cheap and can be recommended. Trace elements are determined if crops show particular deficiency symptoms or if no obvious other reasons for poor growth exist.

The farmer should know the pH of his soil. Nutrient tests consist in attempting to mimic, with the help of extraction solutions, the removal of nutrients from the soil expected to be brought about by plant roots. Only an approximate result can be given in a single analytical figure. In private laboratories a spectrum of results can be obtained including nitrate, ammonia, and various phosphate and potassium fractions. In this case the station employs extraction solutions with varying strengths. This makes interpretation easier and more exact. A more complete program also includes the exchange capacity, the base saturation and in arid areas also the salt content. Nutrient deficiencies can, though may not, be the cause of unsatisfactory growth. Experienced farmers therefore also take into account such possibilities as soil structure, water and air content, etc. Nitrogen availability, as the most important item, is estimated according to total humus or nitrogen content of the soil, taking into account the most recent applications of farm manure and whether legumes have been grown or not (See p. 152f.). The results of bacteriological tests also have to be interpreted. One checks the number of bacteria or counts certain groups. But then the cause for the findings has to be sought, e.g., in the supply of organic matter, in the weather, in the previous year's crops or in cultivation errors.

We have been dealing in this and the previous section with soil properties, fertilizing and farm management. The considerations brought forward are of particular importance when a farmer wants to switch to the bio-dynamic method. At this point we shall therefore go into this with a few remarks.

6 Some Points On Farm Reorganization

The guidelines according to which farms in general are organized today are opposed in a number of ways to those that govern the running of bio-dynamic farms. A bio-dynamic farm is not an organization with one or several specialized production programs. The way it is organized aims at achieving an individual character for the farm as a whole, thus making it a self-sufficient organism. It has already been said that this cannot be totally achieved. Farms, however, will be able to approach the ideal more easily than market gardens, which are on the whole more dependent on bought-in manuring materials.

It is not the case that with an aim of this nature, financial and labor considerations are given insufficient scope. Every farm represents of itself a concrete situation involving ecology, labor and marketing. For bio-dynamic farms there will be a greater variety of marketing possibilities compared with the rest of agriculture, especially where the farmer has succeeded in establishing direct sales to interested customers and where direct marketing absorbs a considerable share of his production. In organizing a bio-dynamic farm what is important is to bring into balance the labor force and the economic situation on the one hand, and the biological conditions for high quality production on the other. No one can afford yield losses and thus income losses even during reorganization. So while there must be a clear goal to aim for, action must be taken step by step only, no measure being put into practice unless it can be seen beforehand that it can be fully realized and more or less what the result will be.

Every experienced farmer knows that his work is a lifelong learning process. A great deal has to be reviewed and rethought if reorganization is planned. It is a good thing to consult one or several farming colleagues who are already practicing bio-dynamics. In some countries, such as the Federal Republic of Germany, farms are organized regionally into working groups. They are assisted by full-time experienced advisers for agriculture and horticulture including private gardens. If such a working group exists in the area, it is an excellent step to join it and so be able to cultivate an on-going exchange of experiences and ideas. In other countries, where bio-dynamic farming is also practiced on a larger scale there exist national organizations that supply addresses and the services of advisers if so requested.

Amelioration

For most farmers there is nothing new in being told that land that needs amelioration does not yield adequate returns. The conclusion to draw from this is that it is better to carry out amelioration first and then intensify or alter cropping. This is right in theory but often difficult to put into practice. In many agricultural areas, particularly those on Pleistocene loams, including sites where these loams provide a thin layer on top of Mesozoic clays and marls, numerous areas needing drainage still exist. Grassland on clays or in low-lying positions tending to waterlog often need tile drainage or ditch drainage as well as outlets before good grass can be produced on it, let alone fodder crops. Recently, deep subsoiling or deep ploughing (2 to 3 feet) on loess and similar substrata has frequently been recommended. This is usually combined with deep fertilizing with lime and phosphates. In fact with these soils their lack of porosity in the subsoil is often a factor limiting the spreading of roots. Yields are not as good as they could be. In addition there is no doubt that intensive cereal cropping over many years coupled with inadequate humus replenishment have made the natural state of affairs even worse. This kind of deep sub-soiling constitutes a considerable interference. (It is different from conventional sub-soiling.) It is therefore usual in such cases to avoid initial yield losses by increasing mineral fertilizer application. Now it should be carefully considered whether the same purpose might not be achieved by planting hay crops, especially as more will be needed anyway during reorganization, and by careful cultivation together with intensification of humus management. With the help of roots, soil fauna and intensified management improvements which are permanent are gradually achieved. The improvement of soil structure by mechanical means is only useful insofar as it can be stabilized by subsequent growth of crops with a rich root system and by manuring.

Testing fertility

The mineral and lime content of the soil must be in order before reorganization takes place. A supply of lime based not only on the soil analysis but also in accordance with an exact knowledge of the soil will often prove useful in the early years. Other materials, such as basalt meal or phosphate, which is particularly important for the cultivatability of clays and clay marls, are not applied immediately or all at once. Instead, additions to farm manure are planned over several years, e.g., 2% ground basalt added at the beginning of composting. If more is to be used, it is best added when the manure is spread. Light soils are visibly improved by this.

In connection with these points, let us mention an experience frequently encountered that should be known to farmers intending to start changeover. Even after several years of bio-dynamic farming the soil continues to improve. This is shown by the fields added to the farm later. After 4-6 years of bio-dynamic farming they are still visibly different from those that have been under bio-dynamic cultivation for 10 or more years. So the stand of plants and their health only become fully satisfactory after a number of years. Long-term changes either of structure or soil microorganisms take place that show in plant population, workability of the soil, and permanence of results. It is difficult to measure the changes directly in the soil, but experience has again and again proved them to be real.

Soil tests are one of the preparatory tasks. There will probably already have been the usual tests for pH, phosphate and potassium. Later these will be repeated every few years. At least for spot checks a more thorough analysis should be undertaken including some orientation with regard to humus content. It is also essential to walk the fields together with an experienced farmer from the same area and the agricultural adviser, testing the subsoil with spade and probes. The result of this investigation finds expression in the fertilizing plan and changes in cultivation methods used.

Tillage

The importance of tillage is increased in bio-dynamic farms. Knowledge of the different methods is among the tools regarded as given for the purposes of this book. Proper tillage is important for growing crops and optimum returns from farm-produced manures. Weed control is also based on crop rotation, the smothering effect of plant population, and working the soil. Tillage is or must be made an essential part of the measures necessary for enhancing soil life. What is meant? If a plough or a spade is used to prepare for a proper seedbed that is free of other plants, or to work in manure, or to bring those substances back that have moved downwards in the top layer, then the ideal structure of natural soils with all their microorganisms is seemingly destroyed. Humus decomposition is increased and soil-inhabiting animals decreased. Deciduous forest soils with plenty of leaf litter are quite different in this respect. The upper layers are well supplied with microorganisms and with digging, burrowing and mixing soil fauna. Capacity for storing rainwater is high yet there is enough coarse pore space to allow excess water to drain off and let air in. Cultivated land, on the contrary, is exposed to rain and extremes of temperature. The even, crumbly structure present just after tilling or planting has become much more compact by harvest time and the surface is probably crusted. But this comparison between forest and cropland does not in itself clarify the importance of tilling the soil. Both the amount and the variety of products grown in the ground in a forest is negligible compared with what the soil in field or garden produces. So tilling means, or should mean, a biological intensification of the soil. If ample organic matter is not supplied or if tilling is not adjusted to biological processes, then all it achieves is the acceleration of humus decomposition and mobilization of the soil's own nitrogen reserves. The interest, care and ability of the farmer determine whether tillage and organic fertilizing achieve a biological optimum in their effectiveness. There are fundamental principles but no ready-made formulae that help to achieve this. Any kind of cultivation, discing, hoeing or harrowing of crops usually also has a fertilizing effect. The soil microorganisms are stimulated to mobilize the nutrients in the effective humus and mineral reserves more intensively. Nevertheless, the aim is to till sparingly, so far as the weeds will permit. Humus-building is promoted if the soil can rest; the beneficial effects of hayfields left for several years are partly due to this. Whether the organic material supplied to the soil can work well depends not least on how it is worked in. Humidity, temperature, air and soil fauna influence the microbial processes in a way similar to those in compost. If the soil is heavy and lacking in air, manures and compost are worked in shallowly. Such soils, which also tend to be too wet at times, decompose farm manure slowly. Zones of anaerobic fermentation are avoided by roots. Lighter soils and those in warmer areas require manure to be worked in to a medium depth. The time at which manure is applied and the depth to which it is worked in influence the factors upon which soil life depends.

With heavier soils the effects of winter frosts should be turned to good use if possible. Autumn ploughing is generally recommended. But silt loams with a poor soil structure are most in need of forcible loosening by mechanical means. In this case spring ploughing is frequently used. An important aim of tillage is to increase the depth of the topsoil layer. But there are limits. The depth of topsoil can increase if the way is paved by roots, soil animals and organic matter, or for instance by means of a good fodder crop with additional farm manure. Deeper ploughing of stubbles for seeding a green manure crop that has itself been fertilized with plenty of liquid manure can also help to improve the structure of the topsoil. Root penetration and earthworm activity are what maintains the connection between topsoil and subsoil. As with topsoil, so also the subsoil penetration with life takes place from above downward. The depth to which it is possible to plough is determined by the depth to which the soil is more intensively penetrated by roots, microorganisms and animals. This point must be taken into account more carefully in the case of poorer soils than where the ground is fertile. The maximum depth for turning the soil with the mouldboard plough is 20-22 cm (8-9"). For deeper layers the subsoiler is used. The chisel plough is especially helpful on loamy soils in the semi-humid and somewhat drier areas. It does not turn the ploughed layer and therefore can work to a depth of 12" and more. All the other relevant points regarding tillage, i.e., timing, optimum humidity, utilization of the best times for mechanical weed control, subsoiling, choice of tools, etc., are to be seen in connection with the view that tillage and humus management must work hand in hand with structure formation and the enhancing of soil life. The changes in crops and crop rotations usually brought about by the changeover to bio-dynamic methods thus also involve alterations in tillage methods that require careful consideration.

194

Planning

Planning the change-over requires a thorough inventory of everything on the farm including all the given conditions for production. The manuring program is a suitable point of reference for the discussion of the intended farm enterprises. These are determined by:

1. the given natural conditions;
2. the interests and particular skills of the farmer;
3. marketing facilities;
4. economic aims, i.e., the necessary net income and also the improvement of the farm and its lasting productivity.

The amount of organic matter recycled and its effect depends on:

1. number and type of livestock;
2. crop ratio and crop rotation;
3. storage and treatment of manure and method of application.

Animal husbandry is aimed at achieving a home-bred herd with a good steady production from feed produced entirely or almost entirely on the farm itself. Density will usually be about 1 CU/ha (0.4 CU/acre). In Germany, where hundreds of bio-dynamic farms are functioning, 1 CU/ha is recommended, the same as for the rest of the country. The average for Germany is 0.94 CU/ha. Bio-dynamic farms do not need more animals than others to maintain lasting fertility. The emphasis will usually be on cattle, though a mixture of manure from different animals is beneficial, particularly of those fed mainly on roughage and those fed on grain.

Land use is on the whole predetermined by the natural suitability of the land or the possibilities of its amelioration. This is not the case, however, as far as the ratio of the crops grown and the crop rotations are concerned. These must take into account the need for marketing produce, feed requirements, distribution of labor, and recycling organic matter. For instance, for the feeding program one must perhaps take into account an intended increase in herd size and/or an intended increase in the use of home-grown feed. The crop rotation should utilize all the possibilities for undersowing, mixed cultivation in its many forms, and catch cropping.

The amount of available manure produced on the farm itself does not depend solely on the number and type of livestock. Storage methods that minimize losses, controlled fermentation according to bio-dynamic principles, and expedient application are all equally important. At the start of the change-over it is best to collect whatever the farm can produce. Sufficient liquid manure, chicken manure, etc., will be needed for the autumn catch crops. Bio-dynamic preparations, possibly compound preparations, must be bought or made. For the first few years of the change-over, roughly the same amount of cash should be earmarked for the purchase of commercial organic fertilizers as was hitherto spent on fertilizers. As soon as the changeover begins to show results, this amount will be reduced. It is useful to use the sprays 500 and 501 several times a year at first. If sufficient manure is available and the soil is in good condition, the change-over can be undertaken in one go, or each field in turn as it comes to be supplied with farm manure. If the soils are poor, the process has to be undertaken in steps.

It is helpful to visit other bio-dynamic farms. This helps in learning to judge the state of one's own fields. Crops under bio-dynamic cultivation often develop slowly in the spring until the stimulating effect of warmth on the soil microorganisms becomes fully effective. The crops catch up on the delay by harvest time. Frequently the ratio of leaf to root mass in beet and carrots and that of grains to straw in cereals is closer than usual.

Chromatograms on round filter papers characterize the condition of the soil; in the original the colors are predominantly brown, yellowish and grey; the patterns as well as the colors are taken into account. Top: Garden soil with good structure and humus content. Bottom: The diffuse outer ring of the chromatogram indicates that the humus is acid and unstable. Photos: Koepf.

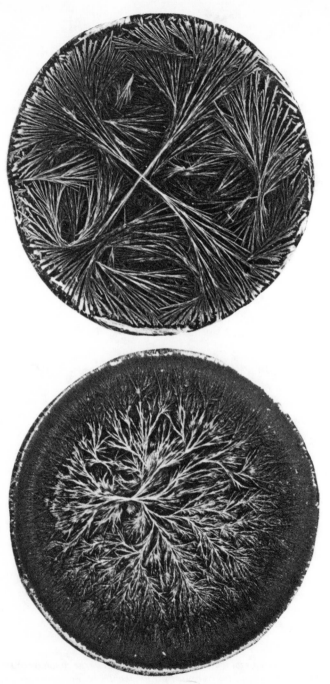

The crystallization of copper chloride is also used in testing plant substances. Top: The crystallization pattern from a seed containing plenty of protein (tares). Bottom: A seed containing starch and protein (wheat). Photos: Koepf.

Chapter Four

Practical Aspects of the Bio-Dynamic Principle

1 General Remarks

The use of the word, principle, in the title of this chapter indicates that the bio-dynamic method is more than just a few suggestions for doing one or two things differently. It is a method covering all aspects of agricultural and horticultural production. This point must be understood! People with the reductionist turn of mind so prevalent today hope to distinguish between bio-dynamic and non-bio-dynamic by the use or nonuse of the preparations. This would be far too narrow a definition, although the preparations are indeed one of the foundations of the method.

On the other hand, bio-dyamic or organic farms do use numerous materials for manuring, pest control, soil conditioning, etc. These are, however, expected to be non-poisonous and inoffensive in their action and to originate from living matter, etc. They are based on traditional methods and also on observation, and the reason why they work as they do has only been partly researched. There is indeed a need for research to help distinguish what is useful from what is not. But the organic movement, which receives no public monies, has so far been

197

unable to finance such projects to any great extent. Not all such materials and measures should be claimed as bio-dynamic, for many are the common heritage of the organic movement as a whole.

The use of the word principle in connection with bio-dynamic work goes back to the former chairman of the Dutch Pest Control Agency, C. J. Briejer (see Heinze 1965 and 1968). The word "principle" is only meaningful if it is based on a particular method. So what is meant?

Conventional research and action seek to find more or less linear cause and effect relations that can also be called mechanisms. Practical measures are then based on these. In bio-dynamic work this attitude is not simply rejected wholesale, but it is regarded as more important to understand every phenomenon in all its connections with nature, the farm and human beings. The kind of questions asked are therefore different. It is well known that the causes of an effect, be it the massive increase of pests or the nutritional quality of products, are never simple and that in agriculture they are always related to what human beings do. The endeavor is therefore made to find thoughts that understand and organize the farm and its environment in a way that makes it possible to grasp the situation with appropriate concepts. Merely to speak of a "totality" is as meaningless as to break up a process of nature into a series of mechanisms.

Many experienced farmers are able to grasp the processes and their factors more rapidly in their feeling than in their thinking. But it is necessary to comprehend with thoughts, for only then can one teach what one knows to others. Nevertheless it is also important to be able to feel one's way into the processes of nature. It is then important to form thoughts that can penetrate into the structure of nature for such thoughts will stimulate rather than banish the appropriate sensitivity. It is the purpose of this book to offer in the different chapters a number of suggestions that can stimulate this kind of thinking and feeling.

Once something is thought of as a mechanism the effect quite unintentionally spreads to everything with which it has a living connection. The "side effects," which are of course unintentional, demonstrate this. These "side effects" are one of the basic problems of environmental conservation, pest control, etc. In contrast, the bio-dynamic principle encourages us to organize a farm as a totality in such a way that high production is achieved while the farm functions according to the inherent "wisdom" of its natural processes. What is this inherent wisdom?

The farmer strives constantly to create optimum conditions in which this wisdom can work, for it is not the human being but the soil and plants and animals that "do the work" in agriculture. The profusion of factors that play their part cannot be fully analyzed nor can they be entirely manipulated. It is a concern close to the heart of the bio-dynamic farmer or gardener to develop the kind of concepts and the kind of relationship to nature that will enable him to do justice to this situation. Neither the mechanisms that are thought out nor the mechanistic thoughts have the power to achieve this.

The attitude of mind and the ideas arising from Goetheanism and anthroposophy do have the necessary quality in the highest sense. Used rightly, they once again gradually unite conscious understanding with surrounding nature. Man as he observes and works learns to act out of nature's wholeness in which he himself is included.

Taking a further step we come upon the question of the forces in living nature. In order to go beyond the obvious explanation of forces as the reason why things happen, we have to depend on our own experience of our organism. Our own exertion is the only means by which we can perceive forces, including physical forces. The only force we experience directly is our will. If we pick up a stone and place it on the table, this is from the point of view of the stone merely a mechanical process carried out physically. But the concepts of "weight" and "exertion" are inner perceptions. The stone would not have reached the table if we had not wished it to do so. Inside our organism the seizing and picking up of the stone is linked to numerous processes that science depicts as separated mechanisms. There is no muscle that does not take part, and metabolic processes commence that in turn intermingle and affect one another. All this is coordinated into a sensible action by the will. The totality of the deed is brought about by the will. It is the will that is the real cause. It is perceived through the inwardly experienced exertion.

How is all this in animals and plants? We must take care not to transfer to them automatically all that we experience in our fellow men and in ourselves. The kingdoms of nature differ from one another. But starting from man's experience of himself we reach different conclusions than we would if our point of departure were lifeless nature. When he picks up a stone, the scientist conducting an experiment does not observe himself. His starting point is a mechanism and he finds such mechanisms everywhere, including in man. A technology arising from this manner of observation cannot but construct technical processes that are nothing more than a series of individual mechanisms.

Living nature, however, is filled with totalities. We can find them if we watch for them. We experience in ourselves how a totality works. Our gaze is turned inward. But in our consciousness we find no material or mechanistic causes, only spiritual ones.

Natural science has as far as possible detached man from knowledge in order to reach objective results. But in the processes described here man works on himself in order to become an ever more complete instrument for understanding nature. In doing so he begins to meet layers of reality that must remain incomprehensible to one who proceeds only by measuring, counting and reckoning. He then experiences more consciously something belonging to the most ancient experiences of mankind: that in natural beings themselves something lives and works that can only be comprehended if he compares it with his own will and indeed grasps it with his own will. Starting from what has just been described, not only feeling but also the will can be exercised in the way it deals with things and thus it can gradually become a kind of organ of knowledge. In this way man can gradually develop an experience of the forces working in life.

This is the path that is likely to suggest itself to the farmer, for he constantly experiences himself as working in nature out of his will. He has to adapt to conditions and yet he can transform them. He also knows how rapidly success recedes if his will weakens. In contrast to the world of urban technology, he daily experiences the limitations of his own will over against things and beings that he has not made and that oppose their own existence to his wishes. If he knows what he is doing, however, he can with his will bring himself into harmony with nature and thus come to a working relationship with nature's forces.

There is no doubt that what average talent and effort can achieve today in this respect is hardly more than a mere beginning. But it is certainly also a real beginning from which wide regions of reality can start to open up. "Science would now be technically in a position to fulfill the demands made (the curbing of pollution and the wise use of natural resources). And yet according to the experiences of the last few years and the manner in which people have reacted to all sensible suggestions and attempts to protect their natural environment and its potential and also to increase it and find new ways of handling it despite the need to tamper with it, it seems that man lacks the moral fibre to take this task upon himself in a wholehearted way" (W. Klausewitz and colleagues 1971). This moral fibre is closely linked to the kind of thoughts man has about himself, his fellow human beings and also nature.

2 Bio-Dynamic Measures

As will have become clear from what has been said so far, applying the bio-dynamic method is a permanent learning process. This alone will lead to real skill. An understanding for the totality of the farm can grow gradually. In R. Steiner's agricultural lectures the basic thoughts and practical measures on which one can lean are described. It seems appropriate, without any claim to completeness, to list them here. The lectures themselves meanwhile are available in print. The experienced farmer will have his own experiences to support him. Blind belief is not required. Work will reveal what can be used. Then an interest in the theoretical background will arise naturally out of daily work.

Central to the bio-dynamic method is the idea of the farm as an organism. This has been dealt with from several angles in the previous chapters.

The yield and quality of crops comes about under the influence of two groups of environmental factors: earthly and cosmic. Light and warmth are the most important of the cosmic factors. Another are the cosmic rhythms. Research into these rhythms over the last 40 - 50 years has provided more exact knowledge about a number of links between the cosmic forces and the phenomena of life on earth. Biological research over the past 50 - 70 years has discovered 107 lunar rhythms in organisms. Many are found in aquatic life (Cloudsley-Thompson 1961, Bünning 1963, and others.) Other phenomena, corresponding to the orbital rhythms of the planets have also been investigated (Schultz 1948 and 1951, Wolber and Vetter 1973). The earthly factors influencing growth emanate chiefly from soil life.

The polarity of earthly and cosmic forces also helps one to understand how the soil provides the basis for growth, and it gives insight into manuring, the treatment of fertilizers including composting, the role of livestock on the farm. New factors with regard to fertilizing are the bio-dynamic preparations. Their use will be described in the next section. Among the more subtle influences on growth is that provided by companion plants.

Over the last 8-10 years it has become obvious how closely the wellbeing of a farm depends on its being surrounded by a healthy landscape. It is also now realized that the quality of the environment depends on the production methods employed. When R. Steiner was giving his lectures in 1924 he pointed in detail to this interrelationship.

Organic soil management and the creation of a varied landscape, etc., are in themselves effective pest control measures. But he also gave specific indications on how weeds, animal and fungal pests can be kept under control. The application of these indications has in a number of instances led to positive results, but there have also been failures. Steiner suggested, for instance, that under certain conditions the ashes of some seeds, pests, etc., be used. It is now obvious, however, that not enough is yet known about the conditions under which this approach would be effective.

Feeding of livestock is today based on the feed tables that have been developed and refined over decades and are now used to compute the rations needed to achieve a specific yield. In the last of his agricultural lectures, R. Steiner gave a morphological and functional description of fodder plants and domestic animals. This adds a number of valuable aspects to what is already known.

This brief enumeration shows that the most important areas of agricultural production are dealt with in the lectures mentioned. Much of what is said in them has proved in practice to be of great use while some of the indications still need working out in more detail.

Rhythms

Recently, interest has been increasingly turned toward the cosmic growth factors. These include more than light and warmth. Research into rhythms is an appropriate tool for gaining additional insight into aspects of the influence of the sun and also of the more subtle relations between plant growth and the moon or planets. Rhythm research has led to the discovery of numerous biological rhythms that oscillate simultaneously or in identical phases with cosmic rhythms. It is important to know these, although to begin with one will find little more than a coincidence of events, while the nature and significance of such rhythms for the growth of plants and man is not yet necessarily clear. A real understanding of them can only be achieved through penetration into the subject with experiments and reasoning in the sense indicated earlier. One might start, for instance, with an attempt to grasp the total phenomenon "moon" including moon-related events in the hydro- and biosphere of the earth. It is a matter of gradually learning about relationships the nature of which is as yet unknown.

Extensive experimental work by M. Thun (1967, 1968, 1969, 1971) and M. Thun and H. Heinze (1973) can be quoted. The work includes studies of a variety of field and garden crops, among others potatoes, cereals, pulses, French beans, radishes, etc. When, for example, radishes were sown daily under experimental conditions, the plants exhibited growth types with regard to root shape, root-leaf ration, flowers, etc., that at first could not be explained. It was possible to distinguish between four morphological tendencies: 1) squat, round radish, not much leaf; 2) longish radish, copious leaves; 3) delicate leaves colored red at the stalk, branched and plentiful flowering, relatively poor seed formation; 4) similar to 3) but stronger in color and better seed yield. A connection with the phases of the moon was unlikely. G. Wachsmuth suggested in a lecture that a relation might exist between plant growth and the sidereal revolution of the moon, i.e., the passing of the moon during one month (27.3 days) in front of the Zodiac. When this pattern was applied to the four morphological types described here it turned out that they were indeed related to the four Zodiacal groups, each of which includes three constellations.

The following diagram No. 8 gives an example. The figures 1 to 12 number the constellations beginning with Pisces and moving counterclockwise around the circle. The yields are marked on the ordinate. One sees that for instance the yields for roots are high in numbers 3, 7 and 11 and for leaves in numbers 1, 5 and 9, etc.

This is what the authors emphasize: In order to obtain results in respect of the moon's effects in the twelve regions of the zodiac the following points must be noted:

- Seed or tubers must be planted when the moon is in the respective constellation;
- It is essential not to work the soil too shallowly on the day of planting or sowing;
- All further cultivation and other measures must be applied when the moon is in the same constellation or in one that belongs to the same group;

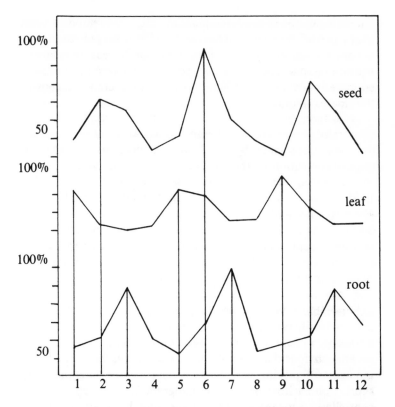

Growth experiment with radishes, 1967. Twelve constellations of one sidereal moon revolution. Yields of roots, leaves and seeds.

- Avoid days and hours when other particular events are expected such as full or new moon, apogee and perigee, nodes, eclipses, occultations, etc.;
- Irrigation was not applied as this seems to interfere;
- Experiments should be conducted on fertile soils with a good organic content.

It looks as though the soil is what mediates the effects. Of course it can be expected that planting, etc., in a temporal series over a period of one month has certain effects on growth that go beyond the seasonal trends of increasing or decreasing light and warmth factors. The experiments so far available showed repeatedly the four types in different crops and the coincidence with the sidereal month.

In a recently published doctoral thesis, accepted by the University of Giessen (West Germany), U. Abelle (1973) took up the suggestions made by M. Thun. Connections between the sidereal revolution of the moon and the growth of crops were found. Spring barley, planted between 21 April and 14 May 1970, yielded 550 kg/ha more when planted on "seed days." The difference to the other sowings is significant at the 5% level. Spring oats in 1971 showed certain relations, however, that were not significant. Carrots planted between 19 April and 14 May produced up to 20% higher yields, i.e., significantly more when planted on "root days." An effect on the shape of the carrots was also demonstrated.

For eighteen years up to the present (1972), A. Fyfe (1967) gathered daily samples of Christmas rose (*Helleborus niger*) and mistletoe, which she analyzed by a chromotographic method. It became obvious that the influences working on the plants at the moments of picking varied widely, even if the intervals were short. Comparisons of the chromatograms taking account of the constellations made it clear that there were definite connections.

3 The Bio-Dynamic Preparations

There are two groups of preparations. The first consists of Nos. 500 and 501, which are applied in spray form. No. 500 consists of cow manure and is sprayed on the soil, and No. 501 consists of ground quartz and is sprayed onto growing plants. During processing, No. 500 is exposed to environmental influences for the duration of one winter and No. 501 is exposed to these influences for one summer. The second group consists of Nos. 502-507, all of which are made from plant substances and are applied to composts and other farm manures. The numbering stems from the time when they were being developed and has no signficance. The actual amount of the substances used is small. The preparations work "dynamically," regulating and stimulating the processes of growth.

The substances to be used for making the preparations should be carefully selected. The cow manure for No. 500 is best collected while the animals are on full pasture or part pasture supplemented by good hay. For No. 501 it is best to use rock crystal if obtainable, or otherwise quartz from the clefts in silicious rocks; feldspar (orthoclase) may also be used. The preparations to be used on manures and composts are produced from the following plant substances:

No. 502: yarrow blossoms (*Achillea millefolium*)
No. 503: camomile blossoms (*Chamomilla officinalis*)
No. 504: stinging nettle (*Urtica dioica*), the whole shoot when the plant is in full bloom
No. 505: oak bark (*Quercus robur*)
No. 506: dandelion flowers (*Taraxacum officinale*)
No. 507: valerian flowers (*Valeriana officinalis*)

Some animal organs are also used: a bovine mesentery; bovine intestine; the skull of any domestic animal; a stag bladder. The details of how to process these substances may be found in the agricultural lectures of R. Steiner.

A great deal of experience and knowledge has been gained over the years in the application and production of these preparations. Originally many bio-dynamic farmers and gardeners made one or all the preparations themselves. This is of course the best way, for the observations made while doing so helps in gaining an understanding of the whole subject and thus also for the application. But the shortage of labor has had its effect here too. Nowadays the preparations for use on manures and composts are usually made by the local adviser in collaboration with individual farms and then distributed to all the farms and gardens in his area. On the one hand this development is a loss, but on the other hand it does mean that the preparations are made each year by the same experienced people.

Using bacteria isolated from the bio-dynamic preparations, E. E. Pfeiffer has devised a formula that he has called the Compost Starter. Originally it was mainly intended for large-scale composting of municipal wastes (Linder 1971). This inoculent, which is applied as a suspension in water, was used in North America.

Understanding how the preparations work

Compared with the weight of soil or of farm manure, which is used in amounts of 15-30 t/ha (6-12 t/acre), the amounts of substance applied as preparations are small. They are not "nutrients."

The use of the word "dynamic" varies. Literally, it means "pertaining to force." If a substance works "dynamically," this then means that it affects the metabolic and growth processes. All substances, including nutrients, which the plant needs for its growth do this. A distinction, however, is usually made between two groups of substances; on the one hand, the substances that are building bricks such as carbon, nitrogen, sulphur, phosphorus, etc., which are contained in plant substances such as fibres, protein, etc.; and on the other hand, the elements that, like most trace or micro-elements, but also for instance potassium, fulfill a regulating function in metabolism. These elements, such as manganese, boron, copper, zinc, are highly effective and appear in amounts of 30-500 g/ha (10.1-7.0 oz/acre) in a good cereal harvest. The more active a substance is, the more critical is also its possibility of switching from a positive to a damaging effect. This is particularly likely if a trace element is applied in excessive amounts as a fertilizer, or if it is present naturally in too great a concentration, as has been observed in the case of selenium. So it is important for balanced growth conditions that the ratio of amounts and the timing of the availability of substances to plants should be regulated by the processes in the soil. These are the biological processes that take place in optimum conditions with regard to the supply of humus, water and air.

Preparations 500 and 501 can be understood out of the polarity of earthly and cosmic influences under which the plant grows. The cosmic and earthly growth conditions in the environment are supported and improved by these preparations. The following table will clarify what is meant. A number of indications were given in the last chapter about the varied effects of silica on plant growth. In the case of these preparations this substance is present in a high dilution.

Yield and quality under the influence of polar opposite growth factors	
earthly influence	cosmic influences
include among others:	
soil life; nutrient content of soil; water supply; average atmospheric humidity.	light, warmth and other climatic conditions, and their seasonal and daily rhythms.
vary locally according to:	
clay, nutrient, humus, lime and nitrogen content of the soil; nutrient and water holding capacity; temperature and precipitation.	sun; cloudiness; rain; geographical latitude; altitude and degree of exposure; aspect of land; annual weather pattern; silica content of soils; etc.
normal influences on growth:	
high yields; protein and ash content.	ripening; flavor; keeping quality; seed quality.
one-sided (unbalanced) effects:	
(when superabundant naturally or from poor management)	
lush growth; susceptibility to diseases and pests; poor keeping quality.	low yields; penetrating or often bitter taste; fibrous woody tissue; hairy fruit; pests and diseases.
managerial measures for optimum effects:	
liberal application of manure and compost treated with bio-dynamic preparations; sufficient legumes in rotation; compensating for deficiencies; irrigation; mulching.	use of mature manure; no over-fertilizing; compensating for deficiencies; suitable spacing of plants; amount of seed used.
use of Preparation 500.	use of Preparation 501

A number of details could be added to these indications. They depict two formative tendencies that can be found in changing ratios in every individual plant but also in every field or landscape. They are influenced by both the preparations, which should always be used in conjunction with one another, though not at the same point in time.

Using the sprays

When bought, Preparation 500 comes in portions of about 80 g (app. 3 oz). 4 portions are used per hectare;* when packed per acre, one set contains 4 oz. Preparation 501 comes in portions of 1.2 g (1/20 oz). 4 portions again are used per hectare, when packed per acre, one set contains 1/5 oz. The amounts used in gardens will be higher. Spraying is done after stirring in water, whereby the ratio of preparation to water can be varied at the discretion of the farmer or gardener, within limits of course. On the farm, 50-60 1/ha (5-6 gal/acre) are usual, whereas for gardens it is practical to use a relatively larger amount of water, i.e, 8-10 1 per 5-10 ar (2-2.5 gal per 5,000-10,000 ft²). Rainwater collected in one or several wooden barrels is used. Spring or pond water are only substitutes if rainwater is not available and care must be taken to ensure that no stream with any kind of pollution flows into the pond in question. If tap water has to be used, it should be allowed to stand for 3-5 days beforehand. Before stirring begins, the water is warmed to about hand temperature, 35-37°C (95-100°F). There must be no traces of detergent in the water. Rainwater, too, is becoming increasingly polluted, particularly in the downwind area of industrial districts. In this case one might have to find suitable water from a well or pond.

The preparation is stirred for one hour by hand or machine into this water. A wooden barrel, earthenware crock or enamel

*The amounts given here are higher than those suggested in the pamphlet "Bio-Dynamic Sprays" (Bio-Dynamics 1971), which reports the applications as made in Great Britain. In most of the countries on the European continent, farmers and gardeners prefer the rates of application reported here. Research and experience have produced positive results in either case. In recent experimental work (Abele 1973) the higher applications have also been used.

bucket are used. The direction of stirring alternates rhythmically. As soon as a deep vortex appears in one direction, the direction is abruptly reversed, the vortex destroyed and built up again in the opposite direction. Thorough mixing is thus achieved.

The size of the vessels used for stirring can vary from a small bucket to barrels with a capacity of 180-200 1 (45-50 gal). The larger containers, however, are only half or two thirds filled, i.e., with about 100-120 1 (25-30 gal). This is about as much as a person can stir by hand, but it is better to stir less in one go if possible. For stirring the larger amounts, the stirring pole, with a few twigs attached besom-fashion, can be suspended from an overhead beam so that it hangs into the water but does not quite touch the bottom. The smaller amounts can be stirred with the bare hand or with a small besom-like brush.

Wooden barrels are preferable for storing the water and for stirring. Metal containers should have a rust-free coating. It is advisable to have a few barrels of water ready for use, though any not used within a reasonable time should be replenished with fresh rainwater when possible. Used barrels and also glazed or metal containers being prepared for use should not be cleaned with a detergent but with a scrubbing brush and hot soda lye. After rinsing, the water in the containers should be replenished several times during the course of a few months before they are finally used for the storage and stirring of the water that is going to be used. The water containers, stirring vessels and spraying equipment should not be used for any other liquids.

If stirring is to be done by machine, farmers usually construct their own or commission a local craftsman. Guidelines can be obtained from the bio-dynamic information services. Care must be taken to ensure that the machine carries out the movements described above as exactly as possible, i.e., formation of vortex, abrupt reversal of direction, formation of new vortex. A promising approach has been developed by bio-dynamic farmers in Australia. More information can be obtained from the Bio-Dynamic Associations in America, Britain and Australia.

The preparations must be carefully stored. After production

or delivery by the supplier, Preparation 500 is best stored in an earthenware crock with a loose lid. The crock is placed in a dark, cool cellar in which the humidity is fairly high. It is enclosed in a wooden box that is lined with sphagnum peat, including the lid. The condition of the preparation should be inspected regularly, since it should neither dry out nor go mouldy. It should be moderately damp and crumbly.

Farms often produce their own Preparation 500, whereas 501 to 507 are obtained from a supplier. Most of these are stored in the same manner as 500. The wooden box can be large enough to contain all the crocks separated by intervening layers of peat. The peat can be prevented from blowing about when the box is opened by covering it with a piece of cardboard or some sacking. The peat, too, should not be allowed to dry out, thus acquiring a dusty consistency. Every now and then it should be sprinkled with water.

Preparation 501 is stored in a glass jar with a screw top on a sunny windowsill. No. 507, which is liquid, is stored in a dark glass bottle with a cork in a room that is not too warm. Small-scale gardeners who do not store the preparations are advised to order shortly before they wish to use them.

In gardens, spraying is usually done with a knapsack spray, though on small beds a whisk broom will do. In agriculture, mechanical sprayers with the nozzles facing downwards or outwards are used. After stirring Preparation 500, the liquid is allowed to settle for about 10 minutes and then poured through a cloth and the hair sieve of the spray into the container. Clogging of the nozzles is thus prevented. Spraying should be finished 1-1½ hours after stirring.

The cow manure Preparation 500 is sprayed onto the damp soil before sowing or latest shortly after emergence. A favorable time is shortly before working with a cultivator, harrow or hoe. Both spring and autumn sowings or plantings are treated and often any land lying fallow is treated at the same time, so that all the fields in the farm in fact receive a treatment in autumn and in spring. The time for spraying is latish afternoon, so that the evening dew can settle afterwards. Spraying should take place when rain is not expected. Spraying is not done during rainfall. Subsequent shallow working of the ground is desirable. Grasses and hayfields are given a spraying in early spring. The ground should have started to warm up or the first fresh shoots be visible. Further applications can be made after grazing or cutting. In horticulture, Preparation 500 is sprayed onto the soil and also in seed boxes, cold frames and green houses in autumn and spring. Whenever possible, every sowing or planting is preceded by a treatment. There are always several areas that can be sprayed at the same time.

Preparation 501 is sprayed on leaves. Number 500 is applied before sowing or after harvesting of crops that continue to grow, but there is no such simple rule for 501. On the other hand, the use of 501 can be more differentiated and subtle during the various stages of plant growth. First, after Preparation 500 has been used at the beginning of growth, a period of growth follows during which Preparation 501 should not be used. This becomes obvious if freshly planted seedlings are treated with 501 too soon. There may be a slight yellowing of leaves, or in the case of lettuce the plants bolt and do not form proper heads. These negative effects thus demonstrate the influence of the extremely small amounts of substance used. The plant must be properly rooted in the earth before any silica is used on it.

Preparation 501 is sprayed in the morning when a sunny or at least partly sunny day is expected. The application is started when the organ later to be harvested is starting to form. Cereals are treated quite early on, when three tillers have been formed and elongation has started. Later treatments, which also help standing ability, can follow. Potatoes are sprayed until they start to blossom. Gardeners continue to spray their potatoes after blossoming, in the afternoon. This helps tuber formation. Clover and clover grass are treated not long after growth has fully commenced. Root vegetables are treated after the root has begun to fill out or, in the case of carrots, when it begins to turn orange. Head-forming leaf vegetables are sprayed when the heart leaves begin to turn inwards. Treatment can be repeated later. Restraint is necessary in the case of lettuce, otherwise bolting and seed formation will be promoted. Peas are sprayed when the first pods begin to form. The spraying of strawberries and tomatoes can continue during the extended period of blossoming. With tomatoes the first treatment can be when the first fruits reach the size of walnuts. Fruit and soft fruit are sprayed when foliage is fully out and later when fruits have started to form. Apples should have reached a diameter of about 2 cm (¾''). Further spraying can be done when ripening starts. Several sprayings are recommended when blossom and fruit formation are important or if the plant is used for its fragrance or other qualities arising out of a maturing process as in the case of medicinal plants and aromatic herbs. Further details may be obtained from the bibliography in this book and from reports in journals that appear regularly (Corrin 1960; Kabisch, Koepf 1971; Remer 1949). Everybody must also make his own observations and gain his own experience; the right intuitive feeling is important for successful treatments.

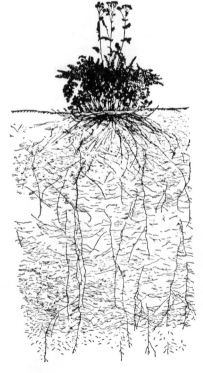

The blossoms of yarrow are used to make Preparation 502.
This plant prefers dry habitats.
Blossom, leaf and stalk are all highly fragrant (After J. Bockemühl).

214

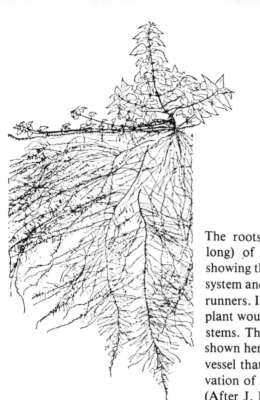

The roots (about 100 cm, 40'' long) of the stinging nettle showing the strong root system and propagation by runners. In a closed stand this plant would form long, erect stems. The single specimen shown here grew in a special vessel that allows observation of the root (After J. Bockemühl).

Preparations for manures and composts

These preparations are used in both solid and liquid farm manures and composts. No. 507 can also be sprayed as a prophylactic measure if there is a threat of frost. This will be discussed further in the section on gardening. Farm manures are added to soil in order to stimulate the life in it. Soil with plenty of active microorganisms is able to supply crops with nutrient elements and trace elements. It is the task of the preparations to regulate these processes. Frequently, watery extracts of these preparations are also used as seed dressings for cereals before sowing. The results observed are fewer attacks by fungal diseases and also earlier and stronger germination. It is well known that valerian (No. 507) stimulates the earthworm population. Let us first discuss the use of the preparations in farm manure.

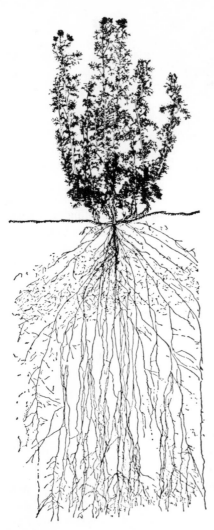

Camomile in a vessel
allowing root observation;
root length about
120 cm (48'')
(After J. Bockemühl 1969).

In compost and manure heaps a set of Preparations 502-506 is applied in such a way that the individual substances are placed 1.5-2 m (5-7 feet) apart. About half way up the side of the heap a hole is poked slanting downwards for about 50 cm (20 inches), at the bottom of which a teaspoonful (not heaped) of the prepara-

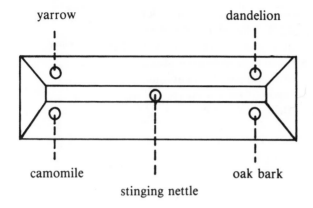

yarrow   dandelion

camomile   oak bark

stinging nettle

Example for the placing of the preparations in a compost heap.

tion is placed (except for the stinging nettle Preparation 504, of which 4 or 5 times that amount is used if available). If necessary it can be mixed with composted earth to form a small ball that is poked to the bottom of the hole. The hole itself is then filled again with the material and closed. The compost heap is approached from both sides. For example, with a pile of garden compost measuring about 3 cubic meters or yards one can go round the heap once using each preparation once. For each set used, 1-2 ml of Preparation 507 in 1 gallon of handwarm rainwater is applied to the surface of the heap with a watering can and rose. If a heap is being added to lengthwise, the preparations are applied every 3-4 meters or yards. In the case of farm manure in rectangular piles, the preparations are also added at distances of 2 meters or yards from each other after a height of 50 cm (20 inches) has been reached. In this case each is scattered over a plate-sized area and covered. The same is done in the loafing or covered yard on every 30 cm (1 foot) layer. Alternatively the preparations can be inserted in the manner described above after longer intervals of time.

For liquid manure small bags can be made from a suitable material to hold the preparations. These are weighted with a

Fully developed valerian plant, root length 110 cm (44''). This plant, which prefers damp, often shady habitats, flowers after 2-3 years when grown from seed. Subsequently propagation continues by means of underground shoots (After J. Bockemühl 1969).

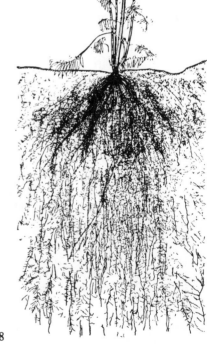

This dwarf variety of dandelion (*Taraxacum laevigatum*) is shown here because it provides a good example of a strongly developed tap root; root length 120 cm (48'') (After J. Bockemühl 1970).

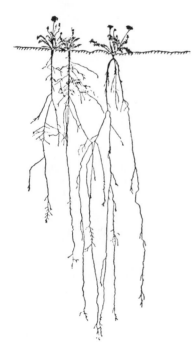

218

stone and suspended from wooden floats so that they hang in the midst of the liquid. One set of preparations is sufficient for 8-10 cubic meters or yards of tank space.

The compound preparation mentioned on p. 169 is made of the solid excrement of cattle (in some instances, the manures of other animals have also been included but this is not a must). One builds a container two thirds of which are in the ground; the walls are lined with thin wooden boards while the bottom remains open (i.e., is formed by the soil). Fruit boxes or bottomless wooden barrels are sometimes used. The size of the container is 60x60x60 cm (24x24x24"). Another handy design is 2 m long with a height and width of 75 cm (80x30x30"). This pit is halved to form two compartments. The container is filled with the manure to which are added 2 or 4 sets of preparations respectively, including Preparation 507. The container is covered with a lid that should not fit tightly but does protect against rainfall. After several weeks the result is a crumbly, brown, friable, pleasant smelling material that can now be used.

As indicated on p. 169 it is spread on the manure before barn cleaning. It can also be spread in loafing barns or yards. Also a shovelful can be spread on 1 cubic meter or yard of compost material and mixed in as the pile is built. Liquid manure and slurry can be treated with good results.

Experimental results

There is a great deal of documentation about experiments on the varied effects of these preparations in manures, seed dressings, etc. Effects on germination, growth, yields, root formation and nodule formation, essential oil content, etc., have been observed. Experiments testing the preferences of earthworms for soils with or without treatment have also been undertaken. E. and L. Kolisko reported on many experiments in 1939 in a book which is no longer available. Similar experiments are reported in E. E. Pfeiffer's book *Die Fruchtbarkeit der Erde* (1969) (*The Fertility of the Earth*), which is available in German and highly recommended. The books by Pfeiffer and Kolisko are available for reference at most centers of bio-dynamic work. At this point we limit ourselves to reporting on two experiments that supply answers to two frequent questions.

It is not easy to measure the effects of small amounts of substances on compost. This is also demonstrated by the fact that numerous bacterial inoculants or starters of various kinds available for compost heaps have no noticeable effect on the temperature curve nor on the content of nitrogen and other nutrient elements in the end product. These characteristics are determined rather by the primary factors of composting, which include the composition of the initial mixture, its water content and aeration. Its microbial activity is an expression of these; in non-pasteurized material it cannot be altered by inoculation. It is a different matter to mix enough (i.e., 10% of the volume and more) ripe compost with a heap that would otherwise be too onesided in its composition. This ripe compost provides shelter for bacteria that will be needed in the later stages of rotting. Meanwhile, they can start to multiply as soon as conditions are right. The external appearance soon shows the beneficial effect of ripe compost.

It must be stressed that the bio-dynamic preparations must not be confused with bacterial inoculants. They work through the substances' own forces. At the same time, however, it is frequently observed that they also have a beneficial effect on the rotting process and the ripening of the compost. The question is, can this be corroborated experimentally?

A compost experiment carried out by the Institute for Bio-Dynamic Research in Germany was described in 1962 by Heinze and Breda. The results are based on three independent series of experiments. In each series the treatments were repeated three times. The results pointed in the same direction and cannot therefore be regarded as coincidental. Out of all these series we shall here show only those comparisons that relate to identically assembled and handled composts when treated or not treated with the bio-dynamic preparations. The three series are:
a) Compost consisting of 50% stable manure, 10% loamy soil, 10% basalt grit, 20% green plants (weeds) and 10% ripe com-

post. Each heap measured 4 m³ (3.2-4 tons). 50 g of preparations were used in the usual way, i.e., approximately 12-15 ppm (1 ppm = 1 g/t).

b) Cow manure in pits 60x60x40 cm (2x2x1.5 ft). Series b)1 contained only cow manure and series b)2 contained cow manure with 20% loam and 15% basalt grit. Bio-dynamic preparations: 25 g (1 oz) per pit.

c) Composted manure, fresh and also temporarily stacked manure. Unlike a) and b), which were situated on the sandy soil of the Institute's premises in Darmstadt, the heaps for this experiment were established on the loamy soil of Dottenfeld Farm at Bad Vilbel. Bio-dynamic preparations as in a).

Among the various tests carried out, the effects of the preparations became apparent in the figures that point to the ripeness of the organic substances. This ripeness can be measured by the ability of the humus to store nutrients. This is a definition that is too complicated for routine checks and the technical details need not be discussed here, since we are simply concerned with a comparison. The unit is T/C_t (Sauerland and Trappmann 1954). T is the storage capacity for nutrients (the so-called exchange capacity) and C_t (total carbon) indicates the organic content of the material.

The influence of the bio-dynamic preparations on the ripening of organic substances in composts (T/C_t values).

Series	without	with
	bio-dynamic treatment	
a) top layer of heap	2.78	4.40
bottom layer of heap	3.03	4.38
b) 1	4.33	4.55
b) 2	3.80	4.16
c) composted stacked manure	8.51	8.81
composted fresh manure	10.34	11.59

The higher figures in series c) are of no significance. Unlike series a) and b), the storage capacity of the inorganic substance was not subtracted. We see that the small amounts with which the heaps were treated have an obvious effect on rotting, which can be registered by chemical analysis.

Manure and soil humus are also physiologically active substances. This gives us a further possibility for registering the effectiveness of preparations. In a laboratory experiment, cow droppings were diluted with water in a ration of 2 : 3. This slurry contained 4.1% dry matter. It was fermented for 3 months under anaerobic conditions at a temperature of just 29°C. Increasing amounts of a mixture of bio-dynamic preparations (mixed at a ratio of 1 : 1 with compost) were added. The additives (laboratory code CPU) stimulated the decomposition of carbon compounds in the liquid manure and there was no loss of nitrogen. The fermented manure was then added in an amount of 1% to a nutrient solution in which germinating wheat plants were raised (Koepf 1966). Some of the findings are shown in the following table:

Influence of diluted cow manure, with and without bio-dynamic treatment, on the growth of wheat seedlings

| treatment | length in mm of | | dry matter from |
	root	shoot	100 roots in mg
control	75.5±0.23	100.0±1.1	364
1% liquid manure	87.9±2.64	102.8±0.33	364
1% liquid manure with 0.08% CPU	99.8±1.6	106.6±0.73	428
1% liquid manure with 2.4% CPU	92.7±0.15	106.2±0.58	410

Fresh manure stored simultaneously in a refrigerator also showed a plant hormone effect. This, however, is not the question here. We see that the preparations affect the length and above all the weight of the roots. The amount of 0.08% CPU in the vessel, half of this amount being the actual preparation, is equivalent to approximately 2 ppm of preparations. This is about the amount used in farming practice, but the effects are much harder to define in the open field since so many other factors play a part. Experience has shown again and again that bio-dynamically treated farm manure favors the development of strong root systems (see p. 230).

Three other much-used substances connected with the bio-dynamic preparations should be mentioned here.

Horsetail tea

The plant in question is the common horsetail (*Equisetum arvense*), which likes habitats that are not too damp. It has barren green stems which continue beyond the last whorl of branches. It is distinguished from other species by the fact that the lowest section of the stalk-like leaves is longer than the corresponding sheath of the main axis. This delicate plant, which is rich in silicic acid, is made into a tea either from the green state or dried. 300 g (10 oz) of the dried leaves are boiled and then diluted with 15 l (4 gallons) of water. Green plants can also be covered with rainwater and either boiled or left to steep for several weeks before diluting. Frequent spraying is done in the garden and under glass to prevent fungal diseases such as mildew, scab and rust. The tea is also often added to sprayings with Preparation 501 or to stinging nettle liquid manure. 1-2% of waterglass (sodium silicate) are sometimes added. For fruit crops particular care must be taken to ascertain how much waterglass can be used without causing burning. This depends on the variety and the weather. Burning is more likely in strong sunshine. In the garden horsetail tea can also be added to watering water. The effect of horsetail tea is preventive, not fungicidal.

Stinging nettle liquid manure

Stinging nettles can be added in quite large amounts to manures and liquid manures. If green plants are left covered with water in a barrel or tub for as little as 24 hours, the result is a liquid, which, though hardly visibly altered, can be used as a spray against aphids when diluted to proportions of between 1:5 and 1:10, though it may take three days for the effects to show. If left to stand for longer periods, the extract begins to color and smell strongly. This can also be diluted and sprayed on leaves in the garden. It is particularly beneficial during a drought. It is either added to watering water and applied by watering can with rose to soil and plants, or it is sprayed together with Preparation 501.

Fruit tree paste

This is used to make the bark of fruit trees smooth, thus affording little shelter to pests. 20 kg (50 lbs) of clay and 10 kg (25 lbs) of cow dung are mixed together and made into a paste with about 15 l (4 gallons) of Preparation 500 and the same amount of horsetail tea. The stickiness can be increased by adding starch. A similar liquid of clay, cow dung and Preparation 500 is used as a root dip for cabbage, etc., before planting out.

4 Biotic Substances

"Biotic" means "through biological action." The biotic substances meant here are a particular group of growth regulating substances.

The harmonious growth and functioning of warm blooded organisms, e.g., domestic animals, is regulated by the secretion of enzymes and the hormones of the endocrine system. Growth, orderly functioning and health depend on the normal working of this system. If there is hypo or hyperfunctioning of one or more of these glands, the animal organism is ill. Plants do not have a comparable glandular system. Their growth is regulated mainly by their environment and only to a lesser extent from within. So plants are an expression of environmental influences to a greater degree than are animals.

In the plant and in the soil there are growth regulators some of which are also called plant hormones. The best-known of these is β-indolyl acetic acid (heteroauxin). This substance stimulates cell division and enlargement. Another group of these substances, the cytokinins, influence the ageing processes in plants. The heteroauxin level is maintained in the plant by an interplay between synthesis and decomposition. The same goes for the soil where the growth regulators of different kinds usually correspond to an activity of 0.02 ppm β-indolyl acetic acid. The relations between plant and soil are not restricted to the intake of water and mineral nutrients by the plant. It was thought by scientists and farmers that this was the case during the period when the theory was held that plants absorb from the soil only inorganic nutrients. We now know that the relationship is more complicated. In agriculture and horticulture not only chemical fertilizers but also such substances as iron chelates are used; others enhance the rooting of cuttings. Gibberellins can bring about a lengthening of the internodes and also influence the dependence of plant development on temperature and light rhythms. Auxins can bring about fruiting without pollination, and so on (West 1962; Hess 1972).

If the manuring, rotation and management of crops is in order, they will be able to resist diseases and pests to a degree that is often astonishing. The reasons are not always clear, but part of the explanation is sure to lie with biotic substances (Koepf 1966). Heteroauxin does not exist only in plants and the soil but is also present in farm-produced and commercial organic manures. Others exist in microorganisms or plant roots whence they are in both cases excreted into the soil. There are also gaseous substances. For instance, the well-known apple gas, ethylene, is excreted by dandelions and is said to speed ripening processes in neighboring plants. Recently it has gained attention among plant physiologists as a factor that influences the maturing of plants. Those humus constituents, root excretions and microorganism excretions, etc., which enter the plant can have varying effects. They can be used as nutrients or they can regulate growth, either retarding or accelerating it. Others can increase plant resistance to disease. Others have no effect or their effect on yield or quality is not known. All in all there is a tremendous mingling of substances, which will only gradually be unravelled by detailed research. In many cases the effects are mild. In others, such as the stimulation of germination, they are transitory. Descriptions of this interesting field of research may be found in the works of Kononova (1961), Scheffer/Ulrich (1960), and Krasilnikov (1961). Domsch (1968) distinguishes between four different ways in which soil microorganisms can stimulate or retard the growth of higher plants. In a euparasitic relationship the latter are damaged by a specific toxin; if the influence is hemiparasitic, there is usually an unspecified toxin that causes disturbances in the plant; symbiosis occurs, for instance, with mycorrhiza; finally there are probiotic relationships such as that with the gibberellins already mentioned. However, the influence of soil microorganisms on the formation of soil structure and on the availability of nutrients from the soil minerals can in many cases be more important than the above.

For the farmer, who is less interested in scientific intricacies, it is important to know that these delicate biological influences exist. He also ought to know what he can do with their help to influence the quantity and quality of his harvest in a positive way. Let us first discuss an example connected with the biotic substances in humus, fertilizers and crop residues.

During World War I, Bottomley in England (1914), and later also others, found that a watery extract of composted peat stimulated the growth of duckweed and also cereals, buckwheat and peas. Composted peat worked more strongly than peat that was not composted. (During a question and answer session in the twenties, Steiner was asked what he thought of these experiments. He drew the attention of those present to the fact that hypertrophic growth could have a negative influence on the quality of fruit and its affects on man. The eating of such products could possibly only show its consequences in the following generation.) Later many experiments were conducted that showed how humus extracts from the soil could stimulate growth. Many reports concerning this are available. For instance Kononova was able to stimulate root formation in young cereal plants with humus extracts obtained from the soil. Christeva

(1948) found that small amounts of humus extract had positive effects on the growth of wheat, barley, oats, millet, maize, rice and buckwheat. The amounts of substance used were small: 0.006 - 0.0001%. Potatoes, tomatoes and sugar beet showed the most marked reaction. The influence was small on peas, beans, lentils and peanuts and hardly noticeable on oil seeds. Other similar experiments have corroborated the evidence that acceleration of growth is greatest in plants that form starch and sugar. This may be related to the fact that some of these humus compounds affect the respiration and thus the net result of photosynthesis in the higher plants.

On the other hand, retardation of germination and growth can also be caused by organic substances. (The toxin theory is among the explanations given for "tired" soils. It is assumed that onesided cropping or the continued planting of a single crop leads to an accumulation of inhibitory substances. Other causes are the build-up of a pest population, exhaustion of a particular nutrient in the soil, and poor soil structure.) There are of course also natural forms of humus that do indeed inhibit growth and germination. Examples of this are raw humus layers in coniferous forest or under heather. Bogs can also have a preserving and tanning affect. This does not contradict Bottomley's findings, since he used composted peat. It is interesting to note that ordinary crop residues such as straw and stubble, and also green manure plants, contain factors that temporarily inhibit germination and growth. This has become known through the work of Winter and colleagues (1952, 1953, 1954), and Schonbeck (1956, 1957). It is not the negative effects of nutrient deficiencies or acid soil reaction that are meant. We are concerned here with the effects of certain organic compounds. McCalla (1948) made an extract of sweet clover with 5-100 parts of water per unit of plant substance. These extracts had inhibiting effects on maize germination ranging from 8% to 62%. The more concentrated the extract, the more marked the effect. Coumarin, a substance contained in sweet clover, was recognized as the inhibitor. Other inhibitors, of which the chemical nature is now known, are found in wilting leaves or stalks. They are related to lignin, which together with cellulose forms the woody parts of plants. These inhibitors, which are early stages in wood formation, are leached out of crop residues, etc., by water. Their effects in the soil can be shown to exist during the months after harvesting. They do not remain for long, however, since they are subject to decomposition and leaching. A good earthworm population accelerates this breakdown.

In actual farming it is not often possible to decide whether and in how far such inhibiting factors are at work if germination is unsatisfactory. In the more westerly regions of the USA, where cereals are the dominating crop, it is essential to use stubble mulch for soil conservation. It is calculated that straw and stubble mulch reduces water erosion after heavy rain by about 80%. Water loss through evaporation can be reduced by half. In dry regions wind erosion is also reduced by 90% (Agr. Res. 1964). Despite these obvious advantages for soil conservation, some farmers are hesitant in their use of stubble mulch. In years with more than average rainfall, yield losses occur that could be caused by shortage of available nitrogen during the early stages of plant growth, the physical condition of the soil, or the toxic substances we have just been discussing, which are leached in greater amounts out of the mulch when there is more rain.

These brief indications regarding the inhibiting and accelerating factors in the soil give at first glance a picture full of contradictions. It is nevertheless possible to order the various observations in a meaningful way.

In many cases it is a matter of influences that are observed during germination and the initial stages of growth and that affect the formation of the roots. Krasilnikov says about the growth enhancing influence of organic compounds: "The strongest reaction to humus substances is observed in young plants. The weight of the root and consequently the growth of the whole plant is increased." The retarding influences in the main also affect the root, though if they are strong enough they also make themselves felt in the shoot. The following table gives an example of this.

The degree of growth retardation depends on the concentra-

The influence of a watery extract from wheat straw on the first five days of pre-germinated wheat seeds (60 plants per treatment)

extract straw:water ratio	Knoop solution	length of root cm	length of shoot cm
control	Knoop sol.	8.4	10.7
1: 20	Knoop sol.	6.1	9.1
1: 50	Knoop sol.	7.7	10.0
1:100	Knoop sol.	8.5	10.3
1:200	Knoop sol.	8.7	10.5
1:400	Knoop sol.	8.9	10.9

tion of the extract and is also more pronounced in the root than in the leaf.

The inhibiting effects originate in certain crop residues that have not yet been sufficiently decomposed or transformed into humus substances by the soil microorganisms. Raw humus layers also form where "digestion" of dead plant residues, etc., is incomplete. In contrast, some extracts from stable humus can promote tillering and growth. These substances favor the formation of a strong root system. Gardeners have of course known for a long time that ripe composts stimulate root formation. Depending on how "ripened" the organic soil substances and manures are, growth is either retarded or the root formation of the plants is greatly stimulated. Klett (1968) has described this, for instance, in connection with a fertilizer experiment using radishes and spinach. The root lengths are shown here.

		radish, cm	spinach, cm
NPK fertilizer	(100:100:140 kg/ha) (90:90:125 lbs/acre)	24.5	17.5
Composted manure	(same amount of N, assuming 50% efficiency in first year)	27.0	23.0

It is interesting to note that these differences only occurred under normal light conditions and not when the plant received reduced amounts of light.

Another point may be added here. On p. 223 an experiment is described in which the effects of the bio-dynamic preparations on slurry from cattle were measured. Used on wheat seedlings, the slurry brought about an increase in the length and weight of the roots — a similar effect, in other words, to that brought about by humus extracts. The transformation of raw waste materials into ripe manure is helped by the preparations. Their regulating effects can be partly understood by the way they help plants utilize soil-born nutrients. This is assisted by a strong root system.

The qualitative nature of the organic substances in the soil determines their inhibiting or stimulating effect on growth. It can be influenced by the management and the handling of manures. Composting is one of the means, and the proper utilization of crop residues and green manures is also important. It is best, once they have been chopped if necessary, to work these lightly into the top layer of the soil so that they are mixed with the soil and its minerals. Materials that are poor in nitrogen need to be supplemented with undersown clover, and liquid or solid manure. Further, depending on the nature of the habitat, enough time must elapse before replanting or resowing. Where the conditions for this kind of sheet composting are achieved, the above-mentioned positive effects on growth can be expected. If these aspects are neglected, then crude organic substances may have transitory negative effects. These will last until the organic substances have been properly "digested" by the soil life.

Apart from those already discussed, other organic compounds also occur in the neighborhood of plant roots. Their importance for the plants can also only partly be assessed. It is appropriate, however, not only to search out the many individual effects that occur but also to discover the ecological conditions and management measures under which the whole complex can attain its optimal effectiveness.

It is usually the case that the organic compounds of plants are also to be found in the soil. Among these are the building bricks of protein, the amino acids, which contain nitrogen. Where humus development is favorable, i.e., in fertile soils or soils that are biologically active, the percentage of α-amino-nitrogen, e.g., nitrogen in amino acids, is higher than where humus development is unfavorable. Also phosphorus, one of the most essential plant nutrients, is present for the most part in organic compounds. The intake by plants of nitrogen and phosphorus compounds can be observed with the help of labelled elements. This intake is increased by the influence of microorganisms in the rhizosphere, as has been shown in comparisons with plants growing under sterile conditions. It also seems that the fate of such nutrients once they have entered the plant is different from that of substances that have been supplied in mineral form. This is shown by the way labelled sulphur is distributed in the plant depending on whether it has been introduced in the amino acid methionine or in an inorganic compound. Also the chelate and chelate-like effects of iron, manganese and phosphorus intake may be mentioned.

In bacterial cultures and also in most cases directly in the soil the vitamins riboflavin, nicotinic acid, pantothenic acid, folic acid, C, K, B_6 and B_{12} have been found. Also provitamin D_2, α and β-carotene. In two different groups of actinomycetes, 90% and 60% respectively were able to produce vitamin B_{12}. The spatial and seasonal incidence of these and other biotic substances follows, on the whole, the distribution of humus and soil microorgansms. There are more in cultivated than in uncultivated soil, and there are more under plants that, like hay crops, stimulate the soil microorganisms. The amount of such biotic substances in the soil decreased with depth, and there are more directly adjacent to plant roots than further away. Favorable humus types such as chernozem contain more vitamins and biotic factors than unfavorable soils such as podsols.

A third group of phenomena should also be mentioned. If microorganisms are grown in the laboratory with two, three or more species or families in the same culture, numerous examples of mutual inhibition or benefit are observed. Krasilnikov is of the opinion that among all the bacteria, actinomycetes, fungi, protozoa, algae, etc., there is not one for which there is not an antagonist that supresses or inhibits it. In the same way the growth of green plants including crops is subject to inhibition by bacteria or other plants.

Soils that harbor many different kinds of microorganisms are also likely to contain a considerable number that have inhibiting effects. Furthermore, the ripened, best types of humus contain many organisms that are antagonistic or inhibitive. The qualitative composition of the microorganism population changes according to the degree of ripeness of the humus. This is shown by an examination of the incidence and behavior of certain groups. The following table summarizes some of the results.

Incidence of antagonistic actinomycetes in different soil groups (1000 per one g of soil, abbreviated, after Krasilnikov, 1961)

soil group	number of actinomycetes	number of antagonists among these
5 chernozems	1500 (1000-2000)	410 (120-800)
3 podsols under forest	920 (360-1200)	250 (0-420)
iron podsol	0.4	0.3
humus podsol	2.0	1.5

Such results furnish an explanation for the incidence of certain plant diseases (e.g., Fusarium, Verticillium and Phthium species). Infected seeds are exposed to the environmental conditions of the soil. It depends on the antagonistic activity of the rest of the microorganisms whether pathogenic organisms can achieve mass proliferation or not. A varied microflora with many antagonists will be able to check the pathogenic organisms. The frequency of root and stem wilt and rot in young vegetable plants and seedlings then decreases. Success has also been achieved in preventing the damping off of pine saplings in tree nurseries by using ripe compost mixed with some forest litter added when the compost heaps were built.

Antibiotics are a well-known case of such antagonistic action.

232

Penicillin, streptomycin, terramycin, aureomycin, griseofulvin and others are products of soil-inhibiting organisms. Their persistence in the soil ranges from days to weeks. Synthesis and breakdown take place simultaneously. Antibiotics are taken in by plants and their presence can be traced in all parts of plants.

The cell sap of plants does not on its own permit the bacteria to grow. It is in fact bacteriocidal. Healthy plants do not contain any bacteria except possibly in the vicinity of a bruised part where the bacteria have been able to enter. The bacteriocidal activity of cell sap, however, is not quite constant. Plants have been experimentally grown in soil to which antibiotic-producing actinomycetes have been added. The bacteriocidal activity of the cell sap was then also found to be higher. This increase also took place after fertilizing with manure and compost, and it was higher in the open than in greenhouses (Krasilnikov 1961).

On the other hand, plant diseases occur even in soils with the best kinds of humus and compost treatment. The protection afforded by antagonistic microorganisms is limited. Nevertheless, this protection is a basis for healthy growth.

The many phenomena connected with biotic substances can be gathered under three headings. There are many exceptions but nevertheless a fundamental pattern is recognizable.

1. Inhibition of germination, growth and root development are often the result of crop residues that have been either not or too little broken down; or accumulations of raw humus; or organic substances rotting in wet, anaerobic conditions; or humus under onesided plant communities (e.g. raw humus layers in coniferous forests).

2. Favorable humus formation with a rich and varied population of microorganisms supports the rooting process of plants; germination, growth and tissue formation are stimulated and increased.

3. Ripe humus developing in soils and composts contains not only growth stimulants but also an increased amount of antagonistic activity; this keeps bacterial and fungal pests within limits and presumably also increases the resistance of plants to disease.

When a suitable habitat and good management lead to the favorable development of humus both in quantity and quality, a number of consequences for plant life can be expected. These include not only the supply of nutrients and the formation of soil structure but also influences regulating growth and protection against parasitic diseases.

The observations of many farmers and gardeners practicing bio-dynamic methods bear this out. Nevertheless, there is still a great deal to be learned. In his lectures on agriculture, Steiner pointed to the relations that have been discussed here, although they were hardly known or appreciated in his day. When he spoke about humus and composting, he pointed to the polarity between forced rampant growth and the controlled effects of matured humus. Compost, being a fertilizer that does not force growth, is pronounced to be a fertilizer for grassland, i.e., for a crop in which strong vegetative growth is typical. There is also a discussion of the role of the earthworm, which promotes the ripening processes in organic substances.

5 Companion Plants

The relationships described in the preceding section also determine in part the interaction between neighboring plants. But there are also other causes for this. Two neighboring plants might be competing for space, light, water or nutrients. The shade cast by one plant over another can either help or hinder its growth. The intake of nitrogen and also of other nutrients such as phosphorus by cereals or other non-leguminous crops is often improved if these are grown as a mixed stand with legumes. In many cases, however, the effects that neighboring plants have on one another are the result of excretions from the roots, or smells. There may also be indirect influences, such as when crop residues bring about changes in the formation of biotic substances via the microorganisms in the soil. So it is not surprising that many observations have been recorded concerning the mutual influence neighboring plants have on one another.

234

Experiments seeking an explanation of these relationships and the resulting statements made on the subject are, however, frequently not convincing. Many indications have proved to be useful and practical and we shall therefore quote a number of examples of these often subtle relationships. More detailed reports are given by Philbrick/Gregg (1966), Scheffer/Ulrich (1960), Krasilnikov (1961), and Wirth (1962).

Trees as companion plants. Beech leaves contain substances that inhibit the germination of ash, spruce, pine, fir, cereals and poppy. Ground flora are often seen to be inhibited under maple, robinia, walnut (*Juglans cinerea*), spruce and beech. Excretions from roots and leaves seem to be the cause of this. Beech and the seeds of other woodland plants germinate well under oak litter. Spruce litter inhibits pine saplings. Inhibitors present in the sward retard the development of the root tips of apple trees. Autoinhibition has been reported in connection with apple and peach trees. According to less recent investigations, negative influences on the growth of young peach trees are oats, potatoes, mustard and rapeseed, while beans and clover do not have this effect.

Field crops as companion plants. Inhibitors in the runners of couch grass affect rapeseed. Rye inhibits a number of annual weeds. Young wheat plants and alfalfa inhibit the germination of field mustard (*Sinapis arvensis*). On the other hand, wild pansy (*Viola tricolor*) germinated better in an experiment in which it was sown together with rye. Wheat and rye inhibit the germination of scentless camomile (*Matricaria inodora*) and mayweed (*Anthemis arvensis*). Dead nettle (*Lamium*) and sainfoin (*Onobrychis viciaefolia*) are beneficial bordering plants for cereals. It is desirable to have some camomile (*Matricaria chamomilla*) and the now rare cornflower (*Centaurea cyanus*) among cereals, but not poppy (*Papaver rhoeas*). Oats grown as a mixed crop with field beans or vetch, and rye grown with peas give good yields. In a field experiment creeping soft grass (*Holcus mollis*) had a negative influence on barley. Watery extracts from field bindweed (*Convolvulus arvensis*) and Canada

235

thistle (*Cirsium arvense*) have an inhibiting effect on wheat and linseed. Leaf extracts from fake flax (*Camelina sativa*) inhibit linseed.

A few yarrow plants (*Achillea millefolium*) on the land have a beneficial influence.

Bordering and companion plants in the garden. A few dead nettle (*Lamium amplexicaule*) and yarrow plants are welcome in the garden along the edges of vegetable beds.

Anise (*Pimpinella anisum*) and coriander (*Coriandrum sativum*) help each other with seed formation and germination, while basil (*Ocimum basilicum*) and rue (*Ruta graveolens*) do not get on well together. Nor do caraway (*Carum carvi*) and fennel flower (*Nigella sativa*). The latter is rarely found in gardens nowadays; it is also not a good companion for beans, tomatoes and kohlrabi. Sage (*Salvia officinalis*) and rosemary (*Rosmarinus officinalis*) are good companion plants. Oil formation in peppermint (*Mentha piperita*) is said to be negatively influenced by camomile. Summer savory (*Satureja hortensis*) is recommended as a bordering plant for onions. Parsley has good effects on neighboring roses and tomatoes. Wormwood (*Artemisia absinthium*) contains inhibitors, among them absinthin, that are soluble in water. This plant inhibits the growth of numerous neighboring herbs, including groundsel (*Senecio vulgaris*). The inhibitors from wormwood stay in the ground for awhile. Many essential oils, for instance those from citrus plants, peppermint, thyme, germander, eucalyptus, coniferous resins, and the poplar, inhibit seed germination.

Stinging nettle (*Urtica dioica*) is used not only for the bio-dynamic Preparation 504 but also fermented extracts and liquid manure. Planted in rows between medicinal herbs, it increases the content of essential oils, particularly in the peppermint.

Planting carrots and onions in alternate rows is supposed to deter both the carrot fly and the onion fly. Some garlic (*Allium sativum*) can be included in the rosebed. Bulgarian peasants used to grow garlic in their rose fields to promote the formation of oil and aroma. Alternating rows of leek and either celery or celeriac are a good idea.

The various cabbage varieties should be accompanied by flowering plants. Dill (*Anethum graveolens*) and camomile are good companions for plants of the cabbage family. It is said of tomato, sage, rosemary, thyme, peppermint and wormwood that they keep the cabbage white butterfly at bay. Cauliflower is said to benefit from the companionship of celery. Cabbage and white mustard (*Sinapis alba*) inhibit one another, while bush beans promote the growth of cabbage varieties.

Carrots flourish with lettuce and chives and in turn promote the growth of lettuce and leek. Rosemary, wormwood, sage and in particular black salsify (*Scorzonera hispanica*) can be used, as well as onions, to keep the carrot fly at bay.

Beans flourish well beside carrots and cauliflower but not with onions, garlic or shallots (*Allium ascalonicum*), which are also not good for peas.

Strawberries react well to borage (*Borago officinalis*). The taste of the fruit is also improved by mulching with spruce and pine needles.

A few garlic plants should grow alongside the potato field or bed. Sainfoin, dead nettle and nasturtium (*Tropaeolum majus*) are also good bordering plants for potatoes. Maize, peas and broad beans are suitable companion plants for potatoes. Hemp (*Cannabis sativa*) is said to prevent potato blight, while this disease is favored by sunflowers (*Helianthus annuus*), tomatoes, apple trees, cherry trees, cucumbers and pumpkins.

Some diseases and pests react specifically to plants. Nasturtiums are planted under apple trees to keep away woolly aphid, and near broccoli to keep away aphids. Tea can also be made from nasturtiums for the same purpose. Spurge (*Euphorbia lathyris*) is said to keep mice away from young fruit trees. Wormwood helps against the flee beetle, and the French marigold (*Tagetes patula*) has a root secretion that works against nematodes in roses, tomatoes and potatoes. A mulch of oak leaves, oak bark or tanning bark is used against slugs and cockchafer grubs. Hazelnut bushes keep flies away from house and pasture. Tansy (*Tanacetum vulgare*) helps to drive away flies and ants.

Too much weight should not be given to a number of these indications, but they are suggestions to be followed by those who are keenly interested and observant.

Chapter Five

Practical Experiences on Bio-Dynamic Farms

This chapter will deal with a number of practical experiences gathered chiefly on arable farms in southwest Germany. The soils there are mainly para-brown earths that have developed on mesozoic sediments and diluvial loams. The series includes all intermediate stages to waterlogged gley soils. There are also clay soils with varying degrees of waterlogging. The measures mentioned are only those that have proved useful or worth considering and that belong to any good crop husbandry.

1 Crop Rotation

Land use on bio-dynamic farms differs on the whole from current practice. In contrast to the tendency to establish specialized cropping systems or even the monoculture of cereals, bio-dynamic farmers usually maintain a 5 or 7-field rotation. This recycles enough organic matter and maintains the life and the structure of the soils. Winter rye and also winter rapeseed are being reintroduced. The rye varieties available today quite often yield 4-4.5 t/ha (61-69 bu/acre) also in bio-dynamic management. There is a secure market for Demeter products. The large amounts of straw are partly dealt with by chopping in fields that

239

had previously been undersown with clover. Larger farms are also starting to use rape in order to extend their rotations. Plant breeders are successfully making efforts to eliminate erucic acid from rape seed, as this damages the heart, linoleic acid, which gives a bad taste to cooking oil, and glucosinolates, which are not wanted in rapeseed cakes. In 1973/74 some European countries had almost completed the transition to rape free of erucic acid.

Another circumstance is also favoring the introduction of better crop rotations in that region. In connection with West Germany's land improvement project (which is causing fields that for many generations have been divided into increasingly small plots to be reallocated among farms) wet and waterlogged ground is also being tile drained. Some grasses can now be ploughed to make more space for arable hay crops (v. Heynitz 1966). Good lucerne grass and clover grass crops produce nutritious fodder, increase the accumulation of humus forming organic substances, and are a help in controlling weeds. Mixed stands of cereals are planted where there is no need on the market for single crop stands of summer barley or oats. An oat-barley mixture with 10-15 kg/ha (9-13 lbs/acre) of peas supplies a satisfactory average over the years of home-produced concentrate. Field beans (*Vicia faba*) enrich heavy soils and they also supply valuable home-grown fodder. They are sown either alone or in a mixture with oats. This depends on the state of the weeds that may start to grow up when later in the season the beans lose their lower leaves. In addition to these crop rotations there is also much undersowing of cereals and also catch cropping. A well-tried techique has been developed. Green manure grown together with the above-mentioned crops also helps to achieve satisfactory yields of root crops that have to be grown for economic reasons. Carrots, beetroot, white cabbage, etc., are being increasingly grown instead of sugar beet for industrial processing. This vegetable production is taking up an increasing proportion of the land that was formerly used for other root crops. Potatoes are grown in sufficient quantities. Marketing possibilities in some cases restrict the acreage.

240

An additional remark about field beans and rape may be inserted here. Field beans (*Vicia faba*) are a valuable carrier of protein. They contain only minor and harmless concentrations of the trypsin inhibitors that are known to exist in soybeans and have to be eliminated by heat treatment before ground soybeans can be used as a feed. As compared with winter field beans, spring beans contain 2-3% more crude protein, the digestibility of which is higher by an additional 3% (Waring and Shannon 1969). As a rule of thumb, the following amount of ground or coarsly ground field beans can be recommended:

For poultry, up to 15% in the ration;

For pigs, up to 10% in the ration (caution should be exercised with young pigs, as beans contain 6-9% fibre);

For ruminants up to 30% in the ration (for calves older than 3 months 80% field beans plus 20% barley is a good mixture).

It has been shown, however, (Wilson and McNab 1972) that in field beans the percentage of sulphur containing amino acids methionin and cystin is low. For adequate performance additional synthetic methionin has therefore been recommended. H. Vogtmann (1974) has repeatedly pointed out that a mixture of field beans and rape seed residues (with a relatively high methionin content) form a complete protein concentrate. One uses the new erucic acid and glucosinolate free varieties. It remains to be established whether up to 15% rapeseed could be used for protein and energy in the rations of monogastric animals. The oil would provide a natural enrichment with the fat-soluble vitamins A, D, E, K (Vogtmann).

2 Green Manure and Hay Crops

The advantages of green manure for soils and also how they are best grown have been studied since the days of Schultz-Lupitz, a leading agriculturist in Germany during the nineteenth century. The knowledge is there and it is only a question of applying it. The use of herbicides is causing undersowing to be increasingly restricted. In general agriculture only rapeseed, mustard and some new breeds are on the whole used for catch-cropping.

In bio-dynamic agriculture, legumes are used as green manure plants. Winter cereals are regularly sown in spring during harrowing. Undersowing of spring grains is frequently undertaken and is recommended where mangles, potatoes or vegetables are to follow. A cheap home-produced red clover can be used, or white or alsike clover (*Trifolium hybridum*). Trefoil (*Medicago lupulina*) is a favorite, since it produces plenty of roots with not too much foliage above ground. It flourishes if there is sufficient lime in the soil and the pH is 6.8 or over. If it becomes uncertain where the pH figure is lower, it is best mixed with white clover, usually 5 kg white clover to 10 kg trefoil per hectare (5 lbs to 10 lbs per acre). White clover is used on lighter more acid soils. It is a beneficial crop with its many runners above ground and its finely-branched root system.

Undersown crops do not usually cause difficulties during harvest. Where cereal stands are strong, the undersown crop usually grows slowly. The combine harvester is simply set higher. The resulting taller stubble is advantageous for the utilization of green manure plants by the soil. Rain at the end of August or the beginning of September usually brings on a lush growth in clover. If this is then used for fodder, its value is reduced by a half or a third compared with its use as a green manure. It is difficult to tell on the farm whether the amount of additional nitrogen gained is 40 or 60 kg/ha (35-55 lbs/acre) but the successful use as green manure frequently causes lodging of subsequently sown barley. Oats can stand more.

Using a large amount of land for clover as a main crop limits the area to be considered for undersown clover. Relatively few problems arise in spite of clovers being grown frequently in rotation. It is important, however, that undersown clover is not left over the winter but is ploughed under in the autumn. Shallow ploughing is preferable to deep. Rotting green manure that has been ploughed in too deeply can have negative effects during the whole of the following growing season.

If the clover is patchy after harvest, or if there are too many

242

weeds, or if summer ploughing is needed, rapeseed, mustard or another brassica crop can be sown. If these are not ploughed under until the following spring, they may start to make new shoots. The use of herbicides in the usual manner is avoided by bio-dynamic farmers, so it is better to plough under in the autumn. Sufficiently large yields are achieved from this group of plants, which need a large amount of nutrients, if they are given plenty of liquid manure, chicken manure, etc.

More frequent applications of small amounts of farm manure
This has proved to be a useful method. Legumes react favorably to bio-dynamic measures. Preparation 500 is particularly good for them. Clover and alfalfa react with dark leaves and good growth. Smaller amounts of composted manure are used in various ways. After harvest, clover grass and lucerne stands are given about 10-15 t/ha (4-6 t/acre) of composted manure in early autumn. This guarantees yields in the following main growing year and continued effects are sometimes observed in the second year. It is an advantage to utilize undersown clover with the help of sheet composting. The clover grows through the high stubble or cover of chopped straw and produces a thick growth. Fast growth is stimulated by an application of 7.5 t/ha (4 t/acre) of farm manure. After this preparation, a good cereal harvest can be expected. If beet, potatoes or other row crops are to follow, another application of 15-20 t/ha (6-8 t/acre) of manure is made in late winter or early spring. This is usually half fermented manure. Spreading this manure is possible on heavy soils if the rough surface from autumn ploughing has been worked.

The use of green manure described here, also the supply of chopped straw to the undersown crop, and the repeated applications of composted manure necessitate corresponding changes in the manner of tillage. The usual shallow ploughing after harvest followed by deep autumn ploughing is hardly ever the method used. The manure and green manure are worked in to a depth of only 12-15 cm (5-6 inches). They thus receive enough air to be broken down. Sekera's idea (1951), that the soil becomes enlivened with microorganisms from the surface downwards, is taken up again. If brassicas are to be sown, summer ploughing is deeper. This contributes to the success of intercropping with rapeseed, mustard, etc.

It should also be pointed out here that the method of frequent applications of smaller amounts of composted manure has also proved favorable for grasses. The growth of the more delicate lower grasses, herbs and clovers is stimulated. The amounts used are 15-20 t/ha (6-8 t/acre). This procedure also makes it possible that the fields and pastures near the farm buildings can be manured with liquid manure without too lush growth or weeds resulting.

3   Soil Analysis and Commercial Manuring Materials

The principles of bio-dynamic manuring should also be applied to the use of commercial fertilizers. The aim is to make it possible for the life processes in the soil to provide a balanced nutrient supply to growing crops. If this is successful, it is not the task of the soil microorganisms to supplement more or less completely what is provided by chemical fertilizers; rather the total nourishment of the plant results from the life of the soil. A number of authors, for instance, Voisin (1966), have pointed out that supplying only the major nutrients (NPK) is extremely onesided.

Soil tests are regularly made in farms and gardens. The advisory service ensures that the resulting information is always available. Experience has shown that a phosphate figure of 200-300 ppm, which is usually considered necessary, is in fact not needed. 100-120 ppm (260-310 lbs/acre) are sufficient, as is shown repeatedly by the most fertile fields. It goes without saying that this applies to well-managed, regularly manured fields. This is understandable when one takes into account the fact that the availability of phosphate is more dependent than that of potassium on the activity of soil microorganisms. The biological activity is hardly reflected in ordinary soil tests. If the activity of the soil life is indeed low, only higher amounts of available phosphate determined by chemical analysis will guarantee an ample supply. Obvious phosphate deficiencies hardly occur at all on well-managed soils that are in good condition organically. Small doses of rock phosphate applied daily to stable manure are all that is required. The usual amount is 50 kg (110 lbs) per CU/year. Many farms do not even require this. Recently, more and more chicken manure, which is relatively high in phosphate, is coming on the market.

A potassium deficiency has so far never been found to occur on loam and heavy loam soils where there is sufficient livestock and where liquid manure is properly collected and utilized. Indeed, the opposite of a deficiency is more likely to occur if it is assumed that the amount wanted is 200-250 ppm. Exceptions occasionally arise in commercial gardens.

Since the researches of J. Utermöhlen (1937) at the beginning of the century in Switzerland, basalt meal has been used to enrich soils. It helps to reduce deficiencies and imbalances, particularly on lighter soils. Amounts of 0.5-1.0 t/ha (450-900 lbs/acre) mixed with corresponding amounts of manure have proved favorable. On lighter garden soils basalt meal is scattered on the soil and lightly worked in.

In recent years, French calcified seaweed has been increasingly under discussion. It is used frequently on grasses and pastures. The amount to use is about 0.4-0.6 t/ha (360-540 lbs/acre). It is also given to the soil via composted manure to which it has been added at an advanced stage of fermentation. Calcified seaweed and ground seaweed have also been introduced into feed mixtures. Experiments with various seaweeds in feed have shown positive results. The use of ground seaweed in agriculture is not least a question of cost.

The term organic commercial fertilizers denotes hair and bristles, horns ground to various grades of fineness, blood and bone meal, meat scrap, chicken manure, etc. Their usual nutrient content is shown in the table on p. 181. There are also a number of blended fertilizers on the market being sold under various brand names. The farms under discussion here most frequently use the substances first mentioned above. If used correctly, they promote lush and healthy growth. It has been found practical to obtain horn waste and bristles in the spring and summer of the previous year, ready to be mixed with farm manure. 400-500 kg/ha (360-450 lbs/acre) are sufficient if winter wheat is to follow sugar beet, maize or carrots. 600-800 kg/ha (540-720 lbs/acre) is applied for root crops following green manure. If applied directly, these fertilizers are lightly worked into the soil before planting or sowing. Their carry-over effect is usually quite distinct in the second year. Winter wheat following potatoes does not require a treatment in the autumn; liquid manure is usually applied in the spring.

Chicken manure from farms with commercial egg production must be specially carefully treated with plenty of bio-dynamic preparations and also nettles. One good way of doing this is to add fresh chicken dung to the slurry or liquid manure tank immediately before application. The way to treat these substances bio-dynamically has been described above. The resulting manure, which tends to have a strong forcing effect, is also suitable in small amounts for spring treatment in order to improve weak stands. Dried commercial chicken manure is applied in amounts of 1000 kg/ha (900 lbs/acre) or more.

Chicken manure can lead relatively quickly to negative effects on the quality, resulting from overfertilizing. It should be remembered that one hen excretes about 1 kg (2.2 lbs) of nitrogen a year in her dung and that a large part of this becomes effective quite rapidly. Recently the use of antibiotics in feedstuffs has been reintroduced on egg farms. These are not absorbed into the blood and are therefore present almost wholesale in the dung. Their life-span and the way they work in the dung varies. Some compounds produced in rotting chicken manure stored in the usual way may cause problems in soils and plants, though small amounts that remain in the top layer are easily digested by an active soil. Chicken manure can be made good use of for catch cropping with green manure crops such as rapeseed, mustard and other fast-growing plants that need plenty of nitrogen.

Composting only functions if organic litter poor in nitrogen is added, e.g., an equal amount in weight of sawdust. Inoculation with good ripened compost is recommended.

Chicken manure can also be added to well-fermented cow manure in amounts of 10-20%. One adds it when the manure is loaded; mixing follows during spreading.

Horn and bone meal and also blood and meat meal are usually included in the blended fertilizers mentioned above and are widely used in gardens. Using them in the seed drill when sowing is another example of how to use small amounts in an economical and yet effective manner. 15 to 20% of the weight of the seed used in the seed drill is advantageous for the development of young plants, particularly for spring cereals. (Merckens 1963).

On farms that are making the transition to the bio-dynamic method the question of the amounts of these fertilizers to be used crops up regularly. The transition has to be carefully planned. Home production of fodder and feed is brought up to the necessary level. A good supply of well-prepared farm manure is accumulated. A useful rule of thumb is to use half as much organic commercial fertilizer as has been used hitherto, so one actually uses smaller amounts of nutrients. For farms that have completed the transition it was usually reckoned before the recent price rises that about 40-80 DM/ha (6.50-13 dollars/acre) had to be spent on organic commercial fertilizers. On the whole the materials can be procured locally. Indeed, the supply may be greater than the demand.

4 Weed Control

Like all others, bio-dynamic farms have been hit by the development in agriculture of an ever increasing acreage with ever decreasing numbers of people willing to work on the land. Herbicides, however, are used only in exceptional circumstances. Bio-dynamic methods do not encourage the increase of common wild oats (*Avena fatua*); foxtail (*Alopecurus myosuroides*); or windgrass (*Apera spica venti*), etc., since these are favored by one-sided grain programs. So the methods available to the bio-dynamic farmer are crop rotation and mechanical control. It is important to build up both to the best advantage.

Crop rotation

Some characteristics of the possible crop rotations have been discussed above. Buttercup (*Ranunculus sp.*), which is a great nuisance on wet fields, disappears rapidly if haycrops are grown after the field has been tile drained. This is also useful against thistle (*Cirsium arvense*), which is averse to being mown several times. Once a good fodder stand has developed, the field will remain clean for several years. The introduction of Persian clover (*Trifolium resupinatum*), which is mown several times during one growing season, has also proved effective against thistles. Strong rye and rape stands, particularly the latter, when mechanical hoeing is done in autumn and spring, contribute to the suppression of weeds. A good green manure of trefoil and white clover has the same effect. After repeated undersowing of clover the common wild oat seems to dwindle.

Couch grass and thistles are suppressed by careful ploughing, rather deeper for this purpose than indicated above. Since good wide mechanical tools are now available for weed control it is difficult to understand why they are not used more. Harrowing and hoeing are, after all, not only weed control measures but also part of general care and manuring of the soil. That no value is attached any longer to the manuring effect of these measures is a consequence of over-fertilizing with nitrogen in general agriculture. Organic farms would not want to do without this means of stimulating soil life. Fortunately, chain harrowing and also machine hoeing of grain crops is enjoying a comeback. It must be assumed that on a well-managed farm there will never be more than a few fields that need special attention at any one time.

Thermal weed control

The positive experiences gained with thermal weed control in other countries led G. Preuschen (1968) to reintroduce this method into Germany. The propane gas used for this leaves no residues. The weeds are destroyed as they emerge, though persistent perennials may need several treatments. Burning the weeds in this way is the more suitable the further the emergence of the weeds is from the emergence of the desired crop. This is the period during which weeds growing in the rows can be burnt. Between the rows, hoeing is done as usual. After three years' experimenting (1969-72) this method of weed control is now worked out for large-scale carrot growing. Experience is still being gathered on how to use it for beetroot, beet and corn. In the USA weed control in corn with an oil flame is used. Handling a flame-thrower run on gas can be dangerous and only recognized and well-tried brands should be used. The hand and knapsack models for gardens are not suitable for agriculture except in exceptional cases. A number of firms have helped to design models suitable for larger fields and recently some industrial firms have also become interested.

Space does not allow for detailed reports on a variety of bio-dynamic farms. But to supplement this description of bio-dynamic measures in a larger group of diversified farms, Ceres Farm, a bio-dynamic dairy farm in New York State, is described in the following paragraphs. This operation is an example of a self-supporting farm that is an economic success, less dependent on input and less subject to economic fluctuations. Then there follows an account of America's oldest bio-dynamic dairy farm, and a report on Meadowbrook Herb Garden.

5 Ceres Farm

Ceres Farm, in central New York, was started in 1959 by a husband and wife team. With their four children, now aged 10 to 15, the farm has begun to evolve into a true family farm. It is 340 acres in size, 248 being tillable land, 3 permanent pasture, 36 wood, 38 an unimproved wooded pasture, 3 farm, garden and orchard, and about 12 acres hedgerows, ditches, ponds and a farm road. The cropland is divided into seven nearly equally large fields of about 35 acres. The seven-year crop rotation includes five successive years of hay and rotated pasture, followed by oats and spring barley, which is then the cover crop for the new grass-legume seeding. The grain acreage provides two thirds of the feed grain needed. The grassland acreage is more than adequate to provide roughage and pasture.

All young stock and every milk cow in the registered Holstein-Friesian herd was born and raised on the farm. Herd replacements were purchased only in the years 1959 and 1960. The number of cattle fluctuates between 70 and 80 head with about one half the number milk cows and the remainder young stock of all ages. This makes the farm a below-average size dairy operation, which is partly compensated for by above-average production. The farm now has 38 cows; the production per cow is over 16000 lbs milk and 600 lbs butterfat.

The gross income derives almost entirely from milk, meat and registered breeding stock. As a sideline, horses are also kept and bred; they presently have four registered Morgan horses.

Soils and tillage

The fields have a rolling, in many areas bedrock-controlled topography, ranging from a little less than 800 feet to nearly 1000 feet elevation above sea level. All fields have drainage problems; the moderately to slowly permeable, silty clay loam contains a variety of rocks and stones. It derives from glacial till. About 80 percent of the cropland has a capability classification of III ("soils with severe limitations that reduce the choice of plants"). The rest is divided about equally into Class II ("soils with some limitations that reduce the choice of plants") and Class IV ("soils with severe limitations that restrict the choice of plants").

Since autumn 1965, no mouldboard plough has been used but only a 12 foot chisel plough. The largest tractor on the place has 70 hp (PTO). The two grain fields are chisel-ploughed in autumn. Spring cultivation is done with a 14 foot double disc harrow and an 18 foot springtooth harrow.

The organic matter content of the soils is medium to high, the nitrogen supply for the crops is more than ample. This is explained by a situation where all plant production is fed, where manure is properly handled and utilized, and where five years of grass-legume are followed by two years of small grains.

In the northeastern areas of the USA, soils are usually deficient in phosphorus. Soil tests show medium to high figures for potassium and calcium, while phosphorus has been low to very low.

Manure and phosphorus

The dairy cows are kept in a stanchion barn, with a chain-type automatic barn cleaner. Young stock is in loose housing. The annual production is about 500 tons of manure, which is treated with the B-D Starter.

The manure from the milk cows is loaded on to a conveyer-type manure spreader, and unloaded at one of two piling sites, such that a four-foot high plateau of manure is accumulated during the barn feeding season. One of the sites is stabilized with stones, gravel and sand so that manure can be piled there at all times of the year.

During the summer the manure turns into compost in the upper layers and especially the lower edges of the manure piles. This compost is spread each fall on some of the hayfields. The remaining half-rotted manure, together with the manure from the young stock, is spread on next year's oats and barley fields. The manure is spread thinly and worked in as soon as possible. One-third of the cropland is treated each year. Ten tons of a colloidal clay, containing 18% phosphate are annually added to the manure.

Grassland management

Five fields of hay and rotated pasture total about 175 acres. The seeding mixture consists of 7 lbs birdsfoot trefoil, ½ lb alsike clover and 6 lbs timothy.

The new seeding is only harvested as first and second cutting hay. In the last (fifth) year it is mostly grazed. In the second, third and fourth years, harvesting for hay and sileage is combined with pasturing; the first cutting is removed and the after-growth is pastured in a rotation system. In this way, enough pasture is available when regrowth subsides during the summer. The pasture season generally extends from the end of May to the middle of October. Miles of fences have been built to make it possible for all 7 fields to be pastured. Access to water was provided by constructing 6 ponds at strategic locations.

The herd

The production, i.e., milk and meat, is being sold into convenient market channels. No higher prices or premiums are received for bio-dynamic quality. Nevertheless, the farm has been successful economically. One of the reasons for this is that, as a result of better quality feed, the animals are healthy and little plagued by metabolic and reproductive disorders. There has been no case of ketosis since 1971, when there was a secondary case. In spite of high production, cases of milk fever occur about one in two years. In all 16 years, there may have been one case of displaced abomasum in 1972. It was not positively diagnosed and the cow is still in the herd.

It is considered quite normal for a farmer to lose 10 percent of his calves to scours and other diseases. At Ceres Farm the losses have been less than 1.5 percent. In a 16-year period, only 4 calves died after having been born alive, and in all cases the cause was accidental.

According to comments from several veterinarians, reproduction is excellent. In some years, it was even possible to have all cows dry at the same time. In the summer of 1970, the whole family spent 30 days in Europe, during which time no cow needed to be milked.

Grain growing

The name Ceres Farm was chosen when it became necessary to register a prefix with the Holstein-Friesian Association of America and later the American Morgan Horse Association. Ceres was the Roman goddess of cereal grains and soil fertility.

In the 1960's the growing of feed grain was considered unprofitable in central and eastern parts of the New York State. Yet the decision was made for two reasons to designate two fields in the crop rotation for feed production. The farm was to be developed into a nearly self-supporting bio-dynamic farm organism and straw was needed for bedding and proper manure consistency.

To produce homegrown feed grain meant not only more expensive feed for the dairy cattle, and more investments in farm machines, but also that the number of milk cows had to be kept lower. The income was lower than it could have been, especially in the early years when the farm carried a high debt load.

But feed grain production was never considered a money-losing branch, because of the need for straw and better food grain quality. What had seemed to be only bio-dynamically right, became also the economically right thing to do when in 1972 all of a sudden feed grain prices rose sharply as a result of world-wide grain shortages. It turned out that diversification for the sake of improving the biological stability of the farm also improved its economic stability in times of shifting demands and economic trends.

From 100 percent debt to 90 percent ownership

The family started their farm with less than 5,000 dollars capital, an old car and lots of enthusiasm and determination.

In January 1959 a "bare" farm with old buildings was bought. This was financed with a first mortgage from a bank and a second mortgage from a relative. Thirty head of dairy cattle were soon thereafter bought with a short-term bank loan. A minimum of second-hand machinery was purchased with a personal credit from a dealer. In 1959 the debt load of the farm was 100 percent. Sixteen years later, in January 1974, the debt load was 10 percent of the total market value of the farm. The original farm was 220 acres in size. In 1968 additional land from a neighboring farm, which had previously been rented, was purchased. During these sixteen years, the farm buildings were modernized and extended, and two new buildings and a new silo were erected. The house was repaired and remodeled.

Especially in more recent years the original old machines were replaced. Today there are three tractors on the farm, which were all bought new. All needed equipment is available for working the land and efficiently harvesting the hay, sileage and grain crops, including a combine and dryers for both grain and all the hay. All grain can be stored and processed for feed on the farm. Also machines and equipment were bought to mechanize the barn work.

In addition to the above-mentioned ponds and fences, a drainage program was carried out that added up to 17,500 feet of ditches and grassed water-ways and 24,300 feet of tile lines.

Productivity

Since all plant production is fed to livestock, no exact yield records are kept. The only exact records available of this farm are milk-production records and the financial records for tax purposes.

In the following economic analysis the year 1959 has been omitted. It was the starting year and not a complete one. The remaining 15 years are grouped into three five-year periods.

The average gross income and the net income were nearly three times and four and a half times respectively higher in the third 5-year period than in the first. This does not, however, give a true picture because of money devaluation. To eliminate the inflation factor the figures for net income and feed expenditures in the following table were expressed in percent of gross income.

The net income increased in direct relationship to the

Economic development of Ceres Farm. Net income and bought-in feed in percent of gross income, 1960-1974.

period	1960-64	1965-69	1970-74
net income	17.7	19.5	26.5
feed bought	18.1	15.5	8.4
totals	35.8	35.0	34.9

period	no. of cows	lbs milk per cow	lbs fat per cow	type of records
1960-64	29.8	10,242	389	OS
1965-69	32.6	12,529	463	DHI
1970-74	36.4	15,080	556	DHI

decreases in feed expense. The combined total is in all three periods around 35 percent. The milk production records show increases in herd averages, and also slight increases in the number of cows. From 1960 to 1974 the annual milk production of Ceres Farm more than doubled. In conventional farming an increase in net income has generally been achieved by increasing the production by raising the input of feed, fertilizer or pesticides. The net income on this bio-dynamic farm was increased by lowering the input of feed bought and by increasing production at the same time. This is a true increase of productivity that has its implications for the farm as a private enterprise, and also for the national economy.

6 Zinniker Farm

America's oldest bio-dynamic dairy farm, still in operation, is located in Elkhorn, Wisconsin. It presents an image of steady growth and economic stability. Located in the sightly rolling landscape formed during the ice ages, the soil of this farm is mainly Miami silt loam. The average rainfall is 35''. To the original 100 acres, 65 more acres have been added. With minor changes, due to the size of the fields, 30 acres are planted in corn (for ear corn and silage), 30 acres in oats and 65 acres in hay crops. Five acres are planted with wheat and rye to provide the family and some customers. The crop rotation for many years has been corn—oats—corn—oats, followed by 4-5 years of alfalfa or clover. The corn gets 10 t/a of composted manure; the other crops are treated with the bio-dynamic sprays. The chisel plough has replaced the mouldboard plough with good results. Corn yields are 90-100 bushels, oats 75 bushels per acre, and hay varies between 2-2.5 t/a.

The farm supports 29-30 Holstein dairy cows, 20 heifers and a bull. In good years the milk production is 12,000 lbs. per cow and over, with an average content of 3.8% butterfat. There is no trouble with reproduction, although 1 or 2 calves may be lost in a year. In 1974 the veterinary bill, including pregnancy tests, was $241.00.

The annual expense for bio-dynamic materials is about $2.00 per acre. In 1974, the feed bill for grinding and concentrate was $18.20 per acre. The concentrate consists of 50% and 25% respectively of corn and oats, with 15% of bought in linseed oil meal, brewers grain and bran. A mineral mixture, including bone meal, is fed. Locally the farm also supplies some bread, eggs and vegetables.

It is important to mention that this rural place is visited by many, especially during the time of the festivals of the year. Cultural and social occasions are offered to neighbors and people from nearby towns.

7 Meadowbrook Herb Garden

Herb growing has been mentioned many times in this book. While most people are aware that herbs are valuable as companion plants in a garden, and as a flavorful complement to food, the true role of herbs in human nutrition has only recently been rediscovered. It is increasingly recognized that these underestimated components of a well-balanced diet have stimulating, soothing and regulating capacities; affecting the nervous system, they influence the healthy functioning of the whole human organism. The nutritional quality of food may be improved by the addition of even the smallest quantities of the right herbs to a well-planned meal.

Private gardeners usually grow some herbs themselves as companion plants and for kitchen use. Their efforts, however, must often be supplemented by a commercial product which, of course, can also make herbs available to those who cannot grow them at all. We are including, therefore, a report on the operation of a successful, commercial bio-dynamic herb garden.

Meadowbrook Herb Garden, in Wyoming, Rhode Island, was established in 1966 by Heinz and Ines Grotzke. Mr. Grotzke had already had seventeen years of experience in this field, both in Europe and in the United States, before he founded Meadowbrook, with the intention of applying bio-dynamic principles to commercial herb-growing. The time and the place were apparently conducive to this effort.

Both the geographic location of the farm and the climate maintain a balance between optimum quality and possible quantity in herb production—a factor of special importance for the relatively small-scale bio-dynamic farm. The average yearly temperature range is from -5°F (-20°C) to 90°F(32°C), with extremes below and above this range occurring infrequently. The holding is five acres, with two to three additional acres of adjacent land currently in use. This area yields a sizeable quantity of a wide variety of herbs. A drying house, a greenhouse, storage and packing facilities and a retail shop, which also offers carefully selected hand-crafted gifts for sale, cover part of the land; an extensive lawn area with ornamental trees, a formal herb garden, and a small orchard complete the picture. An area of woody swamp is being reclaimed and landscaped.

As a rule, due to hand planting, weeding, harvesting, etc., it has been found that two people per acre are needed to cover the whole operation, including greenhouse work, processing and shipping.

1) *Management*

Plant propagation is done in a fiberglass covered greenhouse, measuring 18 by 100 feet. All plant materials are started in the greenhouse from seed or from cuttings. In addition, live herb plants are sold to the public all year round; these are grown bio-dynamically in potting soil made from a blend of 40%manure

257

compost, 40% weed compost, 5% special herb composts, and 15% crushed granite or sand. Insect and disease problems have been minimized by careful sanitation, optimum ventilation, a high quality growing medium, proper temperature and humidity control, and applications of liquid seaweed, nettle and/or the bio-dynamic Preparation 507, as a prophylatic means of disease control.

The soil can be classified as sandy loam and has responded tremendously to bio-dynamic treatment during the past ten years. The most visible improvement is a constant growth of the earthworm population, and the darker color and increased waterholding capacity of the soil. Organic matter content on the average measures 4.5%; the pH is 6.8. Different types of hedges—an area that is still being worked upon—create a micro-climate that is ideal for herb crops as well as. inviting to insect and bird life.

All intensive methods of growing require a special emphasis on fertilization. Herb growing is no exception and successful growing depends on optimum soil fertility, especially since invested labor is high. During the past ten years bio-dynamic manure compost was added to all fields at the rate of twenty tons per acre. Generally, yearly alternations between cow manure and horse manure were possible, but lately the supply has decreased. Fortunately, soil fertility is high enough by now so that even one or two years without continued composting would probably not show much of a decrease in crop production. Alternative solutions to the manure shortage are available but have not yet been attempted.

Manures are composted with bio-dynamic treatment without adding any materials other than an occasional light soil cover. After about six months and two turnings, a small amount of the manure compost—up to ten tons—is kept aside and further composted until it has reached humus soil consistency. This completely finished compost becomes part of the potting soil for the greenhouse. Additional piles are built from accumulated weeds, kitchen garbage, residues from herb processing, leaves, etc. These materials require up to two years before they are ripe for greenhouse use.

2) *Production Program*

The crops grown may be divided into six categories:

I. Annual and perennial herb plants and ornamental plants for sale as live plants from the greenhouse.

II. Outdoor perennial herbs and ornamental plants for sale as live plants.

III. Field crops of herbs grown for culinary use.

IV. Field crops of herbs grown for herb teas of medicinal or beverage use.

V. Field crops of herbs grown for seed.

VI. Wild herbs allowed to grow at random and harvested for use as tea herb.

In the first category, up to one hundred varieties are offered all year round to customers who come and select the plants themselves. The largest selection is available from April to June, this being the regular planting time in the northeastern United States. Some of the varieties available are basil, santolinas, southernwood, rosemary, scented geraniums, all thymes, elfin herb, myrtles, angelica, roman camomile, all mints, plume poppy, korean mint, and many others.

The hardy perennial herbs are kept outdoors in beds and sold directly from there during the growing season. Around sixty varieties are available, among them tansy, broad-leaved chives, oriental garlic, agrimony, false indigo, roman wormwood, germander, hyssop, and yellow yarrow.

The growing of culinary herbs has always received major attention, and the following choice is now available. Basil, calendula petals, chervil, celery leaves, chives, comfrey leaves, coriander leaves, cress, dill leaves, fennel leaves, leek, lemonbalm, lemon verbena, lovage, marjoram, nettle, oregano, parsley, peppermint, rosemary, rue, sage, salad burnet, savory, spearmint, tarragon, thyme. In addition, the following culinary herb blends are offered: salad herbs, poultry seasoning, fish herbs, hamburger seasoning, soup herbs, heart's delight. Almost equally important are the herb crops used for various kinds of herb teas, and cultivation includes an average of twenty-five varieties. In this category are apple mint, camomile, catnip, comfrey, gold melissa, heart's ease, horehound, hyssop,

259

kentucky mint, korean mint, lemonbalm, lemon verbena, mugwort, nettle, oswego tea, spice mint, and many others.

The growing of herb crops for seed is limited mainly to those that would be hard to purchase or not available at all in the seed trade. The list includes agrimony, ambrosia, sacred basil, belladonna, burdock, yellow camomile, german camomile, catmint, celandine, chicory, clove root, foxglove, garlic chives, korean mint, mullein, heart's ease, sweet cicely, etc. Altogether about 180 varieties of herb seeds are offered. Some herbs are native to the geographical area and are encouraged to grow wild in waste places; at the proper time they are harvested and dried. These are St. John's wort, celandine, some nettle, yarrow, and mullein. Their seeds are collected and sold.

The treatment of all these crops, whether in greenhouse or field, follows the established rules of bio-dynamic farming. Preparations 500 and 501 are routinely and repeatedly applied. Direct means of disease and insect control appear to be unnecessary, and whenever an outbreak seems likely, the crop can generally be harvested prior to such timely nuisances as, for example, the Japanese beetle on basil. In general, the health of all crops has been excellent and consistent from year to year, probably due to the active soil life. The emphasis on a quality herb product does not seem to allow even the slightest use of sprays other than 500 and 501 on field crops. All others would to some extent leave a residue on the leaf and alter the pure flavor of the dried material.

3) *Harvesting and processing*

The various crops are harvested from May on, some three or four times during the season, until heavier night frosts bring this activity to a halt. The earliest crops are chives and nettle, closely followed by comfrey and camomile, while some of the latest cuttings can be made of parsley, celery and thyme. At the right moment, the herbs are cut by hand, high enough above ground to allow new shoots to develop for a second growth. Yellow leaves, leaves soiled by bird droppings or used by insects to pupate or hibernate, spears of grass, and weeds are removed by hand before the herbs are placed into baskets with stems lined up in the same direction. Only then can the herbs be chopped to reach a uniform particle size. The possibility of milling the herbs with hammermill or steel grinder has not been considered at all, because precious essential oils would escape in the process and thereby lower the quality of fragrance.

Directly after harvesting and chopping, herbs are spread on screens in the drying house, and checked once again for impurities. The immediate drying is ideal because herbs like marjoram tend to heat up in the basket rather quickly, and should this happen, they lose color and fragrance. The loaded screens are placed in racks in the drying room at 100°F (38°C) for as long as it takes the specific herb to become brittle enough for either rubbing, immediate storage, or further processing. The maximum load of the drying house is about 1800 lbs. of fresh material on 180 screens measuring 36" by 72". The average weight ratio between fresh and dried material is 7:1. In the case of the finished culinary herbs, this ratio increases to about 14:1 because, with the culinary herbs, only the leaves are used, while the stem part is rubbed out.

All of the herb seasonings are rubbed, which is a process that separates the leaves from the stems by pressing the former with a circular hand motion through a fine screen. The stems stay on top of the screen and are discarded and composted, while the broken-up leaves are rubbed once again and then dedusted before being put into jars.

All these bio-dynamic crops—whether used for herb seasonings or teas—are packaged on the premises, with an eye toward maximum freshness for the consumer. None are sold in bulk. Wholesale orders from stores and co-ops make up 85% of Meadowbrook Herb Garden's business; another 6% consists of retail orders from individual customers; a final 9% consists of all that is sold in Meadowbrook's own retail shop.

Chapter 6

Animal Husbandry and Feed Production

1 General Remarks

The animal and the soil

In nature, plant growth and animal life are always closely linked. Since ancient times man has kept the large herbivores in herds and has wandered with them from pasture to pasture. In other regions plant cultivation came into being making it necessary for man to settle.

If the number and type of animals kept are suitable for the habitat, animal husbandry will lead to an enhancement and strengthening of plant growth and soil formation, since the nutrients taken in and transformed by the animals for the most part return to the soil, leading to a strong stimulus for microbiotic soil life, humus formation and root formation. This applies chiefly to cattle. Their organism needs fodder plants, which are for the most part in a state of the highest vegetative development. The ruminant digestive system needs roughage to stimulate inner movement and cud chewing.*

When the fodder plants are quite large but have not yet

---

*Rumen or paunch in the following paragraphs frequently stand for rumen, reticulum and omasum, the three compartments that do not secrete digestive juices.

reached the dying stages of full flower and ripe seeds they have also made the maximum amount of root and have thus reached their maximum depth in the soil. They reach to a considerable depth in arable hayfields, greater than in permanent grassland. The former is more suitable for dryer regions, the latter for humid climates because of the thickness of its turf. At this stage in their development the plants are able to grow rapidly again after grazing or mowing so long as there is sufficient humidity. Thus the very nature of the way plants grow under good grazing and mowing techniques brings about an enlivening of the soil coupled with increasing fertility.

The animal lives by transforming its food into nourishment. A large part of the carbon compounds is exhaled, or used to build up body susbstance or excreted in the milk. A great deal, however, is also excreted in a metamorphosed form in dung and urine. Thus animal husbandry leads to an increase in substance turnover in agriculture. This applies to almost the total amount of the minerals as well as to the nitrogen in fodder. These are contained in a large variety of organic substances that, left over from the food, have been mixed by the animal's organism with digestive ferments, slime, waste intestinal epithelia and secretions from the blood and intestinal microorganisms. The nature of these excretions depends on the farm-produced fodder crops and bought-in concentrates, and on the type of animal in question. Appearance, smell and constituents show clearly how much the different animal species differ in this respect. Site conditions and animal metabolism adjust to each other.

Fertilizing with fermented manure or composted manure is the measure by which the soil is so enlivened that the plant's growth is quickened and balanced and the soil's fertility is built up. At every stage of their growth the plants then create the corresponding balance between vegetative development and the limiting nature of the formative factor, flowering and ripeness. In this way, even with high yields, taste, keeping qualities and wholesomeness can be maintained.

Because of its influence on soil development and the constitution of fodder plants, animal manure is a prerequisite for biodynamic farming in the humid temperate zone.

The exercising of livestock

Exercise is important for animal health. It stimulates metabolism, circulation, the formation of muscles, sinews and bones, digestion and many other functions. Research conducted in connection with space travel has shown that when muscles are not used, the bones quickly lose calcium. This even occurs if a person has to spend some time in bed or has a limb in plaster. The additional stimulus of changing weather activates the organism's capacity to react and thus remain healthy. An animal also has to use its sense organs in quite a different way if it moves about. Its movement is governed by its sense perceptions. Inwardly also then the functions of the nerve-sense organization and those of metabolism work more intensively together in a rhythmic way. Even in the lower evolutionary stages of animal life, movement is the typical characteristic of the animal. It is through movement that it develops itself (Portmann 1960). Not only its performance but also its health has to be recreated daily and this requires exercise.

Exercise is particularly important for the raising of young stock. All animals show a natural urge for exercise if they are given the opportunity and this is particularly the case with calves and heifers. They also need to be able to exercise in winter, so their stables must be planned accordingly.

When out at pasture, young animals should be encouraged to exercise more than they would of their own accord. If possible, they should therefore be put out to graze on the steeper slopes. This is also the best way of utilizing such slopes. Horse breeding is worthwhile on larger farms even if only because of the extra exercise young cattle are forced into by the more playful foals, provided the breed kept does not chase the cattle unduly. Dry situations are better for young cattle than damp ones. Care must be taken to ensure that the pastures are free from parasites. Additional significant advantages of exercise in the case of cows are stronger growth, more easily detectable heat, and easier parturition.

Exercise for sows, particularly between the weaning of one

litter and the birth of the next, is an indispensable factor for healthy breeding without too many problems. It was observed that weaned piglets, bred from mothers who were kept in a run on farms where most of the fodder was home grown, were the only pigs over many years that did not become infected by the local cough in a large fattening stall.

To be able to grub and take in good earth is important for piglets.

Light

Light also plays an important and frequently insufficiently recognized part. Ultraviolet light, which does not penetrate through glass, brings about the development of vitamin D_3 in the skin and this in turn is needed for the proper utilization of calcium and phosphorus in metabolism. It is an important preventative factor against rickets.

Light works into almost all metabolic functions through the eyes. The ophthalmic researcher Hollwich found that the hypophysis, which is essential for hormonal regulation, remains a third smaller than normal if there is no intake of light through the eyes. In an experiment with chicks, the formation of hemoglobin was dependent on light entering through open eyes. Steiner in his agricultural lectures described and emphasized the importance of breathing and sense activity for the building up of the body of domesticated animals.

Specialization in animal husbandry

Specialization for the sake of reducing labor, concentrating expert knowledge, saving machinery, producing large amounts to ease transport to market, and other reasons, results in farms limiting themselves to one type of animal or even to a particular period in the life of a particular animal (e.g., piglet production, calf fattening, etc.). But the loss of variety leads to the loss of a number of hygienic factors. For instance, a chicken run remains free of parasites if it is used simultaneously or alternately for other livestock. Furthermore, where only one type of livestock is kept, circumstances often force the farmer to take the big step of embarking on factory farming where life is maintained, under conditions of isolation from the environment, by disinfection, vaccination, medicated feeds, etc. To have healthy animals it is better to keep several kinds on each farm or farm unit.

In agricultural, not yet industrial, situations and where there is sufficient biological understanding and suitable measures are taken, a certain amount of specialization can be biologically balanced and managed on healthy lines. This is particularly so with regard to dairy farming combined with breeding. In such cases, however, the manure must be well fermented. This removes from the excrement some of the too dominant characteristics of the animals in question, which if crude manure were used would repel livestock of the same or similar kinds. Fodder from freshly manured fields is made into hay since the "ripening" processes that take place during drying remove some of the remaining onesided characteristics. The odor of ammonia and other manure smells are quickly absorbed when the manure heap is permeated by fungal growth. Fermentation can be controlled from the beginning by compound preparations (see p. 206). Remer has developed a formula, from the bio-dynamic preparations, which enhances fungal and bacterial growth. Intensive fermentation also kills disease germs and parasite eggs (ascaris eggs in about 14 days, and salmonella in 5-8 days). This is brought about not only by high temperatures but also through the biological processes that occur.

It is rather more difficult to achieve this with slurry, which should be allowed to ferment for as long as possible. (See also above about experiments to aerate slurry.) The start of fermentation in the floating manure canal can be fairly easily controlled since this offers a large enough surface. Apart from the bio-dynamic preparations or compound preparations, also well-rotted manure scattered on the canal or stirred into water and applied with a watering can may be used to start the process off in the right direction. This is hygienically important on farms specializing in one kind of livestock, i.e., where the cycle soil-plant-animal-soil is rather short.

Many farms also tend to specialize in pig breeding. This is virtually impossible to manage without the availability of alternate enclosures. Sows should be regularly vetted for parasites. The same applies to horses on stud farms. Fresh dung samples are sent to the veterinary testing service.

Specialized livestock keeping can also be practiced in cooperation with neighboring farms willing to accept some of the manure. Seen from the point of view of natural processes, this combination is like one large mixed farm or biological unit. Such cooperation is a positive step.

In present-day large-scale livestock farming, most of the problems with regard to buildings, feeding, lighting, ventilation and work routine have been solved, at least well enough to enable such farming to be carried out. The unsolved problem is what to do with the excrements that accumulate in such masses that they cannot all be utilized on the spot. If collected in pits or disposed of in lagoons they make the neighborhood uninhabitable by their smell. They also endanger both the surface and ground water. Chicken dung can be dried but this is expensive and the product is difficult to sell. So far as we know, no solution has yet been found as regards the manure from large-scale pig farms, especially as it has become evident that lagoons are not the final answer. Maybe it would be possible to compost it together with municipal waste.

Because the dung cannot be used, valuable substances are being lost on such operations. Instead of being useful they become a threat to man who has come to regard agriculture merely as a matter of capital, labor and the market, i.e., an industry. He loses the personal relationship with individual animals and is simply managing a business. He has to force himself to disregard the impressions that would otherwise overwhelm him, impressions such as the stink of massed livestock and the unhappy, panicky behavior of animals kept in this way. In the past the farmer's main concern was for the welfare of his livestock. Today there is often a quite unconcealed lack of feeling resulting from an extremely commercial attitude. Anyone who has not experienced this would do well to visit, for instance, a battery hen farm and then read a good description of traditional peasant life. There is no call for sentimentality about the good old romantic days, but it should not go unnoticed how much of his humanity man must be prepared to lose if he wants to be a slave to his computed profits. Factory farming produces a cheaper product and so family farms are forced by prices to adopt an industrial attitude in their turn. In the present phase of overproduction of animal products in the western world, the existence of a farm depends on how cheaply it can produce. In consequence, animal husbandry is becoming increasingly detached from agriculture, i.e., from the soil, and more and more it acquires the nature of an industry.

Meanwhile people's deep longing is revealed in the increasing number of pet animals kept and in the way city dwellers stream out into the countryside for weekends and holidays. This "holiday mentality," however, is frequently onesided, superficial and sentimental and does not meet man's real needs. People today so often lack the necessary basis for knowledgeable, objective and yet at the same time feeling relationship to nature. That they have the need for this is, nevertheless, just as much part of their inner selves as is their striving for money, power and comfort.

It should not really be necessary to ask whether it also matters to nature what kind of inner attitude people have who work on the land. Those who do, know from experience, which is born out again and again, that it does indeed matter.

Another viewpoint should also be mentioned. It has now become necessary to set aside recreational areas, which for economic reasons can no longer be maintained by agriculture. This leads to a peculiar split in man's relationship with nature, for now nature as the provider of food is separated off from nature as a place where recreation can be found. The money saved by more economical ways of producing food is used to maintain those areas that offer food to man's soul because of their particular beauty. Problems arise in both fields, difficulties connected with maintaining the landscape and the quality of foodstuffs, and also those arising from pest control and soil conservation. Animal husbandry has been disentangled from its intimate relationship with the land, and the ecological requirements of nature have been disregarded.

2 Feeding

As far as possible, home-produced feeding stuffs should be used. Wherever livestock is kept it becomes obvious that the animals adapt themselves to the local habitat. The many difficulties that arise with bought-in animals, difficulties familiar to vets and breeders alike, are an obvious indication. Fodder crops that have been grown carefully for their quality bring advantages not only of yield but also of good health. It has frequently been observed that during the change-over to bio-dynamic methods, the animals gradually become more choosy about the fodder they like. This is not always convenient, but it does show that their instinct becomes more certain, i.e., their connection with nature stronger, while at the same time they gain a comfortable familiarity with man. Feeding can therefore become a better yardstick for the quality of the fodder, and observation of the animals while they eat provides an important basis on which to judge whether the measures undertaken in the growing of the fodder crops have been worthwhile or not. It is essential for a farm to do its own breeding; then feeding can start to play its part even at the embryonic stage of development. The very body structure of the animals improves. It has been observed, for instance, in cattle that the horns are firmer and more pointed and erect.

For satisfactory production the requirements for carbohydrates, protein, fat, minerals, trace elements and vitamins must of course be satisfied. Foodstuff tables giving the composition and digestibility of the various feeds can be used as guidelines for compiling rations. Such indications, however, can only give a rough average. The yields of the best cows nowadays can be higher than the amount of energy they can possibly eat. Composition of feedstuffs varies and so do the requirements of individual animals. Even if fodder is analyzed every three months and supplemented according to the results of the analysis, it is still not necessarily possible to improve reproduction in cattle. This shows that chemical analysis is not sufficient to determine all the factors at work in fodder with regard to health and nutrition. A comparison with medicinal plants may make this clearer. In their case, no one would think of determining their value on the basis of their protein content. Quite other factors are what matters. Usually one goes by the pharmacological substances they contain, but experience has shown that even this does not specify the effects sufficiently accurately. The plant, or the parts of it that are fed to the animal, works as a whole, whereby the substances it contains do not simply add up to a sum total but actually enhance or inhibit each other. The structure of the plant works by releasing its substances. The enzymes also work. It is not merely a matter of the substances taken in but of the processes by which they come into being and fade away again. The same applies to fodder plants. The animal's organism receives from fodder not only substances that give it energy and build up its body but also numerous impulses that stimulate or inhibit its different functions. Fodder beets, for example, have proved valuable for the yield and health of dairy cows. It has been found that they cannot simply be replaced by maize silage, though going by the chemical analysis, one would have thought this would be a good substitute. Roots, leaves, flowers and fruits have quite specific relationships to the organic systems of the animal (see p. 291f.).

Feeding cattle

The organization of the ruminant enables it to eat large amounts of roughage that is subsequently given a second thorough chewing. Digestion in the rumen is carried out by microorganisms, some of which are taken in with the fodder, while others are passed on from one generation to the next by mutual licking. These microorganisms are even able slowly to break down the cellulose. This releases the cell substances and makes them more digestible. The bacteria in the rumen form short-chain fatty acids (acetic, butyric, propionic acids) that are for the most part absorbed into the animal's blood. Carbon dioxide, methane, ammonia, hydrogen and other gasses also arise. These are partly released by the gullet and partly absorbed (e.g., ammonia).

The animal must have enough roughage for this activity in the rumen, otherwise the reflex to regurgitate cud does not function. Chewing the cud is not only vital to the animal because of the activity of chewing but also because of the enormous amounts of saliva thus secreted: about 90-120 L/day (22-30 gal/day). This contains sodium hydrocarbonate, which neutralizes some of the acid created in the rumen. When not enough saliva is secreted, e.g., if there is not enough cud chewing, the rumen content becomes too acid. Digestion in the rumen then comes to a standstill and severe metabolic disease results. Roughage is important for the formation of acetic acid in the rumen and this is in turn important for the formation of fat in the milk. This need of cows for stalks and leaves is, as has been shown above, an important fertility factor for soil life and consequently for the whole farm.

The quality of stalk and leaf depends on the soil and the way it is manured. When large doses of soluble nutrients are given, too many of these, in relation to the sugar formed, reach the upper parts of the plants either untransformed or too little transformed. They are only metabolized properly by the plant in its leaves. There are then great differences in the content of leaf stalk and leaf blade. In the leaf stalk, for instance, hardly any protein has yet been synthesized. When overfertilizing takes place, the nature of the plant typical for the root and the stalk is pushed up into the leaves. Seeds and fruit are less affected by this. This is why the value of green fodder, hay and silage is particularly dependent on the type of fertilizer used, in addition to the habitat and the weather. (The situation is similar with regard to vegetables for the human being.) On permanent grasses these factors also determine the type and quantity of plants.

In the effort to achieve high milk yields that make it necessary for the cows to have enough easily digestible foods, more and more farmers have been giving increasing amounts of grains and protein concentrates. As consumers of ground oil-seed residues and grains, cattle are deprived of playing their special role in agriculture, which is to be first and foremost consumers of roughage, thus serving not only the production of milk but also the enlivening of the soil. Agricultural use of the land in regions where high precipitation makes crop growing impossible can only take place through cattle and other ruminants.

The ruminant has the unusual faculty of being able to reabsorb urea from its own protein metabolism, i.e., nitrogen, into the blood before it leaves the kidney and thus build up bacterial protein in the rumen. Thus nitrogen is utilized again for nutrition. This knowledge has so far only led to the idea that it is a good thing to add synthetic urea to fodder concentrates. On the farm, however, the cow is a useful animal if it can produce a good yield while remaining fertile and healthy for its proper life-span while being fed on farm-produced fodder, particularly roughage, which is usually relatively poor in nitrogen. Cows like this can only be bred in herds fed an appropriate amount of roughage. Restraining the amount of protein in the ration makes it possible to recognize such animals.

Feeding cattle in spring and summer

More than other animals, the ruminant needs a ration that remains more or less even in its composition. The vital fermentation in the rumen is adapted to the daily intake and cannot readapt to a completely new mixture from one day to the next or even within several days. For high yielding animals, however, few options remain with respect to the ration. Substantial diet changes call for extra metabolic activity and consume additional fodder.

Young grass contains a high proportion of nitrogen, so to give concentrates while the grass is young is a waste and puts stress on the animals' digestion. The fibre content is small, but it contains much water, potassium and phosphorus. Grass with plenty of water and comparatively low in carbohydrates also grows again in the autumn when the days, and thus the amount of sunshine, are short and there is still strong growth from the relatively warm earth.

Grazing should start if possible with the cattle going out to pasture for half of each day. Some additonal hay should be fed. Straw can also be given, or dried sugar beet pulp or silage if the fodder was sufficiently ripe when it was harvested.

Young plants are at first heavily dependent on soil fertility. They have to build their leaves from what they find in the soil. Thus if young grass or young fodder crops are fed to livestock, every fertilizer imbalance is particularly dangerous. Nitrogen is used in the leaf to build protein with the help of sunlight. If there is too much nitrogen in the soil—and young plants tend to absorb much nitrogen—then protein will not be synthesized quickly enough. There will then be more than the normal content of free amino acids, urea and nitrate, etc., in the leaves. This incomplete synthesis leads to rapid breaking down of the fodder in the rumen. Large amounts of ammonia and possibly also nitrite are produced. Bloating can arise (carbon dioxide, methane, ammonia, etc.). A large amount of ammonia is taken into the blood and is then changed into urea by the liver and excreted by the kidneys. This puts an extra strain on the liver, which can lead to disturbances such as ketosis, particularly if the liver is already burdened by liver fluke or something else. Ammonia and potassium both have an alkaline reaction and shift the reaction of the blood in this direction as well. In extreme cases this leads to grass stagger. It has been noted for some time that in intensive fattening units, almost all the livers of slaughtered cattle have to be discarded. This is not the case under bio-dynamic management as Cockleburr Farm in central Illinois demonstrates.

The presence of too much intermediary nitrogen compounds in leaves can also occur after heavy fertilizing with liquid manure and slurry shortly before the animals are put out to pasture. This happens particularly on the grasses near to farm buildings, which tend to get more than their share.

High potassium content in fodder makes it necessary to give salt. The sodium in the salt helps the animals to excrete potassium. Feedgrade salt can lead to diarrhea if there is a high water and low fibre content in the fodder. In the early part of the year it is therefore good to give mineral supplements with, in addition to salt, phosphate, ground lime, magnesium, bolus alba or kaolin, and charcoal.

The most important measures to prevent difficulties at the beginning of the grazing period are those involving fertilizing. As always, the aim is to achieve the right composition of the plant with the help of an active soil life. If fertilizing has been carried out in autumn or winter with compost treated with bio-dynamic preparations and also with fermented liquid manure, difficulties will be much less likely. Bio-dynamic sprays should be used before grazing starts.

It is important not to start grazing on too young grass unless straw or hay are given as well. This applies to the whole of the grazing period. When plants are a little older, the ratio of substances they contain is more balanced. The plant needs sunlight for all its processes of synthesis. Apart from the weather, this is also a question of the total leaf surface and of time. The yield harvested increases while the water content decreases. As they increasingly absorb calcium, the plants create a better P:K:Ca ratio for the animals. The formation of sugar brings about a balance between carbohydrate and protein, and the formation of fibrous substances improves the consistency of the fodder and helps digestion in rumen and intestines. More acetic acid is formed in the rumen so that the fat content of the milk and also the hardness of the butter fat is maintained after grazing has started.

Older grass makes deeper roots, so plants have a larger root space at their disposal. The second and third growth after grazing are more rapid as there are now deeper roots in which the plants have stored reserves. Accessible water reserves are also correspondingly greater, which is decisive for yields if the weather is dry. As the now taller grass is more trampled by the animals, an important asset to successful grazing is a forward fence (usually electric), which is moved once or preferably twice daily. This also allows for longer periods for regrowth. We are speaking about one of the most important measures for achieving high yields, high cattle density, and high yields per animal coupled with positive effects on the soil through the use of farm-produced fodder and manure.

The composition of the turf in permanent grassland depends on habitat, fertilizing, cultivation and grazing techniques. In short leys, composition can also be affected by sowing. The more fertilizer used, the shorter the growing periods between grazing or mowing, the weaker the insolation, e.g., on north-facing slopes or because of damp weather, and the more onesided the soil, the fewer will be the number of permanent plant species. So the composition of the fodder will be less varied. In extreme cases permanent pastures have been known to turn virtually into monocultures. With high fertilizer applications these pastures can be made to produce high yields. But the health and fertility of the livestock becomes a problem, and even to maintain milk production extra amounts of concentrates have to be fed in addition. This development need not be feared when well-rotted manure is used, preferably including some pig manure, and composted with some soil, or slurry as described. The bio-dynamic preparations have a great deal to offer.

Permanent grass should, whenever possible, be influenced only by means of manuring and grazing. Ploughing the old turf always involves breaking down humus, which depletes the possibilities of future growth. Sowing of new varieties can change the composition for a few years. But frequently undesirable plants increase, for instance, the buttercup (*Ranunculus acer*) or the dock (*Rumex obtusifolius*). Seeding in existing grassland is not usually successful unless the turf is first broken by heavy harrowing or some similar technique. This is risky, however, as masses of weeds are also likely to spring up. Only in rare cases, for instance if there is excessive tufted hair grass (*Deschampsia cespitosa*), is it justifiable to plough and resow permanent grass. With most weeds it is better to remove the conditions which encourage them to spread. Otherwise they will soon reappear after ploughing, like for instance rushes.

Grazing of clover grass in an arable crop rotation is easily undertaken provided the cows do not have to walk too far or if milking is done in the field. An electric fence can be used but these are not always sufficient. A hayfield for grazing as part of a rotation is most useful from the point of view of cropping. From the point of view of the livestock there is no likelihood that they will become infested with parasites.

This method is common in Denmark. Seeding of herbs is worthwhile if the hayfield is only used for 2-3 years, since they usually last for this period. In addition to clovergrass the following can be sown: ribwort plantain (*Plantago lanceolata*); yarrow (*Achillea millefolium*); great and lesser burnet (*Sanguisorba officinalis* and *S. minor*); *Pimpinella saxifraga, P. major, P. anisum*; caraway (*Carum carvi*); chicory (*Cichorium intybus*); sage (*Salvia pratensis*); and parsley (*Petroselinum crispum*). Other herbs are also suitable. They are there to enrich the diet and need not be present in amounts that would satisfy on their own. Though the seeds are of different sizes, as a rule of thumb 0.5 kg/ha (0.5 lbs/acre) can be sown of each kind except for the ribwort plantain, of which 4 kg/ha (3.5 lbs/acre) can be used. But the ribwort should only be sown if there is none already present. Whether it will flourish depends on the habitat and fertilizing. Some firms supply clovergrass mixtures with added herbs. To be quite certain of what one is sowing one can also order the seeds separately and thoroughly mix them on the farm.

Permanent pastures with plenty of herbs can be established on dry, sunny slopes or in similar positions. If only ripe compost is used for fertilizing, a suitable flora will come about of its own accord. Shallow dry slopes can stand more manuring without suffering appreciable changes in their plant community. When reseeding, the main varieties already present should be taken into account. On dry, warm slopes, for instance, bird's foot trefoil (*Lotus corniculatus*) and lucerne (*Medicago varia*) can be sown among others. The plant population will adapt to conditions within a few years. Naturally, these pastures cannot be used as main grazing lands. They offer a health-giving supplement to the diet and are grazed for a few hours or half a day at a time, or by pregnant animals just before calving. Above all they are for haymaking.

Manuring grasses and hayfields

An increase in yields does not necessarily mean an increase in biological value. A hayfield in a crop rotation is all the better for the subsequent crop if it grows strongly itself. It is therefore well worth fertilizing it with composted manure. Another advantage is that the manure spreader used on ground containing plenty of roots does not impair soil structure. On the whole the best time for applying manure is in the late summer so long as the soil microorganisms still have time to work it in and the plants can still utilize some of it. Another good time is early spring just as vegetation starts to grow. Fermented manure can also be applied in summer and is particularly effective then so long as it is not allowed to dry out on the surface during a long dry period. Since it gives shade and retains humidity it promotes the second and third growths. Slurry and liquid manure penetrate immediately, but the animals often reject the subsequent growth. It is probably mainly the smell of the ground that repels them since they are quite happy to eat lush grass cut from spots where droppings have been so long as it is offered elsewhere. The organic substance in slurry is rapidly decomposed in the soil. Therefore its growth-promoting effects also dwindle rapidly, and it does not necessarily enrich the humus. Instead, there can be leaching of calcium and magnesium. Patches of lush growth resulting from the fresh urine of the cattle are usually eaten without objection.

On farms where cattle are strictly fed on home-grown fodder only, the cycle of nutrients is almost entirely closed. If 5000 kg/ha (4500 lbs/acre) of milk is sold per year, 4-5 kg/ha (3.5-4.5 lbs/acre) of phosphorus and 6-7 kg/ha (5.4-6.3 lbs/acre) of potassium are sold off. This is almost negligible in comparison with the reserves in the soil and the turnover in fodder and dung, even if the sale of animals is taken into account. Regular buying-in of many times this amount, which is the custom except on bio-dynamic farms, can easily lead to excessive amounts of these substances in plants and thus in fodder. Of course, deficiencies can occur in poor, inactive soils, so if in doubt a soil test should be made. To reinstate the life processes, supplemental applications may then be necessary for a few years. Such conditions are unlikely to be found in central Europe, since so many minerals have been applied during the last decades that imbalances resulting from excesses are far more likely. Nutrient deficiencies are more likely to occur on newly established pastures for wildlife in woodlands. Farmers have demonstrated that the regular use of bio-dynamic sprays alone can promote varied growth on grassland. More distant meadows that receive less manure have thus been able to produce valuable additional fodder.

Problems arising from excessive supplies of potassium are well-known. Similar situations with regard to phosphate are building up on farms that have been intensively fertilized for long periods. If the variety of different plants in pastures decreases, for instance, if there is mainly rye grass, the calcium-phosphorus ratio in the fodder may fall below 1.5:1. Problems of sterility will then arise. When this happens, farmers are afraid to use the manures from their animals because too many mineral nutrients might be returned to the soil. In accordance with Liebig's ideas of nutrient replacement, it is easy to work out the figures, but it took 100 years before anyone troubled to do so. The complete omission of mineral fertilizers coupled with smallish applications of well-fermented manure will restore a more favorable situation.

Small applications of manure stirred in liquid manure just before the latter is hauled out can bring about a considerable stimulation of growth. Frequent small applications are better than less frequent larger amounts.

Calcified seaweed (Algomin) can be a great help, particularly at the beginning of a change to bio-dynamics. In amounts of 300-1000 kg/ha (270-900 lbs/acre) it stimulates the soil micro-organisms and plant growth and promotes a variety of plants all of which grow equally well and are tasty as fodder. The doses should be repeated a few times during the first few years.

In damp climates aggravated by lack of frosts in winter, there is a tendency for acid humus to form. This leads to a loss of nutrients in the soil, particularly alkaline substances such as potassium and also calcium. The greater the population of microorganisms, the less this need be feared. Permanent grass is least subject to leaching. Since in thickly populated European countries it is probable that sulphur dioxide will fall in rain and snow in yearly amounts of 200-300 kg/ha (180-270 lbs/acre) or in places even up to 1000 kg/ha (900 lbs/acre), forming sulphurous acids in the soil, a watch must be kept on the pH figures and the calcium situation so that if necessary lime can be given. If calcified seaweed is not desirable for some reason, dolomite, which consists of calcium and magnesium, can be recommended. The pH figures should, however, remain at a level suitable to the soil type. Repeated large applications of lime can lead to disturbances.

Distinct imbalances resulting either from the nature of the soils or from cultivation should be redressed. Imbalances can be discerned by means of soil tests covering the whole farm. Such tests should include information on the more slowly accessible nutrient reserves. In Europe, with the exception of extreme habitats, serious imbalances have only been discovered with regard to phosphorus on cattle intensive farms. Rock phosphate scattered on the manure while still in the barn is included in the organic process of fermentation. It accelerates good fermentation of the manure, which leads to a rapid rise in grass growth and also stimulates considerably the growth of legumes. For instance, in the Black Forest poor growth in clover, which revealed that there was a deficiency, changed to strong growth with plenty of blossom after the first application of composted manure enriched with rock phosphate. After several years' application leading to the development of grassland with thick turf and a greater variety of plants the addition could be omitted. The expense was small since 50 kg/cow/year (110 lbs/cow/year) given in 200 days in the stable proved the right amount. That is 250 g (10 oz) per cow per day during the winter. This is 1% if it is reckoned that each cow produces about 25 kg (60 lbs) solid excrement per day.

3 Preserving Roughage

Growth during the summer half of the year does not take

place evenly. So the size of the pasture must be sufficient for the number of animals even when growth slows down during the second half of the season. Correspondingly, the main hay harvest must take place during the earlier, rising part of the year. Growth is at its highest until mid-June. At the end of June or the beginning of July there is usually a slowing down until it picks up speed again in August. Finally, growth gradually comes to a standstill during the autumn. This rhythm varies according to the weather but it is basically quite regular.

Plant organs, which in themselves have reached a certain termination of activity, e.g., grains, are not on the whole too difficult to store. The situation is quite different with roughage that has to be harvested when it is at the height of vegetative development.

When a green plant is cut, the water supply is interrupted thus stopping synthesis. But life continues. What constantly takes place in the plant in the root, and at night also in the leaf, now occurs in the whole plant, that is, respiration, the opposite of photosynthesis. The plant begins to use up itself. Then the bacteria gain the upper hand and break down the remains.

In order to preserve roughage, one of the three elements of water, air or warmth must be removed as quickly and radically as possible. Then the life processes are brought to a standstill and the plant is preserved more or less in the condition in which it is cut.

Haymaking

In hay it is the water that is removed. The steamy warm atmosphere under freshly mown, thick green fodder on to which the sun is shining is the best place for respiration to take place. A great deal of what we would like to feed our livestock escapes into the air unnoticed, for carbon dioxide and water vapor are invisible and give off no smell. So an important first step is to lay the hay as loosely and airily as possible. It is best when the same machine does the mowing and the tedding. A crimper breaks open the plant structure of the stalks thus speeding up the drying process. This is particularly important for clovergrass. Rapid drying inhibits respiration so that bacterial activity cannot develop. The sap of the plants, which now starts to dry, contains mineral and organic substances in solution. These remain and dry up, rather like salt or sugar that remain in the bowl when a solution is allowed to evaporate. This process takes place particularly at the surface of the plant. For this reason rain falling onto mown hay leads easily and rapidly to the loss of nutrients.

In this drying process, the warmth of the sun causes in the plant another change of substance, the formation of fragrance. This is the opposite of the precipitating and solidifying processes. Solid or liquid substances become volatile as aromas. In the plant this usually takes place in the flower and in some cases also the fruit. Though the processes are not identical, it could be said that haymaking continues the process of ripening and even goes beyond it before the plant has completed its stages of growth. We use the same means as the plant world: air warmed by the sun. By cutting the plant we have already brought to an end the working of the earth and of the water rising with its substances in solution. We so to speak enhance a "fruiting" process in the plants.

Haymaking can be speeded up by ventilation under cover. This has proved quite useful. It enables the more delicate parts of the plants to be brought in before they become brittle and while they still contain most of their goodness. Short periods of good weather can also be better utilized, and there is far greater certainty that the hay harvest will be satisfactory. The quantitative and qualitative losses undergone by the hay as a result of too much warming up in the rick can be almost eliminated by this means.

A further development of artificial ventilation is hot-air drying. Warm and hot-air drying is particularly important in regions where relative air humidity is high. If it is done properly, it rapidly halts the life processes in the plants, so that there is hardly any decomposition, e.g., of carotin. Problems may arise, however, because of the cost of the plant including heating.

There may be residues from the exhaust fumes and also some losses from overheating.

Silage

Silage making is quite a different process, for it requires the exclusion of air. The water in the plant remains. The finished product shows that silage making has nothing to do with hardening but has rather more to do with dissolving. If the fodder is cut with plenty of moisture there will be a silage effluent containing a great deal of substances in solution.

Immediately after cutting respiration starts. It continues after the grass has been placed in a silo. Respiration releases carbon dioxide, which is heavy and remains in the container. It is invisible and odorless but it suffocates human beings! The more oxygen there is, the more heat is produced in the plants. This heat accelerates decomposition and leads to losses. It also stimulates other changes that make the fodder less digestible. As the temperature rises the color of the material darkens and it begins to give off a pleasant smell like something roasting, or somewhat like caramel. In order to prevent losses of nutrients and digestibility air must therefore be excluded as rapidly and completely as possible. Respiration then ceases and anaerobic fermentation continues.

All substances are constantly in contact with bacteria, fungal spores and yeasts. These are present almost everywhere. The fodder plants cut for silage are also covered with bacteria. So long as the plant is alive it resists decomposition by microorganisms. But once it wilts and juices start to emerge from the weakening tissues the number of bacteria increases rapidly and they become extremely active. For silage it is important that the bacteria that produce lactic acid are favored. This is the case if enough sugar is present in the plant saps; it should be at least 2%.

The sugar content of fresh green plants varies from 0.8 to 6.5%. Grasses and ripe grains contain relatively large amounts while legumes have less than 2%. Autumn catch crops such as rape have only a little. The concentration of sugar also varies according to age, weather, fertilizer used, and the time of day. Once flowering starts, the amount of easily fermented sugar decreases rapidly. High rainfall, lack of sunshine and the short days of autumn all lead to watery plants with poor sugar content. Rain, especially if it falls just before or during cutting, causes low amounts. When the sun is shining, sugar contents rise until the early afternoon and then decrease again toward evening. During the course of a day, the content can double and then be halved. Large amounts of soluble nitrogen also decrease the sugar content. It is probable, although no exact experimental results are available as yet, that the bio-dynamic sprays increase the suitability of fodder plants for silage making.

The best way of starting a good silage process is to allow the plants to wilt sufficiently. This is certainly the case in climates where this can be achieved in a single day at the most. The procedure is the same as for haymaking; straight after cutting, the swaths are fluffed up. The sugar remains in the plant tissue and its concentration increases, particularly at the surface of the plant. The bacteria that produce lactic acid can then proliferate rapidly. For low moisture silage the final pH figures need not be as low as for moist silage. The formation of butyric acid is greater when there is more moisture.

If conditions are unfavorable, apart from wilting the addition of sugar, molasses and ground grains may be needed to help raise the sugar content to 2%. It is important to distribute these evenly. They can be added either during chopping in the field or when the silos are being loaded. The amounts in percentages of the total of green fodder are 0.5-2.0% sugar, 1-5% ground grain, and 1-4% molasses. To stimulate rapid fermentation and ensure lactic acid formation some roasted malt can be added to the grains. This accelerates the transformation of starch into sugar. Ground grains also have a certain drying effect. This procedure was worked out by E. Rasmussen in the damp cool climate of Denmark. In Finland Virtanen has developed a procedure involving the addition of inorganic acids. This, however, leads to quite a different situation with regard to the minerals.

The added acids have to be balanced by additions of calcium and magnesium given when feeding.

In addition to lactic acid, other acids also arise, in particular butyric and acetic acids and sometimes also alcohol. The formation of butyric acid increases if low pH values are not achieved quickly. The tendency to form acetic acid arises when a great deal of fermentable sugar is present and also if the silage is rather wet.

To achieve a more rapid settling, which excludes the air, it is useful to chop the fodder. This also enables the saps to exude better so that they can be more quickly attacked by lactic acid producing bacteria. Where air pockets arise during filling, mould usually forms. Around the edges of the silo the stuffing should be done particularly carefully. The mould itself is not usually dangerous, but near it butyric acid and rotting often occur and these can cause disturbances.

The formation of acid causes the plant cells to die and seeds to lose their viability, which is not the case with haymaking. The eggs of many livestock parasites also die after awhile. Acid formation also plays a part in the preserving of food for man, as in sour dough or leaven, and also sauerkraut in which the juices of cabbage leaves are released with the help of salt. Some feedgrade salt can also be added to silage.

As the plant cells die, the contents of the silo consolidate. Substances in the cells are released. As a result NH_3 arises, which has an alkaline effect, so the pH figure rises again. There must at this point still be enough sugar left to enable the lactic acid forming bacteria to go into action again. If this is not the case, butyric acid can still start to form.

A third way in which the silage can be spoilt is by yeasts, which live on the lactic acid itself. This process can still start at quite a late stage. The necessary low pH figure is then lost and wrong fermentations can start.

The higher the outdoor temperature, the more rapid is the fermentation and the greater the danger of overheating, which can lead to losses of digestible nutrients. After the fodder has settled a kind of "air pump" effect can occur as a result of the changing differences of temperature between the gases in the silo and the air outside. The slightest leak will release internal pressure and let in air as soon as the outside air pressure has increased again due to barometric pressure changes. The same occurs if there are leaks in silos that are heated by the summer sunshine during the day and cool off at night. At best the pressure differences may not lead to wrong fermentation but there is nevertheless a loss of feeding value, which tends to go unnoticed. The mistakes of the summer do not become apparent until the winter when the silage does not satisfy the cattle sufficiently to maintain their milk yield. Usually the spots where the air has entered can be discovered because of the mould that has formed. Rapid, continuous filling is important to ensure the prevention of losses.

When the silo is opened for feeding to start, the chemical changes begin once again, to a greater extent of course in the summer than in the winter. The daily ration removed must be sufficient to include silage that has not been touched by the extra fermentation brought about by the opening of the container on the previous day. The surface area inside the silo must be the right size to allow for this. In winter it is usually sufficient to remove a layer of about 10 cm (4 inches) but in summer it will have to be more like 20 cm (8 inches). The volume weight of silage varies between 500 and 1000 kg/m^3 (32-62 lbs/ft^3) depending on the type of plants, their condition, the degree of preliminary wilting, and the pressure applied.

After mastication and salivation, digestion in all vertebrates continues with a stage involving strong acid action in the stomach. It is hydrochloric acid, which as the digestive secretion in the stomach starts to penetrate, kills and dissolves the food collecting there. In the rumen of the ruminant, however, there is no such secretion. Instead of being killed, its contents undergo a strong enlivening process. In man and animals with simple stomachs this would cause serious illness, but for ruminants it is vital. This elivening process takes place in completely anaerobic conditions. Other acids, however, do arise in the rumen such as carbonic, propionic, acetic and butyric acids. Strangely enough,

285

hardly any lactic acid is found. When the animals are fed on silage, the lactic acid is rapidly absorbed and also broken down in the rumen.

The saliva of ruminants is an almost pure solution of sodium hydrocarbonate. This alkaline salt prevents the development of too low a pH figure in the rumen. It neutralizes some of the acids. The pH figures of the juices in the rumen vary between 5.6 and 7.0. After the intake of fodder, the fermentation processes increase, producing acid. The thorough permeation of the fodder with saliva during cud chewing then gradually neutralizes it again. Almost everywhere where acids are found, we detect a breaking-down action of bacteria. In this way, silage making is a kind of predigestion, an initial breaking down, which must remain held in the acid stage so that the food value is retained. Fodder intake, chewing, fermentation and breaking-down can then take place easily and rapidly because the processes have already been started. Good silage is thus a suitable fodder for producing yields.

This is also the case because the method makes it possible to include young green plants without loss. Some of the goodness is lost in curing and drying for hay, especially that of the leaves of herbs and legumes. Various active substances and their high magnesium content can be depleted. Silage, on the other hand, allows the preservation of leafy young fodder. It also preserves the disadvantages of young plants: their high water, nitrogen, potassium and phosphorus contents, and their lack of fiber and calcium.

On the whole cattle are partial to silage, but if it smells musty owing to mould or if the acetic acid content is too high, they may reject it. Presumably silage tastes much like the contents of the rumen though it does not have the typical smell of propionic acid. So taking in the fodder is for the animal similar to chewing the cud. This is probably the reason why the animals are not repelled by the strong smell and taste of butyric acid.

Butyric acid is one of the normal products of fermentation in the rumen. If it is eaten in addition, it can lead to metabolic disturbances, e.g., ketosis. If it is present in silage, it is a sign that considerable nutrient losses have taken place, because it arises when the pH figure is not low enough and therefore fermentation did not cease soon enough. There is then a danger that other unfavorable products of fermentation and rotting might be present, such as substances produced when protein is broken down. The animals' senses of taste and smell do not protect them from eating bad silage.

Since the pH figure in the rumen is not so low it is not surprising that butyric acid should arise there. Protein is also broken down there. The resulting large amounts of ammonia can affect the health of the animals if they are fed on young, heavily fertilized fodder. If the breaking down is not too rapid, the bacteria in the rumen can ensure that protein is constituted. Whether this is successful depends on the composition of the ration and on the health of the animals.

While silage contains valuable nutrients, it nevertheless differs from hay. In silage the aromatic substances of the ripening plant, which come about in haymaking, either do not arise or are virtually lost during fermentation. The occurrence of vitamins A and D also shows the difference between silage and hay.

Vitamin A is formed in the organism from carotins, the yellow pigments, which are present in large quantities in fodder plants. It is needed by those tissues in which regeneration is particularly intense. The mucous membranes are the inner barriers between the body and what is taken in from outside and they have the capacity for rapid renewal (lung passages, intestinal canal, etc.). Vitamin A deficiency leads to absorption disturbances, hardening (eyelids, tongue of newborn piglets) and susceptibility to infection. Fertility also depends on sufficient carotin in the diet. Deficiencies lead to weak heat, stillbirths and small litters of pigs.

Fresh green plants usually contain plenty of carotin, though the nature and amount depends on the conditions and methods of cultivation. It arises in plants that are growing strongly. Breaking down starts immediately after cutting. Rapid artificial drying and also the silage process halt this breaking down. Thus there is usually plenty of carotin in alfalfa and other green meals and in silage, though the actual amounts depend in large part on the method of curing and also on the weather. Breaking down still continues slowly during storage. It is more rapid if the concentration is higher, so varying amounts at the start are gradually evened out. Airproof storage still delays breaking down, even at this late stage.

With vitamin D (D_2) the situation is rather different, since it does not arise in the plant until the sun shines on it after cutting for hay. With vitamin D it is rather a ripening process in which valuable substances are created, as is also the case with the aromatic substances mentioned above. Vitamin D promotes the absorption of calcium and phosphorus from the intestine and their building in the bones. It is one of the factors that promotes strength and form in the organism. People and animals form vitamin D (D_3) in their skins if they have sufficient light. Indeed, vitamin D might be called a "light substance" whether it is obtained from plants or produced by the organism itself.

For the sake of completeness, the preservation of fodder by removing the warmth factor should also be mentioned. As the removal of water leads to drying and the removal of air to lactic acid fermentation, so the removal of warmth leads to cooling and finally freezing. But freezing is far too expensive a process to be used on fodder. During cooling certain enzymes remain active and changes in the concentrations of nutrients in the juices take place. Germs that flourish in the cold might increase and continue with their metabolism. Cold storage is practical for keeping fodder beet either in the cellar or in clamps. A cool temperature together with high air humidity is maintained and the qualitative characteristics of the beet only change slowly.

Every method of preservation has its own laws, processes and effects. Basic fodder is sometimes given exclusively as silage, for instance, on so-called Harvestore farms. However good the quality, the total intake of dry matter is thus reduced. Cud chewing may then also be reduced and this in turn makes it necessary to give too much ground grain and protein. The final result is a feeding program that does not conform to the nature of the animal. Some hay is an indispensable part of a good ration.

4 Fodder and Feeding of Cattle in Winter

Rations are calculated for maintenance and production. The higher the milk yield is, the greater is the importance of finding the right nutrient balance. The actual nutrient content of fodder varies widely. Apart from the kind of fodder used (hay, silage, grains, beet, oil-seed cakes, etc.), variations also occur through many factors: soil, manuring, weather, plant varieties, stage of development of plants when harvested, method of harvesting, preservation, storage, duration of storage. Onesidedness on a large scale almost always harms the livestock, on occasions even after a great deal of time has elapsed. Variety is always a safety factor.

Exact nutrient contents can only be ascertained by tests that in practice are made only in exceptional cases. We can learn a great deal, however, if we make good use of our five senses. The look, feel, smell and taste of fodder all tell us about its quality in addition to calculations of protein content, nitrogen-free extractives (carbohydrates), fat, fiber, minerals, trace elements and vitamins.

Digestibility can considerably alter the actual value of the nutrients. For high production, good digestibility is asked for. Nevertheless, cattle need roughage. For the ruminant, the source of the most digestible protein is the protein forming bacteria in its own rumen. These organisms produce the even flow of short chain fatty acids that supply the ruminant with energy in accordance with its nature. The more easily digestible the fodder is, the more rapid are the processes of fermentation. Nitrogen is then lost in the form of ammonia, which burdens liver, circulation and kidneys. Ratio changes of the different acids or their irregular formation during the course of a day can lead to metabolic disturbances, susceptibility to disease and actual disease. This is particularly so in the long term. There is therefore no point in making the ration too easily digistible merely in order to achieve even higher yields. The value of fodder for the health of the animals depends on other factors besides those that can be assessed by the nutrient contents. Among these other factors is the variety of different plants in the hay, i.e., the grasses, legumes and herbs and their proportion of the whole.

The profitability of a ration cannot be measured by taking only the average annual yield into account. Other parts of the calculation are the yield of a whole lifetime, and the number, health and yields of offspring. Easily digested foods such as sugar, molasses, ground grains, young plants and oil-seed cakes are usually expensive and onesided. As regards the farm itself, the ratio of fodder crop yields to soil fertility must be taken into account. Expenses incurred because of disease can also damage profitability. Disturbances of digestibility caused by overheating of the haystack or silage can, of course, be avoided.

Hay is the most important basis for winter feeding. If possible, hay should be fed to satisfaction. How much is eaten depends partly on its taste. When presented in large amounts, cattle usually eat less keenly then when they are given frequent smaller amounts. Probably they enjoy the smell of fresh hay but dislike the effect their breathing has on it. The amount of hay given must depend on the total ration. A cow should not have less than 2 kg (4.4 lbs) per day, though hay-free feeding is also possible. Actually, considerably higher amounts are desirable. If storage permits, a mixture of first and second cuttings should be given, or one in the morning and the other in the evening. Hay from the first cutting contains more stems and from the second more leaves. Many herbs are fully developed either in the spring or in the summer, and often either only the first or only the second cutting has been cured without rain. A disadvantage of hay towers and also of silos is that there is no way of selecting which part to give the cows next. If possible the harvest from qualitatively valuable fields such as herb pastures should be stored separately so that portions can be taken at any time. Long hay is more easily eaten than chopped hay.

Silage is also a valuable component of the ration. For

intensive dairying the minimum amount of silage needed is 4 cubic meters or yards per cow. This assumes a daily ration of about 16 kg (35 lbs) per cow, not including the amount required for young animals (1 m³ = app. 800 kg or 1760 lbs). Double this amount is still reasonable. Low moisture silage from varied permanent grassland or clover-grass with sown herbs is best. Many reasons have already been discussed for recommending clover sowing only in a mixture with grass: weed control, guaranteed yield, humus formation, suitability for silage, nutrition.

The following remarks about silage and fodder beet apply to bio-dynamic farms in Europe (See also the report on Ceres Farm). In Europe silage maize has become exceptionally popular in recent years because of its high yields, the possibility of using agricultural chemicals to save labor, and its suitability for silage. However, plant diseases and some unfavorable dietetic effects of a large proportion of maize in the daily ration show that its use has been exaggerated. Since the effects of herbicides at this point cannot be foreseen in the long term, the chemicals are not used on bio-dynamic farms. Consequently, in these cases maize is not such a favorable crop from the point of view of the labor requirement. Maize, however, can be grown quite well on good land, for instance after a winter forage mixture (winter vetch, ryegrass and crimson clover). In limited amounts, e.g., 10 kg (22 lbs) daily per cow, corn silage is a valuable part of the ration.

Silage is often the only way of preserving catch crops, since in autumn the days are short, with mist often lasting well into the morning, and the sun too weak for proper drying. The most important fast plants for catch crops are brassicas. This together with the weather conditions in autumn means that catch crops have on the whole too little sugar, too much water and poor quality protein. Their feeding value is therefore limited and silage can hardly be made from them without additives such as ground grains, molasses or sugar. Silage effluents form during fermentation, so empty silos must be used to prevent spoiling lower layers of silage made earlier. On the whole it is preferable to plough in catch crops that cannot be fed fresh during the autumn.

The amount of labor involved in growing fodder beet has meant that many farms have had to give this up during the last decade. Experience with livestock on mixed farms, however, has also encouraged numerous farmers to take it up again. A reasonable minimum per cow per day is 5-10 kg (11-22 lbs). Assuming 200 indoor feeding days in the year, this amounts to 1-4 ar per cattle unit, i.e., at this rate one acre supplies 10-40 cows. The keeping properties of bio-dynamically grown fodder beet are considerably better than usual, provided they are properly stored either in clamps covered with straw and soil or in a cellar. Even so the supply should be used up by the time fresh green fodder becomes available. Silage can be stored longer without difficulty. The smaller the amount of beet included in the ration, the more it is worthwhile growing carrots instead of beet, particularly for the second half of the winter and for cows in late pregnancy as well as calves.

During the second half of the winter, cattle often suffer from loss of fur round the eyes, on the neck and at the base of the tail; there may also be sterility, and diseases in newborn calves, etc. Apart from lack of exercise and light, the main reason for this is the gradual deterioration of the fodder during storage. Vitamin A and the carotin from which it arises have been particularly studied. They are broken down through contact with air, and the consequences are increases of mites, susceptibility to infection, and sterility.

Carrots contain a specially large amount of carotin and at the same time they are a typical root fodder with a high content of essential oils. They are highly suitable for young animals, helping their development, preventing worm infestation and curing diarrhea.

The leaves of beet and sugar beet are a nourishing fodder. On farms growing much sugar beet there is also a danger that cattle may be given too much, to the detriment of their health. After late harvests when the ground is wet the animals can often not be prevented from eating considerable amounts of earth (about 2 kg or 4 lbs of sand a day or more). The oxalic acid content of the leaves fixes the calcium in the alimentary tract, so it is necessary to give proper amounts of feedgrade lime, if possible in combination with some mildly constipating substance such as charcoal, aluminum hydroxide, bolus alba, etc. Care must be taken with silage from sugar beet leaves and the way it is fed because it can contain relatively large amounts of butyric acid.

The production of cows of proper genetic constitution can easily be increased by feeding grain and protein. The advantage of bio-dynamic growing, however, is that the roughages can have the same effect. Cattle are suited to utilizing roughage. As ruminants they are able to be sparing with nitrogen in their metabolism. These capacities should be preserved and enhanced. This is a matter of breeding that can only be achieved by those who decide to work along such lines, as production and reproduction can then be combined in the same animals. When large amounts of concentrates are fed, the genetic capacity can be fully utilized but it is not possible to discern which cattle are also good users of roughage. It is also easy to overstep the limits with concentrates, achieving ever higher yields at the expense of reproduction.

Farm produced grains, particularly rolled oats, are suitable. Many observers have confirmed the value of oats as a concentrate. Apart from bolstering yields they also help preserve health and fertility. Mixed sowing with field beans (*Vicia faba*) or peas has proved useful. Newer varieties of peas are adapted to combine harvesting. Rolled oats are also useful mixed with ground barley.

In addition to crops from the farm, protein feed can be bought in. It should be taken into account in this connection that mixed concentrates should contain mineral additives to match the protein. Phosphate requirements increase when more protein is eaten. Even if the components, such as wheat bran, ground linseed and soya bean meal, are bought separately, they should be mixed before feeding. Protein-rich oil cakes are expensive and may be found to be adulterated with fillers. They can also go bad, or may already be bad when purchased. If this happens, moulds can cause the formation of toxic substances, as, for example, aflatoxin in peanuts. Protein feeds including the farm grown legumes should not account for more than one third in the mixture with ground grain. High milk yields can be reached through feeding but this easily puts too much strain on the cows so that sooner or later they will fall ill or remain infertile. Udder diseases are closely linked with feeding.

The metabolic disturbances described above, udder complaints, and hoof diseases show, if they occur in any numbers, that there is something wrong with management and feeding on the one hand and yield on the other. All these phenomena show a weakening of the formative processes. To counteract this, feeding with additional herbs can be recommended. The variety of stimulus obtainable from these specialized plants often brings about rapid improvement or does not allow the disturbances to arise in the first place. They do not grow where fodder is grown intensively but need special habitats with plenty of sunshine, a good supply of soil-born mineral elements, or ripe compost. It has been found useful to establish a herb garden in which the herbs can be tended and harvested before being carefully dried and stored. The greatest need for herbs is from January onwards. Stinging nettles have proved particularly useful and a good supply should be dried, possibly with forced aeration. If they are not properly dried they easily go mouldy. They should be harvested before the lower leaves begin to drop. Even small amounts are useful for special cases. 10-20 g (½ oz) per cow per day, i.e., 2-4 kg (4-9 lbs) per cow for the winter half of the year, is a good amount. For many centuries nettles have also proved excellent as first fodder for young animals. Many other herbs that can be grown in the garden are also suitable, such as fennel, caraway, anise, coriander, dill, chervil, lovage, angelica, marsh mallow, hog's fennel, balm, peppermint, marjoram, summer savory, hyssop, basil, fenugreek, common goat's rue, camomile, a little wormwood, ladies love and rue. The greater the variety the better, but there is no need to grow the whole arsenal. Only small amounts are needed. The best way is to grind the plants after drying in shade, ready to be added later to concentrates. Keep them airtight; they must not go damp again.

Every farmer who grazes his cows knows how greedy they are for the foliage of trees and shrubs. Hedges have many uses on a farm, and one of them is to provide supplemental feeds that stimulate the metabolic processes in cattle. Only small amounts of foliage are needed; it is not supposed to replace any other fodder, though it is rich in nutrients. Labor conditions being what they are at present, it is rarely possible to harvest and store "leaf hay" although mechanization does not seem out of the question. But it is easy enough to allow the animals to forage a a little in hedges, provided these contain the right shrubs. In Europe in earlier centuries the ash was considered especially important for the health of livestock and trees were kept from which young branches were regularly cut, giving them rather the appearance of osiers. If work conditions on the farm make it impossible to supply livestock with supplements of this kind, there are on the market a number of proven herb mixtures either with or without seaweed.

In summer, feeding according to yield is not easy to carry out. In winter it is easier with the help of concentrates. It is then possible to squeeze the last possible ounce of production out of the herd. This aim is easily exaggerated, however, for the animals with the highest yield resulting from special feeding are also the ones most likely to suffer in health. It must be remembered that if the milking capacity of every cow is utilized individually with the help of concentrates, it becomes impossible to detect which animals are good utilizers of roughage. Yet it is from these that breeding should be done. Both the above points might persuade a farmer to refrain from feeding each cow individually for maximum yield and he might well find that such a course of action would not be in the least unprofitable.

Finally let us remember that apart from making milk, every cow also has to produce a calf each year. Feeding adjusted to the nature of the animal is also important for the growth and subsequent health of the calf. Steiner pointed out that a pregnant cow fed on clover would subsequently produce more milk. Experience gained during the changeover of farms to bio-dynamic methods has proved this to be so. There is of course no question that the feeding of the mothers has a direct bearing on the health and vitality of their newborn offspring.

5 Minerals

The high yields of which cows are capable today, and which are for the most part necessary for economic viability, do place a great strain on the metabolic system. Daily yields of over 40 liters (80 lbs and more) can hardly be sustained however much the cow eats. This is even more the case with individual nutrients whose relative amounts in the ration never remain the same. If there is something lacking the animal supplies it from its own body. This is particularly the case with minerals, of which the amounts in the milk are kept remarkably constant. A bottle neck is most likely to occur with phosphate, and this first affects reproduction; the cow fails to come into heat, or suffers a reduction in the size of her ovaries, or calves prematurely without having any infection, or contracts an udder inflammation, etc. This has been observed repeatedly. Only a rough calculation can be made of the relative intake and output because no account can be taken of the processes within the animal, for instance, the amount of phosphorus absorbed from the fodder into the blood, or its effects in the metabolic processes between blood and the organs, etc. What is certain is that all these processes, in addition to depending on genetic factors, are much influenced by the type and nature of the feed. This is so not only of the minerals taken in with the plants but also of known and unknown active substances. If the amount of protein in the fodder increases, the need for phosphate not only rises, it also rises relative to the amount of calcium in the ration. Young plants contain much phosphorus and potassium and not much calcium. As the plants grow older this is reversed. Herbs usually supply plenty of calcium and magnesium. Ground grains and particularly bran contain relatively large amounts of phosphate. After many years of intensive fertilizing with phosphorus and nitrogen the Ca:P ratio can become too close in pure green pastures. It has been known to drop below 1.5:1, i.e., there is relatively too much P. Before this intensive fertilizing starts, this ratio is usually at its highest value of 3.5:1 or more. There can, however, also be a high phosphorus content in soils that have received no phosphorus fertilizer at all. In a series of hay tests in the jurassic limestone region of Southern Germany the highest phosphorus figure for 1970 was from a bio-dynamic farm that had used no phosphates for forty years.

In short, general indications must be supplemented by an exact knowledge of the farm before a proper supply of minerals for the cattle can be worked out. The actual ratios in the soil and the fodder can be ascertained through analysis, but all other facts and experiences about the farm must also be taken into account, as well as the health of the animals. This is possible. The offering of mineral mixtures with medium amounts of phosphate (around 20%, Ca:P ratio 1.5:1), magnesium and block salt are a means of achieving some guarantee of good health. In many cases where yields are high this is a must. It is good to combine this with herb fodder. The amounts suggested by the makers are often higher than necessary.

If the change to young pasture has been rather sudden because of a lack of any more hay, straw, sugar beet pulp etc., there may be a calcium deficiency in the fodder. There is also a danger of diarrhea. Warm damp weather increases the likelihood of this. Mixtures with plenty of calcium and some means of preventing diarrhea are useful in this transitional period. If there is a magnesium deficiency, which may have arisen from many years of fertilizing with potassium, grass stagger can easily occur. Sufficient additional magnesium must therefore be given. No sickness of this kind has so far been noted on a bio-dynamic farm.

Onesided effects similar to those on young pastures can occur if the amount of sugar beet leaves or maize silage (maise in its vegetative state) is too high in the ration. Sugar beet leaves, which are actually rich in calcium, can cause the fixation of calcium in the intestines because of their equally high oxalic acid content. If diarrhea lasts too long, absorption disturbances can lead to an acute failure of the mineral supply. Maize, if fertilized with too much phosphate, can come to have too narrow a Ca:P ratio.

Silica is a mineral in fodder that is almost totally overlooked. In quite small amounts it is important for organ formation and metabolism. In the human and animal organism silica has an important part to play in connection with body fluids, the formation of strong connective tissues, and the functioning of the nervous system and the senses. It should be remembered that one of the reasons why ripe grass and straw are important in fodder is because of their high silica content.

6   Reproduction in Cattle

The healthy rhythm for reproduction in cattle is one calf a year. As the average gestation period is 284 days, cows should be serviced again 12 weeks after the birth of a calf. Repeated earlier insemination can lead to sterility after a few years and thus an unnecessarily early end of a cow's reproductive period. Cows should therefore not go for service sooner than 10 weeks after parturition. If they are well fed on a good ration with mineral and possibly herb supplements coupled with daily exercise, servicing is more likely to take. It is a necessity to keep exact records of each cow's calving, heat, and service dates and of the results of pregnancy tests. The dry period should start 8 weeks before calving. If cows are still giving a great deal of milk, they should not be given anything to drink before milking for two days and possibly also receive less feed (straw fodder). The udder must be in perfect health before the dry period starts. If inflammation is thought likely after the start of the dry period, the cow can be given 500 g (15 oz) Glauber's salt dissolved in a drink two days before the commencement of the dry period.

Once the cow is dry she should nevertheless still have daily exercise till the calf is due. Her fodder should be of good quality in amounts suitable for a cow giving 15 l (30 lbs) milk a day. The bulky farm-produced fodder is probably enough for this. Concentrates should not be given until two weeks before calving. The best is ground oats and a few days before the birth possibly soaked linseed. The supply of minerals must be good as the cow stores reserves ready for the demands the calf will make on her.

Calf rearing

The cow should be allowed to lick her new-born calf dry. As soon as it can stand it can drink from the mother. The first drops of milk contain protective substances. In addition the sucking of the calf helps the womb to shrink and stimulates expulsion of the afterbirth. During the first few days the calf should drink four times a day and it should be allowed to drink from its mother for at least four days because it is essential that it should take all the colostrum.

In smaller and larger herds, if the stanchions are large enough, it is a good idea to tie up the calves for 2 to 4 days at the front end. Then they can be allowed to roam freely in the stable. They will return only to their own mothers and are free to do so when they please. Milking takes place as usual. The calves are thus fed according to their needs and they can also take as much exercise as they want.

Another possibility is to keep the calves in pens while work is going on in the stable. A two liter (4 lbs) container can be attached to the milking machine to collect the mother's milk, which can then be given straight to the calf while the mother is being milked. Another good method is to use calving pens in which cows, either tied or untied, can spend a few days or weeks with their calves.

For replacement calves that are to remain in the herd it is good to give whole milk for six weeks, gradually increasing the dilution with water. Particularly at first the milk should have a temperature of 30-35°C (85-95°F). Daily rations at first are 3 liters given in 4 meals of ¾ l (6.3 US pints in 4 meals of 1.6 pints). This is gradually increased to 7 liters (15 US pints) at the end of the second week. By diluting, the amount of whole milk is then reduced to about 4 l (8.4 pints) (3rd-6th week) and later 2 l (4.2 pints) can be given until the 12th week. Skim milk, which often has to be transported, is better not used unless butter is made on the farm. Good hay is given from the second week on ad lib. Ground oats possibly with field beans, peas and linseed is also given from the second week. The herb mixture described above can also be given from the start. Carrots are useful, especially toward the end of the winter. In good weather the calves can be let out on dry pasture so long as there are absolutely no parasites. They must be carefully observed and always returned to the barn at night. If this is not possible, they should be kept under cover. Full grazing is not possible until the 7th month.

In their second year, young cattle are robust enough to thrive on roughage alone. They can also be allowed to graze over the pastures just vacated by the cows so long as the latter are quite free of parasites because the young animals are still rather susceptible. Parasites to be avoided are parasitic worms such as liver fluke, husk, stomach worms and parasitic gastro-enteritis, etc.

7 Pig Rearing

In Europe, the factory farming of cattle is only gradually becoming established. The need of these animals for roughage is to a certain extent a deterrent. With poultry and pigs, however, the methods have gained a good deal of ground. But the diseases contracted by the animals are causing a corresponding amount of concern. We must learn what we can from these efforts and at the same time not forget that yield depends on health and the future of a breed on robust animals that do not have to be carefully cosseted and nurtured until they are ready for slaughter. The health of future generations depends to a great extent on natural conditions being provided for the mothers, though selection for health is also important.

Exercise is particularly important for sows, but conditions detrimental to health can easily arise in outdoor runs used only for pigs. Intestinal parasites may become particularly prevalent. It is therefore good either to alternate pastures with different animals, or to apply manure from different animals, or to alternate with crops every 2-3 years. Pigs with nose rings (clips are not sufficient) do not dig, so properly organized grazing becomes possible. Restricting grazing to 2-3 hours a day is another way to prevent pigs somewhat from breaking up the ground. Since rooting is part of the animal's character, however, it is probably better to create conditions where it is possible, i.e., by alternating with crops. Jerusalem artichokes and comfrey are particularly favored by pigs. They produce new growth out of any remains in the soil, so they are useful in plots where pigs are allowed to root.

Sows that have had sufficient activity during their youth and during pregnancy can then easily be confined in farrowing crates when their litters are due. These allow the sow to turn or change from one side to the other without crushing the piglets. Moreover, sows that have had sufficient exercise are less clumsy and more careful than those kept without exercise. Allowing the sows out of their pens for feeding has proved a useful method, since it gives them a certain amount of exercise while making feeding easier; they also then leave most of their dung outside the pens.

For the first few days, piglets need some kind of extra heat. This is also useful because it makes them settle together in a suitable spot. Plenty of bedding with an overhanging bale of straw also provides good shelter. From the second week, when their own ability to regulate their temperature is more stable, the piglets can easily become spoilt. The heater is gradually moved further away and then removed altogether.

So far as is known, every mammal is born with a large supply of haemoglobin iron. After birth, some haemoglobin is broken down and iron is excreted. The milk of mammals contains little iron and the pig is no exception. The rapid growing capacity developed by breeding is one of the reasons why piglets can easily become anaemic during the first 2-4 weeks. Against this they are usually injected between their 1st and 3rd days with organic iron compounds (iron dextran), which help build up the blood, thus preventing anaemia and its possible consequent diseases. This does, however, amount to an involuntary negative selection because it makes it impossible to determine which animals have good blood building capacities of their own and which have not. It has been observed that piglets with enough exercise and the possibility for taking in earth regenerate their blood as rapidly as those that are given the injections. From their first day, i.e., long before they can take in solid food, which is not until the 10th to 14th day, they root busily in earth that smells of humus and also take some in. They do the same in compost. Earth can be taken from molehills, from cow pastures or fertile fields, but not from a pig run. Special composts can also be prepared, including composted cow manure with its high vitamin content (B-complex). This is possible even for large herds if it has been made possible to store the earth in a dry place so that it is available even in frosty weather. A special run for piglets should be included in any plans for stables. Remer (1968) has worked out a means of controlling manure fermentation with the help of the bio-dynamic preparations, which produces a particularly favorable compost. Even garbage compost of sufficient quality has proved highly successful (Hauri 1968). Composts must have a good earthy aroma. If they contain too much nitrate, they can damage the digestion. A considerable advantage of this method is that the piglets start earlier than otherwise to take in food other than mother's milk. In addition, light penetrating through the eyes via the hypothalamus to the hypophysis has important effects on metabolism and particularly the building of blood. If pig breeding is not to become increasingly severed from a natural environment but brought instead consciously into closer contact with it, which is the aim on bio-dynamic farms, there must be proper conditions for the piglets. Such conditions are an important source for health in the long term and it is quite possible to create them within the framework of planned farm developments.

Selection for breeding on the basis of good health has become important with pigs now that the rapid spread of special breeds giving a high percentage of lean meat has brought about an almost total disregard for the health aspect. Extreme measures such as birth by caesarean operation and rearing without sows so that no virus infections can be passed on by mothers show how warped the attitude to health has become. (One speaks here of "special pathogen-free farms.") Of course the prevention of infection is important for some diseases, but it is far more important to breed animals that are so healthy that diseases do not spread. Suitable measures and particularly selection for good health can be carried out systematically with pigs.

Feeding pigs

In comparison with cattle, pigs are rather more fastidious with regard to the digestibility of their feed as well as its protein content. Usually this means buying in protein, which must be of high quality. The feeding of root crops (potatoes) has been greatly reduced as a result of the unavoidable rationalization of labor. Ground farm-produced grains form the basis of the ration. Oats are particularly suitable for breeding sows and also as the first food for piglets, but they must be dehulled. It can be worthwhile to grow hull-less oats for this purpose. Some cultivars are available.

Green meal in the summer and silage and dried green meal in winter are important for the feeding of breeding sows. This must be made from young plants, as pigs do not digest crude fiber at all well. Breeding sows should not grow fat, but neither should they be too thin while suckling. Giving the piglets supplementary feeds as early as possible spares the strength of the sows.

The herb mixture already mentioned in connection with feeding cattle in winter has also proved useful both for sows and piglets. Even stinging nettles alone are valuable and seaweed products have proved a useful supplement.

In large herds it is important to weigh the animals. Increased intake must keep pace with increases in weight. This cannot be achieved without proper calculation and weighing. Fattening does not aim to achieve the quickest possible weight gain. What matters is weight put on per unit of feed and the least possible increase of fat. Pigs that are fed to maximum gains in weight are susceptible to erysipelas (swine fever), heart failure, and other maladies. This is not so much the case with pigs fed optimum amounts.

303

Young breeding stock should be kept out of doors for as long as possible after weaning, during which period they should grow rapidly but not be fattened. Dehulled oats are an important fodder for this. These animals must have exercise and plenty of opportunity to move about.

8 Poultry

The non-industrial keeping of poultry by farmers has almost disappeared. It should therefore be stated that chicken runs for 1000 or more hens can be kept healthy provided they are shared or alternated with other animals, such as calves or pigs, or even horses or cows. If the turf is killed here and there in the run, straw or other litter can be strewn and this will combine with the chicken manure to activate intense soil life. The whole area becomes a kind of compost and should therefore preferably not be trodden by people or larger animals. There must always be enough litter to prevent the surface from becoming soggy. A light application of lime will promote fermenting. Calcified seaweed is good for this. Bio-dynamic compost preparations should also be added. Strong activity of bacteria, soil fungi and soil microorganisms destroy the larvae of parasites. In this way it is possible to achieve a relatively extensive healthy stock of free range hens on the farm, an undertaking that can be quite profitable. Hens can be kept successfully for several years. Keeping layers in this way, however, does require some interest and attention of one of the members of the family or farm staff.

Composting is of course also possible where broilers or hens are kept indoors on deep litter (see p. 169). Deep litter houses can be relatively small (up to 6 hens per square meter or yard) if the hens are let out regularly. The litter to use is wood shavings from chemically untreated wood, and chopped straw (20 cm or 8 inches). Some peat and the bio-dynamic preparations are added regularly where the dung accumulates under the roosts.

Some of the grain is scattered in the house or the run. Wheat, oats, barley and maize (also millet) can be used whole, mixed in equal proportions. Oats and barley can be dehulled. The hulls burden the organism with crude fiber that is difficult to digest and limits the amount of food that can be taken in; on the other hand, they stimulate the digestion and supply the important silicic acid. Larger breeds can do with a larger proportion of oats. Alfalfa meal is a good supplement (approximately 5 g or one sixth of an ounce per hen per day). It must, however, be made from alfalfa cut very young so that it contains the necessary amount of carotin and not too much crude fiber. If it is young and also bio-dynamically grown more can be given.. Weeds and vegetable trimmings are eaten with relish. Like pigs, hens kept indoors take in quite a bit of compost (about 3 kg or 6-7 lbs per 100 hens per week). Skim milk either fresh or sour is an excellent fodder (5-10 L = liter or 1.3-2.6 gal per 100 hens per day). Plant and animal proteins (about 15%, i.e., 1.8 kg or 4 lbs per 100 hens per day) and also minerals will have to be bought in together with vitamins. Feeding along these lines has brought favorable results with regard to health, egg yield and prize-winning breeding champions. Supplementary herbs (25 g or 1 oz dry substance per 100 hens per day) have also proved useful. The mixture is stinging nettle leaves (50%); seeds of caraway, fennel, anise; leaves of birch, haxel, bilberry, rosemary, coltsfoot, St. John's wort; tips of spruce; bark of oak; root of gentian. This mixture is available on the market also in combination with minerals and calcified seaweed. Given in larger amounts, over 50 g or 2 oz, the mixture stimulates moulting, which then takes its course rapidly.

If kept in a healthy way, battery hens due for slaughter can be selected for transfer to bio-dynamic management. After a transitional period these hens can still become good layers. If all the grain is grown on the farm, 90 g (3 oz) per hen per day must be reckoned with. 100 hens (= 1.7-2.5 cattle units) therefore need 3.28 t per year. This means a grain harvest of 0.6-1.0 ha (1.5-2.5 acres). Of this, almost 100 kg (220 lbs) of nitrogen and almost the same amount of phosphate is returned to the soil in an effective calcium-rich manure. This will be correspondingly less the more the hens are free range. In this situation a hen gives 200 eggs per year. If hens are kept for several years, the eggs are fewer but larger.

9 Housing Livestock

This is not the place for detailed instructions on the building of stabling, but a few points relevant to the views expressed in this book and based on experience are worth mentioning.

When buildings are being planned, the following points should be examined separately and then combined in the plan:

1. Protection of livestock against extreme weather conditions to maintain high yields and health and also to accustom the animals to the presence of humans.
2. Storage of high quality fodder; ease of access.
3. Protection against winter weather for farm workers and the greatest possible reduction in daily labor requirements.
4. Optimal conditions of hygiene in obtaining the animals' products, especially milk.
5. The biological effect the stabling of animals has on the farm (availability of solid and liquid manure, the ratio of different livestock, the route taken by the animals on their way to pasture).
6. The aesthetic aspects of the buildings in the landscape.

These six aspects of livestock buildings cover a variety of often conflicting needs. To harmonize these may be more complicated than the building of dwellings for human beings. In recent years ideas for stabling have been under the influence of the need to save labor. This is certainly one of the main aspects but it should not be virtually the only point considered. As a result of the onesidedness arising from this, many of the new designs are already no longer being built, while scientists are now turning their attention instead to a more thorough study of the behavior and actual needs of the animals. From the bio-dynamic point of view this recent development is most welcome. In 1959 Bielenberg summed up this aspect in the rather extreme formulation: "The animal is the architect." We shall not discuss types of livestock buildings but rather go in more detail into some points connected with #1. above. A variety of practical solutions can be found to conform with each of the objectives put forward.

306

The high yields we have to demand from our livestock make it necessary to protect them from the vagaries of the weather and provide them with suitable housing and first class fodder. Health is one form of "production" that has to be maintained and renewed continuously. Too much protection simply softens animals so that important body building functions lapse and the animals weaken.

Animals are creatures that need to move. In summer when they are out at pasture they have plenty of opportunity for this. It has been found that even for larger herds it helps fertility if the animals are given opportunities for exercise in winter as well. There are various possibilities for this: the loafing barn, free stalls, or cow barns with standings and appropriate tyings. This point is particularly important on farms where the cattle are kept indoors the year round. If they can by some means be made to walk just a short distance uphill and downhill, this alone can stimulate weak senses and help with muscle and bone building and the maintenance of proper circulation.

Domestic animals have a great need for light. Cowhouses should let in as much light as possible during winter. In summer, though, they need to be shady and cool. To achieve this the barns should be oriented east to west so that one long side faces south. If the edge of the roof overhang on this side is at an angle of 50-60° to the bottom to the bottom edge of the windows, the midday sun will not shine in between the end of April and the end of August. (At 50° latitude north the midday sun arises 40° above the horizon on 21 March and 23 September, 63.5° on 21 June, and 16.5° on 21 December. The corresponding figures for 40° latitude north are: 50°, 73.5° and 26.5°.) During the winter months, on the other hand, every available sunbeam will shine right in, so long as there are no tall buildings, trees or hills in the way. The higher the wall the greater the overhang must be in order to achieve the necessary angle (J. Ober 1954). In the east and the west the sun is of course low in the summer as well as the winter, so ordinary shutters can be used at the ends of the buildings. Sloping window sills and edges also increase the amount of light that can be let in.

Ventilation ducts do not work in summer and in winter they can turn into channels of ice. Forced ventilation by means of a ventilator in the wall works perfectly and cheaply. In this way, light winter and cool summer stabling can be achieved with plenty of ventilation while avoiding draughts.

Cattle and pigs are herd animals, which if left to themselves function largely according to a common daily rhythm. This is important for their health, and most noticeable at feeding time with pigs though it is less obvious with cattle. The investment costs of automated feeding have meant that feed banks have been made to accommodate only a third of the herd at a time. This means that one third of the herd is always eating, so the daily rhythm of the herd as a totality is broken. Fights to dominate at the feeding bank become more ferocious. Enough fodder has to be given in the bays for fattening pigs to completely satisfy them. In fact, this leads to overfeeding of some while the weaker ones still do not get enough. The same applies to young cattle. So it is important that there should be enough room for the whole herd to feed at once. An exception may be feed banks where more than one kind of fodder are offered at the same time.

It is also important for the fertility and health of the animals that they should be kept sufficiently under observation by humans. There should be an intercom linking barn and dwelling house. Automatic feeding and milking parlors mean that there is only brief contact with the animals during milking when the farmer's attention is concentrated mainly on the milking itself.

Concentrates given in the milking parlor attract the animals' attention to the fodder. With the cowhouse system many animals stop eating during milking. The resulting peace and quiet cannot possibly be achieved in the milking parlor. A further advantage of the cowhouse system is that during the whole milking period the cowman is in contact with all the animals. The problems of work requirements and kind of work coupled with hygiene have to date been solved better for milking than for any other process on the farm.

If given the choice, animals avoid grids and slatted floors. Without any doubt the best bed is one with plenty of straw. In the age of the combine harvester, the collection of straw for bedding does add to the amount of labor necessary in the field. Although plenty of straw for bedding cannot really be replaced by anything else, there are a number of other possible solutions. Short standings with some sawdust have proved useful in stables with floating manure and dung channels. Where little or no bedding is used, the floor must be well insulated. Wood paving in the standings or cubicles is the warmest. Perhaps loose gravel in free stalls will turn out to be a proper solution. Poor bedding, standing and lying on grids, sharp edges or cold cement, slipping on slippery floors, etc., have been the cause of many injuries, diseased joints and udder infections. When trying to get up on slippery floors cows often tread on their own teats. If they cannot quite reach the fodder in the trough or manger they push and slip with their back legs. Troughs should always make it easy to reach the fodder but the cows should not see it while it is placed in the feeding passage.

Tyings and troughs must leave enough space for the animals to stand, lie, and above all get up again. In the cowhouse system yokes for catching a group of animals at once are important for saving time and labor when letting the animals out to graze.

The routes the animals have to take (inside the stable, in the loafing yard and on the way out) should not have any bottle necks that could lead to fights. Otherwise dehorning may become unavoidable, though with the right handling even for the loafing barn animals do not need to be dehorned.

If possible, the breeding bull should share the barn with the cows. The calf rearing methods to be used must be taken into account before the cowhouse is built because this will determine what is possible later. Yearlings can most easily be moved to other premises. Different animal species should not be mixed in the same stabling.

Feeding and its labor requirements must be properly planned for. The possible necessity of summer feeding indoors should not be ignored (e.g., because of exceptionally muddy pastures or sickness). It should be possible to drive along the feeding passage with a trolley or trailer.

Wood has so far not been surpassed from the point of view of animal health as the building material for barns, floors, walls and roofs, but of course other materials can be used. Faraday cages (iron cages brought about by iron enforced concrete in floors, walls, pillars and ceilings when metal is connected) should not be used since they alter the electrostatic field. Cold places near walls and corners can be avoided by using building materials with a different thermal conductivity. Primitive huts made from straw bales and beams are highly suitable for yearlings and pigs.

All stabling should allow for alterations without too much difficulty, for instance, changes in feeding (e.g., return to fodder beet), in the ratio of young stock to cows, in the method of barn cleaning, and any alterations that might be necessary.

The method of barn cleaning used has its effect on the whole farm. For the soil a mixture of solid or liquid manure with straw is far and away the best. But it is a matter of the work involved and the available labor. Whatever method is used it must be planned in advance. Since manure has to be dealt with daily the decision is important. Solid manure is kept best in low, long or wide heaps with new matter gradually being added to the already better fermented section of the pile. In high heaps anaerobic zones are unavoidable.

It is important that farm buildings should harmonize with the landscape. Rural districts with well-built farmsteads have their own stamp of beauty. Hedges, which offer cattle additional fodder, should not be forgotten when planning the paths to their grazing grounds.

The negative effects coming from the earth, which can be discerned by some people with the help of a divining rod, do also affect domestic animals. Though little is known about these, it does not do to deny their existence. A serious diviner should be consulted if possible before building so that at least the places where the animals will lie are placed outside the more active zones. Since there are always so many considerations to be taken into account, this may not be possible. It is certainly worth knowing the most active spots in existing buildings. The healthier the animals are and the more often they are allowed to change their places, in winter as well as in summer, the less likely are they to be affected by such influences.

Chapter Seven

The Sick Animal

The shortage of farm workers and the consequent overwork of those who are available, and also the mechanization of farm work, both mean that it is becoming less and less possible to observe the animals during daily work on the farm. There has been a substanial decrease in the amount of individual care that can be given to each animal. The working day is so full that almost anything extra is too much. This is why an increasing number of sick animals are treated by the vet. In some ways this is quite right, since training and experience are on his side. On the other hand, for cost reasons many farmers seek to carry out routine treatments themselves, such as prevention of anaemia in piglets. Animals are regarded as commercial objects much more than was formerly the case, so every treatment has to be costed out first. Vets are thus often forced to use medicaments that will lead to a quick conclusion: either the animal is well tomorrow or it must be slaughtered. Thus the number of treatments to choose from are greatly reduced. One advantage is that a certain selection on the basis of health and fertility does take place even though this is not the actual motivation. Like pest control, chemotherapy, which is almost always the method used, brings with it legislation enforcing safety intervals after treatment in order to protect the consumer against residues.

As herd sizes increase it is becoming increasingly necessary to treat the whole herd instead of one animal. This means adding medication to fodder or drinking water, which in turn brings new and difficult problems of agricultural practice, veterinary practice, pharmaceutical production and legislation.

As a result of mass animal husbandry a great deal has been learned about feeding. It is known how important a good and complete ration is and what can be controlled via the diet. Research is concerned with the specific effects of additives. These effects, however, almost always extend beyond the specific purpose. Then dietary constituents start coming into conflict with medication. Side effects become a problem and so do residues.

With few exceptions it has been overlooked that there is a connection between the health of livestock and how their feed is produced. Mass animal husbandry is dependent for the most part on industrially mixed fodders purchased on the world market. But here is an open field for a biological and ecological approach; if the soil and the crops are healthy then man and animals will also be healthy. This was a demand made in 1959 by Voisin. For Steiner and bio-dynamic farmers it has been the basis of their research and action since 1924 and has led to a great deal of practical experience.

This question is particularly important for the treatment of bulls at AI stations since male fertility reacts rapidly to environmental conditions and bulls usually obtain the bulk of their feed direct from farms (Aehnelt and Hahn 1969).

Quite apart from its medical treatment, therefore, the sick animal always poses questions for the farmer. Is the animal integrated in the proper way in the farm organism? Is feeding taking enough account not only of yields but also of health? Is the growth of the feed the result of the life in the soil or of higher applications of fertilizers? Have deficiencies in the soil been overcome? Is animal manure being returned to the soil in a way that quickens the soil or are there short cycles back to the animals? (Schaumann 1963 a and b).

313

These questions must be asked when an animal falls ill, though at that moment they cannot be much help. A change in the diet can nevertheless bring about immediate relief: the amount of feed is reduced and it must be good and easy to digest with a reduced share of protein. The person in charge of the animals has an advantage over the vet in that he can notice illnesses right at the beginning before they have become acute. He can often help by simple means if he knows what he is doing. He can use medicines that presuppose a therapeutic approach much suited to agricultural attitudes, namely, that the animal has within itself the possibility of creating its own healing process. Nothing can be achieved without this possibility within the animal itself. But the usual attitude is rather to counter directly the pathogenic factor or give a substitute for substances lacking or seemingly lacking, even if these are substances that the organism is quite capable of producing on its own.

A farm should have a supply of medicines on hand so that they can be used without delay should it become necessary. The stable needs a medicine chest just as much as the household does. But the stable medicine chest should not be kept in the stable since the air there is likely to make labels illegible and even spoil some medicines, both of which can lead to trouble and mistakes. A small wall cupboard kept in a cool dry place somewhere near the stable is ideal.

The following pages present some practical experience in the use of veterinary medicaments that have been developed from indications given by Steiner. These medicaments are not available in all countries. Where this is the case, the matter may nevertheless be interesting as information.

Abbreviations

V.M.	Veterinary Medicament
Dil.	Dilution, i.e., in liquid form
Trit.	Trituration (Powder)
Ungt.	Ointment
Prep.	Medicament, special method of preparation
aa	equal parts
Ø	Undiluted, Mother Tincture.

How to administer the veterinary medicaments

Undiluted liquids: These are best given in a small glass tube (about 4" long) or egg cup or small bottle direct onto the animal's tongue or between the lips. Alternatively, give the required dosage on bread.

Diluted liquids: Best given as a drench from a large bottle. Both liquids and powders can be given with concentrates. Powders can also be given between bread.

Ointments: Ensure skin is dry and rub on against the grain. Bandages can obviously only be used on head and limbs, but where they are used put ointment on gauze, apply to wound, cover with cotton wool and fasten with elastic bandage.

Lotions: Soak piece of suitable material in the lotion and apply several layers. Cover with waterproof material (plastic or rubber) and fasten with elastic bandage. When applied to a foot, place bandaged foot in sack and secure with string and Stockholm tar.

Much ill health in domestic animals can be overcome if dealt with in the early stages. In all doubtful cases the veterinary surgeon must be called.

Sterility diseases (especially in cattle)

1. In all cases check first whether the ration is as it should be, e.g., amounts and ratios of mineral supplements; herbs added to fodder can stimulate different functions; ensure plenty of exercise.

2. *Cow does not settle.* The cow is obviously in season. Although she has been served, season keeps recurring. The best time for serving or insemination is 10-12 weeks after calving, not sooner! Have the vet make sure that the cow is not in calf despite the recurring season.
   Pulsatilla pratensis D4/Sepia D6 aa Dil. V.M.
   15 drops 2-3 times daily for two weeks.

3. *Not coming into season.* A clear mucous discharge without any other signs shows that the cow is in fact in season. Possibly it should be served immediately. If the mucous is cloudy

or flaky, treat in the same way as 1 and 2. Pregnancy tests by the vet ensure the early recognition of cows that are neither in season nor in calf.

Ungt. caeruleum V.M. (blue ointment). First wash the outer skin with warm water and soap. Then apply a piece of ointment the size of a hazelnut to the mucous of the vagina, having possibly first dried the part with a paper tissue. This should be done three times with one day's interval in between.

Aristolochia clematitis D1/Apis D3 aa Dil.

15 drops 2-3 times daily for 2 weeks.

The ointment and the drops complement each other.

4. *Disturbances after calving.* If the after-birth is retained, have it removed by the vet. Cow has a foul-smelling, pussy secretion, eats poorly, milk production is slow to start:

Marjoram/Melissa V.M. (Marjoram-Melissa tea).

Irrigation of vagina and womb with ½-2 litres (1-4 pints).

Add 2 heaped spoonfuls to 1 litre (2 pints) boiling water. Remove from heat immediately and allow to steep for 5 minutes. Sieve through a cloth. Wash around the vagina thoroughly with warm water and soap. Insert the hose, which has a rounded end and side opening, pushing upwards gently for about 20 cm (8 inches). The liquid introduced should have a temperature of 40-42°C (104-108°F). Cooler irrigation is less effective and hotter will burn the mucous membranes. Keep the animal, the hands and the apparatus absolutely clean! Once daily, later once every 2-3 days, for 3-20 days.

These irrigations have given excellent results for decades. The condition of the animal worsens only on rare occasions. It is good if the animal reacts by pressing. The irrigation stimulates the discharge so that purification takes place. The foul smell will soon disappear. If the animal's condition does not improve, cease the treatment. Try instead an irrigation with camomile tea, reducing the temperature gradually to 38°C (101°F).

Lachesis D8/Pyrogenium D15/Echinacea D3 aa Dil. V.M.

15 drops twice (up to 5 times daily in acute cases) daily for 8 days. This can also be given to stimulate uterine contraction. Treatment as under 1 and 2 can be given subsequently.

Illnesses of the udder

1. *Acute retention of milk, acute inflammation.* The udder is hard, hot and painful, little milk is secreted, and that is yellowish or reddish brown in color. Milk every hour. Do not massage udder.
   Apply udder balm to the whole udder.
   Phosphorus D6 Dil. V.M.
   15 drops 2-5 times daily.
   If general state of health is disturbed, give in addition:
   Lachesis D8/Pyrogenium D15, Echinacea ang. D3 aa Dil. V.M.
   20 drops 3-5 times daily.

2. *Chronic mastitis* without general health being affected.
   Marjoram-Melissa Ungt. V.M. (or udder balm). Rub on thinly twice daily and massage for a quarter of an hour. Milk as often as possible.

3. *Clots in the milk without apparent reason.*
   Pulsatilla pratensis D5/Phytolacca decandra D1 aa Dil. V.M.
   15 drops 3 times daily. This can also be given when milk secretion does not start properly, particularly also for pigs.

4. *Hardened quarter after mastitis.* Massage firmly twice daily with Massage Oil.

5. *Cowpox on the udder.* Apply Combudoron ointment or jelly instead of milk fat as soon as the condition occurs. Milk the cows affected with udder trouble last. Do not milk onto floor or bedding. Observe strictest cleanliness, particularly of the milking machine. Test the vacuum of the milking machine. This should be as low as possible. Feed affected milk to pigs only if heated. Keep the udder warm. Protect from draughts.

Indigestion

With all digestive troubles, including colic in horses, give Coffea prep. Ø warmed to blood temperature, up to 100 ml (4 oz) depending on the size of the animal.

1. *Diarrhea.* Fast the animal and remove drinking water. Give a saltspoonful to a teaspoonful of Scour Powder 2-3 times an hour.

Additional suggestions: camomile tea, linseed slime, thin porridge. Possibly change diet.

2. *Constipation*. Give a saltspoonful to a teaspoonful of Plumbum prep. D6 Trit. V.M. 3 times daily, depending on the size of the animal.

3. *Loss of appetite, cessation of rumination*. In addition to Coffea, give 20 drops Nux vomica D4 Dil. V.M. 3 times daily. Give a drench of wormwood or caraway tea (for a cow, several pints daily). Give the sick animal the cud of a healthy cow (see Illnesses in calves 2).
Did the animal eat a foreign body? Usually it will hang its head, arch its back and groan. Give no food nor medicaments. Raise the forepart of the animal by 20 or 30 cm (8 inches to 1 foot). Call the vet.

4. *Sudden bloat*.
a) If a foreign body (apple, beet, etc.) is stuck in the animal's throat, saliva will be dribbling from its mouth. Call the vet, and meanwhile do not give anything.

b) If the bloating is caused by fodder, give 100 ml (4 oz) of Coffea prep. Pull a hay rope through the mouth and tie behind the ears. Raise the animal's forepart as high as possible. Massage the left side of the paunch. If the animal is fighting for breath, puncture the rumen by inserting a trocar in the middle of the left paunch. Wait till there are regular breathing movements of the rumen at the spot where the case of the trocar is inserted and then pull out with a sharp tug. Have several cases in stock, as several animals may bloat at once if the food is unsuitable.

Diseases in calves

1. *General weakness, retarded development*. Arsenicum album D4 Dil. V.M. 5 drops 3 times daily; Phosphorus D5 Dil. V.M. 5 drops in the morning for about a week. Parasites are often the cause if young animals are slow to develop. Have their dung tested! Give extra herbs (see p. 299).

2. *Diarrhea*. Give camomile tea instead of milk, at least for one

day. Give Coffea prep. Ø V.M. 1-2 dessertspoonfuls several times daily warmed to blood temperature (or a strong cup of mocha). Give Scour Powder: Carbo Betulae 10%, Quarts D4/Argentum met. D4/Arsenicum album D4 aa Trit. V.M. A saltspoonful every 2-3 hours together with Coffea Ø.

For calves of weaning age: First give Coffea and Scour Powder, and then a handful of cud from a cow whose dung is well formed. Take the cud the moment she has brought it up by pulling out her tongue with one hand and taking the cud with the other. Take care, the cow's molars are sharp!

3. *Weak bones, swollen joints* (*rickets*).
Phosphorus D6 Dil. V.M. 10 drops in the morning for about 8 days.
Conchae verae D3 Trit. VM. A saltspoonful twice daily.
Plenty of sun and exercise!

4. *Inflammation of the navel, lameness of joints.*
Arsenicum album D4 Dil. V.M. about 5 drops 3 times daily.
Lachesis D8, Pyrogenium D15, Echinacea angustifolia D3 aa Dil. V.M.
Direct treatment of navel.
Give a drench of Coffea prep. Ø V.M., half a bottle twice daily at blood temperature. Do not massage!
Prophylaxis: Plenty of exercise from the first day on. Milk from the calf's own mother. Plenty of sun. Lime powder concentrated V.M., a saltspoonful twice daily.
Arsenicum album D4 Dil. V.M. 5-8 drops twice daily.

Diseases in piglets

1. *Diarrhea.* As a prophylactic give the piglets some sod, or earth from fresh molehills, or good compost from their first day. Stir some Scour Powder to a paste (see Calves) with a little flour and Coffea prep. ØV.M. and put a saltspoonful on the tongue of each piglet with a wooden spatula or spoon handle.
Watch the sow's diet and possibly change it. Are her teats healthy? If there is a suspicion of post-natal disturbances give

the sow intestinal enemas with Marjoram-Melissa tea. Give older piglets thin porridge or roasted barley. As a prophylactic, give every piglet ½ teaspoonful of dry nettle powder mixed with the first feed.

2. *Black knees and tail.* Quartz D10 Dil. V.M. 10 drops twice daily. This often arises by wrong feeding or keeping. Are intestinal parasites present? Herb supplement should be fed.

Wounds and sprains

1. *Unclean and festering wounds.* Apply honey directly on the wound. No dressing. After some days if necessary apply a dressing soaked with Calendula 20%; diluted 1:10. Moisten the bandage repeatedly with this. Or use Calendula ointment V.M. If the place is swollen, use Arnica instead of Calendula.

2. *Painful subcutaneous inflammation or inflamed swellings* (e.g., incipient ulcers between claws). Apply Calendula 20% externally, diluted 1:3. Wet the bandage repeatedly with this. Lacheses D8/Pyrogenium D15/Echinacea ang. D3 aa Dil. V.M. 15-20 drops 3 times daily taken internally.
   For all subcutaneous injuries and damage to muscles and soft parts: Arnica D3 Dil. V.M. 15 drops 3 times daily.
   For damage to harder parts, bones, sinews, tendons, it is probably better in northern regions to use Symphytum officinale D1 Dil. V.M. 15 drops 3 times daily.

3. *Sprains.* Symphytum officinale D1 Dil. V.M. and Rhus toxicodendron D3 V.M. 3 times daily on alternate days.

Chapter Eight

The Bio-Dynamic Method in Garden, Orchard and Vineyard

The domestic garden, with which we shall be mainly concerned here, should be a source of pleasure and recreation, a place where the children can play and a source of fresh vegetables, herbs and fruit for the family. Gardening is the most intensive form of food production demanding a high degree of care for the soil and the environment. The goal is to achieve this with a reasonable amount of work.

The way the available land is divided up will depend on the individual needs of the family. Usually the lawn and play area surrounded by flower beds and ornamental shrubs will be closest to the house. The vegetable garden should if possible be a single plot evenly open to the light and big enough if possible to include the cold frame and the herb patch. The irregular corners and sloping parts of the garden are best used for fruit trees and soft fruit canes. It would be wrong to plant fruit trees in the vegetable patch since they would cast too much shade and their roots would compete with the vegetables. It is, on the other hand, useful to have a lawn near the vegetable beds so that the trimmings can be used as a mulch with as little trouble as possible. Finally, there must be a compost heap on which to collect all garden and kitchen waste to make the basic fertilizer for the garden. The best place is in semi-shade between the ornamental and functional parts of the garden up against the hedge that is likely to surround the whole property.

Low shrubs, 1-3 m (3-9 ft)		Fruit for
Japanese Rose	(*Rosa rugosa*)	jam, syrup, jelly
Sweet Briar	(*Rosa rubiginosa*)	
Blackberry	(*Rubus fructicosus*)	juice, jam, jelly, eating raw
	(*Amelanchier laevis*)	stewing, jelly
Raspberry	(*Rubus idaeus*)	juice, eating raw, jam
Japonica	(*Chaenomeles japonica*)	jelly, juice, jam

Medium to tall shrubs, 3-5 m (9-15 ft)		
Hazel nut	(*Corylus avellana*)	eating raw
Elder	(*Sambucus nigra*)	juice from berries syrup from flowers
Sea buckthorn	(*Hippophae rhamnoides*)	juice
Blackthorn, sloe	(*Prunus spinosa*)	juice
Azarole	(*Crataegus azarolus*)	jam

Tall shrubs, 5-10 m (15-30 ft)		
Siberian crab	(*Malus baccata*)	eating raw, puree
White Mulberry	(*Morus alba*)	eating raw with sugar
Medlar	(*Mespilus germanica*)	stewing, jam
White beam	(*Sorbus aria*)	stewing, jelly, juice
Rowan	(*Sorbus moravica*)	jam
Service tree	(*Sorbus domestica*)	edible
Quince	(*Cydonia oblonga*)	jelly, juice

The garden hedge provides privacy, protection against wind and also the retention of warmth, quite apart from offering shelter to all sorts of beneficial creatures. Birds, hedgehogs and insects all help control pests. The hedge can contain not only flowering shrubs but also some that bear fruit (see table, Scheerer 1965). Hedges that need a lot of cutting can be avoided by using shrubs of varying heights loosely aligned. These will only need occasional thinning below or trimming on top. Small twigs can be composted while larger branches when burned supply ash for all sorts of fertilizing purposes.

1 The Vegetable Garden

Enough vegetables can be grown on 70-100 sq. m. (750-1000 sq. ft.) to feed 3-4 people, depending somewhat of course on local soil conditions and climate. It is quite practical to divide the available space into beds of 1.20 m (4 ft) in width separated by paths of 30 cm (1 ft) in width. The beds can run the full length of the plot. The orientation of the beds on soil that does not warm up easily should be north to south so that the sun can shine right on the soil during the hottest part of the day. On soil that dries out easily east-west orientation is better so that shade is cast by the rows on to the soil when the sun is high.

The number of beds needed is determined by crop rotation. Vegetation if left to itself, e.g., grassland, regulates the processes in the soil by producing a community of plants that includes grasses, clovers and aromatic herbs. They follow the law of succession until a climax state is achieved that establishes a lasting natural fertility of the permanent grass. A gardener, however, seeks to meet the requirements of the soil by working with a suitable rotation of vegetables. Usually it is thought right to use a course of strong feeders, light feeders and legumes in which the strong feeders are given high fertilizing. There are all sorts of "proven" crop rotations for every area and every gardener's aims. Some interesting points with regard to crop rotations have emerged from Thun's researches (1963) into the more subtle connections between natural processes. Garden vegetables have on the whole been developed with the emphasis on one particular organ of the plant. Thus carrots, radishes, white radishes, black salsify, potatoes and celeriac are quite clearly roots, while lettuce, spinach and cabbages have been developed for their leaves. There are no vegetables that can be regarded as flowers since we eat cauliflower, broccoli and artichokes when they are still buds. On the other hand, seeds and fruits such as peas, beans, tomatoes, cucumbers and peppers are familiar.

Experience gained over more than a decade has shown that the division of vegetables into roots, leaves and fruits is a good basis for establishing a crop rotation thus giving us a useful guideline when planning a vegetable garden. In successive years the soil is enabled to produce the different "organs" of the plant and in this way fertility is maintained instead of being weakened by, for instance, repeated demands to form leaf vegetables. In a four-year cycle each plot can produce first leaf vegetables, then roots, then fruits and seeds, and finally flowering plants.

A vegetable garden should therefore be divided into four plots one of which is used for annual flowers so that the soil also has the opportunity to form flowers. If there is no use for flowers, potatoes can be planted instead. Most gardeners like to grow strawberries, which unlike other soft fruit are usually moved from one bed to another more frequently. In a bio-dynamic garden, however, there can be a separate bed for strawberries, since with bio-dynamic treatment they can remain on the same bed for several years, which also saves labor.

Plan of a "dynamic" crop rotation in the garden

	Plot 1	Plot 2	Plot 3	Plot 4	Plot 5
1st year	root veg.	leaf veg.	annuals or potatoes	fruit veg.	strawberries
2nd year	fruit veg.	root veg.	leaf veg.	annuals or potatoes	strawberries
3rd year	annuals or potatoes	fruit veg.	root veg.	leaf veg.	strawberries
4th year	leaf veg.	annuals or potatoes	fruit vet.	root veg.	strawberries

After four years the strawberries should be planted on another bed. They do best after root vegetables. If they are to be planted in July-August they can be preceded by early carrots. Otherwise, a winter catch crop of rye-vetch on the root plot can precede their planting in spring after the plot has been dug over early enough with compost. We now have a jump in the rotation, taking the old strawberry plot for leaf vegetables, most of which are brassicas and therefore most likely to grow healthily and without pests on a plot that has been without them for several years.

	Plot 1	Plot 2	Plot 3	Plot 4	Plot 5
5th year	root veg.	fruit veg.	annuals or potatoes	strawberries	leaf veg.
6th year	fruit veg.	annuals or potatoes	leaf veg.	strawberries	root veg.
7th year	annauls or potatoes	leaf veg.	root veg.	strawberries	fruit veg.
8th year	leaf veg.	root veg.	fruit veg.	strawberries	annuals or potatoes

Now again the strawberries are planted in the root plot and the next leaf vegetables are planted on the old strawberry bed. This change can of course also be made after the third year on the strawberry bed. This rotation is useful because it is so flexible so long as the general principle is adhered to. The weather conditions of a particular year or the gardener's own preferences can easily be accommodated. The table indicates the rotation for the main crops only. Early and late crops on all the plots can be for example:
spinach
perpetual lettuce, cabbage lettuce, endive
chicory "Sugar Loaf"
lamb's lettuce (*Valerianella olitoria*)
These can be considered neutral in this particular rotation. In

this way all the beds are well used and variety can be achieved even in small gardens without making onesided demands on the soil. This way of adhering to a dynamic principle that takes account of the demands made on the forces of the earth also makes the plants more resistant to damage from fungi or insects, which are in the first place expressions of disharmony in the soil.

The herb garden

The use of herbs and spices in cooking dates back to ancient times. They not only enhance the flavor but also increase the digestibility of dishes. Herbs for teas, seasonings and medicinal purposes as well as those we keep for their aroma not only enrich the garden in summer but can be dried for winter use when they continue to give of their beneficial effects. Chives, wormwood, horse radish, garlic, thyme can also be allowed to ferment in water or used for making a tea which, when applied to plants, helps check insect pests and fungi. Since most garden herbs are perennials, the only work required by the herb bed is planting at first, weeding, occasional manuring and harvesting. The benefits derived from this small input are great. The most commonly used herbs are:

Common garden herbs		
perennial	perennial	annual
wormwood	burnet	anise
summer savory	rosemary	basil
garlic	chives	borage*
caraway*	sage	dill*
lovage	thyme	fennel
lavender	rue	chervil*
horse radish	hyssop	coriander
peppermint	lemon balm	garden cress
tarragon		camomile*
		parsley

*These usually seed themselves in subsequent years once they have been established.

326

The herb garden should be in a warm, sunny position. The soil should be kept manured well to ensure rapid growth and the greatest amount of aroma. A few square yards are generally all that is required. Some herbs not needed daily can also be grown among the perennials in the flower garden. More information about herbs and their uses can be found in the works of Lippert (1953), Pfeiffer-Riese (1970), Pelikan (1962), Geutter (1962), and Grotzke (1975).

Gardening under glass

For domestic gardening a cold frame for transplanting from a seed box and raising young plants of early lettuce and radishes followed by cucumber is usually all that is required. It should be within easy reach of the house in a position where it is sunny all day. The sides are of strong wood impregnated with linseed oil. Cement walls cause large differences of temperature. Two or three windows are a minimum. It is useful to have the kind with automatic temperature regulation. The main purpose of a cold frame is to achieve an early warming of the soil. This can be helped with horse manure or extra heating. The frame is filled with a good mixture of garden soil and compost in a ratio of 1:1. A few spadefuls of basalt meal will help raise the temperature.

The earth in the cold frame is treated with the bio-dynamic preparations for compost. Before sowing it is sprayed with Preparation 500. Preparation 501 is not used in the sowing bed. Equisetum tea is frequently used.

It is often a problem to obtain seeds that have been bio-dynamically cultivated since gardeners usually only succeed in growing their own seeds for some of their crops. There is a need for more bio-dynamic seed growing; at the moment it is only done in a couple of places.

The compost site in the home garden

Both garden and kitchen supply a variety of waste materials that can be composted to make good garden fertilizer. The best position for the compost heap is in a semi-shady spot as near as possible to the kitchen and all parts of the garden, for the various waste materials first have to be carried to the heap and the heap later spread about the garden.

The compost heap is the mainstay with regard to manuring the garden and thus enlivening the garden soil. Any organic waste from kitchen and garden can be used: weeds, litter from harvested beds, lawn trimmings, stalks, wilted flowers, hedge clippings, vegetable trimmings, etc. There should be as far as is possible a good mixture of everything. The heap is 1.5-2 m (5-7 ft) wide and as long as necessary.

It is a good idea to start the heap with a pyramid and then continue adding to one of the sloping sides. Each time a new load is added, 2-3 handfuls of horn-bone meal can be sprinkled on to help start fermentation.

A box the height of a chair or table can be used to collect the wastes. When the box is full it contains a worthwhile amount to be added to the heap. In summer most of the waste is green matter together with kitchen trimmings and the remains of soft fruits. For these it is good to add a light sprinkling of lime to each new layer, which is then covered with a spadeful of earth. The ratio is approximately 3-4 handfuls of lime and one spadeful of earth to 5-8 forkfuls of compost material. If there are weeds among the collected material there is usually no need for further soil. Care must be taken to ensure that there is sufficient moisture, though green materials usually have enough of their own. Dry litter is thoroughly moistened a few days before the pile is built. Quicklime or ordinary agricultural lime can be used. Calcified seaweed (Algomin) has also been introduced in recent years. It is also possible to use basalt meal instead of lime.

When a heap of the appropriate length has been built it is then treated with bio-dynamic compost preparations.

About 2 g = 1 level teaspoonful is used of each preparation, including the liquid valerian Preparation 507. These amounts are sufficient for 2-5m^3 (70-170 ft^3). Since it is best if the preparations can work on the materials immediately, they can also be added continuously as the heap grows longer.

When the heap has been built and supplied with preparations

a shovel full of compost from an earlier heap with plenty of worms is placed at the base every half to one meter to establish a good population of worms in the new heap. At the base the earth remains damp and cool whereas the heap itself builds up heat that could kill the worms. As the temperature decreases, the worms can penetrate the heap from the cooler moist area beneath. They are a great help in creating good compost because they eat earth and organic matter. These are formed in their intestines into the fertile worm castings. The heap must be covered with grass trimmings, the greenery from peas, peat or other kinds of dry litter.

Should compost be sieved? It is of course useful to do so for use in seed drills or seed boxes for small seeds. But it is hardly ever necessary for ordinary garden beds. When compost no longer contains many worms it is a sign that it is ready for use. When there is not much more to work through, the worms move on to the next heap or withdraw into the soil beneath the heap.

If the above method is used, the compost will be ready for use in 3 to 12 months. For tomatoes, potatoes and various cabbages it can be used after 3 to 6 months when the particles still show what materials it consists of. For seed drills ripened twelve-month compost is best. Thus, after one year at most a usable compost is produced.

In the autumn, everything can be collected near the heap and made ready in larger amounts for composting with the use of the following mixture. This will induce temperatures up to 50°C (122°F) even in cooler weather. The weed seeds will first germinate and later be destroyed by the fermentation processes. The heap must be well covered so that the heat can reach right to the outer layers. The mixture that will induce this heating in 0.5- 1 cubic meter or yard of compost material is: 10 l (2½ gal) warmish water (about 40°C or 104°F) with 300-400 g (10-14 oz) molasses and one cube of brewer's yeast dissolved in it. This is sprinkled evenly over everything with a sprinkling can while the heap is being made. Of course other additives can be used too. 2-4 kg (4.5-9 lbs) of quicklime or up to 10 kg (20 lbs) of calcified seaweed meal can be sprinkled on every cubic meter or yard of green compost material. Other additives, depending on requirements, are any one or two of the following at a time:

	5-8	kg	(10-18 lbs)	bone meal
	3-5	kg	(7-10 lbs)	wood ash
up to	8	kg	(18 lbs)	horn-bone-blood meal
approx.	5	kg	(10 lbs)	seaweed meal
up to	20	kg	(45 lbs)	manure from pigeons, chickens or other small animals
up to	15	kg	(33 lbs)	basalt meal

per cubic meter or yard. In every case the five compost preparations are applied and Preparation 507 sprayed on before covering.

Leaf compost: Some gardens have large amounts of leaves to be dealt with in the autumn and apart from preparing some leaf mould also a good compost can be made from these if the following indications are noted. The leaves should be raked up while damp or made quite damp a few days before the pile is to be built. Oak, beech and pear leaves are mixed with lime, chestnut and apple leaves. 5 kg (10 lbs) quick lime or up to 10 kg (20 lbs) calcified seaweed meal are mixed with every cubic meter or yard of leaves. About 2 buckets of fine earth, or better still of residues from sieving compost, will help start the fermentation and maintain humidity. It is also good to add green trimmings and the last green stinging nettles of the season. Everything should be well mixed and not laid in layers. The preparations have to be added while the heap is being made because holes cannot be poked into a pile of leaves. More leaves or similar waste materials are used to give a warm covering, the whole being held in place by twigs or wire netting. Turning is not usually necessary for garden composts, but for leaves it is unavoidable. A warm day in spring or early summer of the following year is the best time and a new dose of the preparations is added during the process. A mixture of

1000 g (2 lbs) matured compost

20 g (1 oz)   blood meal or guano
30 g (1 oz)   clay, powdered
50 cm³ (50 ml) liquid seaweed

soaked in 10 l (2½ gal) water and sprayed on 10 ar (a quarter of an acre) of fallen leaves brings about more rapid fermentation in leaf compost heaps.

Of the many silos and special containers on the market for garden composting only those should be chosen that allow plenty of air to penetrate. New materials always tend to be added to the heaps in these containers in horizontal layers and this makes it more difficult for the earthworms to penetrate unless great care is taken to add thin layers. If one of these containers is used, especially when it is new, it is advisable to scatter some handfuls of good garden earth or mature compost on each layer. A compost bin always looks tidy and is a good though expensive solution if neighbors object to compost heaps.

Fertilizing

"We must know that to manure the earth is to make it alive so that the plant may not be brought into a dead earth and find it difficult, out of its own vitality, to achieve all that is necessary up to the fruiting process." This ideal was stated by Steiner in 1924. The best way of achieving it is to use compost as a fertilizer.

A liquid manure can also be made out of stinging nettles, shepherd's purse (*Capsella bursa-pastoris*) and other weeds and also herbs: 1-1.5 kg (2-3 lbs) of green plants are steeped in 10 l (2½ gal) water together with some ripe compost. After two weeks, if the weather is warm, a strong smelling fermentation will have taken place. Diluted 5 to 10 times, this is a strong juice that can be poured on plants or sprayed on leaves up to the middle stage of growth.

For green manure younger green plants are dug in and give the earth additional organic substances. A mixture of field beans, peas and tares (spring vetch) in equal amounts together with some oats is particularly suitable for reclaiming land. 2 kg (5 lbs) of seeds per 100 m² (1000 sq ft) gives good root penetration and the closely growing plants smother weeds. Though this means doing without vegetables for a month or two, their growth will be all the better subsequently.

In the autumn a winter cover of rye and winter vetch can be sown. This is lightly dug in during spring. Spinach, mustard and Persian clover (see p. 338) can be sown between the rows of growing plants. These are cut low down and left on the ground where they are eventually pulled underground by the earthworms.

Autumn is the main season for manuring. After the soil in the garden has been suitably worked according to its type, about half a bucket of compost per square meter or yard is spread on the surface. It does not matter at all if there are still some lumps in it. These might be broken up by frost during the winter. Unrotted twigs and stalks from perennials also do not matter since they will be raked up in the spring and added to the new compost heap.

To the basic treatment with compost are added the various commercial organic manuring materials that are used according to the state of the ground and the crops planned for the following year. About 60-100 g/m² (2-3 oz/10 ft²) equalling 1-2 handfuls are used. Apart from the materials mentioned below, also feather meal, dried chicken or cow manure, guano, and castor bean meal are used.

Basalt meal. Fertile soils arise where basalt is weathered. Basalt meal used in the garden supplies trace elements and minerals that are transformed into clay. 80-300 g/m² (3-10 oz/10 ft²) is spread. This must be done evenly as otherwise sticky patches might form because the meal is usually extremely fine.

Calcified seaweed (Algomin) supplies trace elements collected by corals from the sea. It has a favorable calcium-magnesium ratio. Though its effectiveness is not only due to the 80% portion of lime it contains, it should be sparingly used for strawberries on calcareous soils. 20-100 g/m² (1-4 oz/10 ft²) are used annually for general soil improvement.

Wood ash contains on average about 10% potash and can be important for root vegetables and cabbages and also for strawberries, currants and raspberries, particularly on sandy soils. 50-100 g/m² (2-3 oz/10 ft²) is usually used. Beech charcoal has the same effects and in addition improves soil structure.

Clay such as bentonite used together with compost can greatly improve the coherence and water holding capacity of light, sandy soils. Depending on the permeability of the soil, 100 to 300 g/m² (3-10 oz/10 ft²) are scattered evenly on the compost (Thiess 1958).

Seaweed fertilizer in amounts of 50-100 g/m² (2-3 oz/ft²) is either lightly worked into the top layer of the soil or scattered on the soil before compost is added in autumn.

Bone meal contains phosphate and also a small amount of nitrogen, which speeds up the transformation of the bone substance in the soil. 50-100 g/m² (2-3 oz/10 ft²) is the amount usually used.

Of course, one can make suitable mixtures from all the above or treat individual beds differently with one or other. If there is enough compost, however, no additional fertilizers are needed. These additional fertilizers are there to fall back on if there is not enough compost.

Apart from the main application in the autumn, repeated applications of compost may be needed during the course of the year, perhaps before the second or third crop on beds that have been harvested. For a smallish bed, one bucket of compost with 4 handfuls of horn-bone meal will suffice.

It is sometimes asked whether sphagnum peat moss should be used in the garden. Peat consists of mosses (sphagnum) that are only slightly rotted and then preserved in a wet condition in a cool climate. This material is worked into the garden soil to loosen it. But one should not forget that peat moss is taken from a situation in which the natural cycle that prevails in living soils, that is "growth and decomposition," has been halted. Instead dead organic matter has been accumulated due to lack of oxygen in the elevated fen from which the peat comes. It has insulating and absorbing effects. This is why bio-dynamic preparations are stored in peat, which retains their effectiveness until they are applied to the compost or manure heap. The mechanical effect of peat, which loosens and aerates the soil, is better achieved by compost that in addition enlivens the soil. Peat, however, does have its uses in connection with compost, even though it is not made into compost. Plant life continues to work inside a compost heap and the insulating properties of peat can be used to retain these forces by covering the heap with peat. Furthermore, about 5% peat can be mixed in with compost when it is a matter of preventing the evaporation of nitrogen from poultry or stable manure, etc.

Decorative shrubs are often planted in peat. They do not produce food, so compost is better saved for crops that do. Shrubs and trees can be planted in a peat-earth mixture, particularly if, like rhododendrons, they need an acid reaction.

Bio-dynamic preparations in the garden

As nearly as possible before sowing or planting out (in the case of lawns, after cutting), i.e., at the beginning of a growth cycle, Preparation 500 is sprayed on the ground in large droplets so that it can penetrate a little. About 5 g are used for 1 ar (one sixth of an ounce for 100 ft^2 or often a bit more). Comparison with an unsprayed plot will prove that spraying improves root formation. The soil microorganisms are also affected. Spraying is best done in the evening. This preparation is also useful sprayed finely on green leaves during times of drought when there is a danger of premature ripening.

In a domestic garden there is constantly something being sown, planted or planted out. It would be too much work to stir the preparations anew for every small bed. So during the main sowing periods all the appropriate beds can be sprayed in advance so that each receives 3-4 treatments a year. Peparation 501 is used in a similar way. In the green plant it promotes the formation of substance under the influence of light. A knife tip full is used for 10 l (2.5 gal) water. As it is sprayed on the leaves in a fine mist, this amount is quite sufficient for 10 ar (10,000 ft^2).

Whether to spray or not depends on the stage of development

the plant has reached (for details see above, p. 212f.). The time of day can also vary, but normally it is done in the morning. The earlier in the day, the more definite will be the results. Fruit vegetables react particularly. Thus 7-8 applications in a rather dull year have made it possible to harvest ripe tomatoes from the garden. In experiments covering several years, beans sprayed in this way yielded 27% more than those on control plots sprayed only with water (Leihenseder 1961). Afternoon spraying has proved useful when vegetables are ripening, e.g., for potatoes. Ripening blackberries are sprayed in the afternoon with Preparation 501 as well; this helps also the growth of side shoots and they are then better able to resist fungal infestation. Afternoon spraying also helps delay the bolting of lettuce.

Some hints for the gardening year

Spraying with horsetail tea (*Equisetum arvense*) (see p. 223) or a fermented (cold water) extract in spring and autumn is a good prevention against fungal diseases. In threatened places and for seed boxes spraying is recommended before every new sowing or planting. If the fermented extract is made, this can be stored ready for use at any time.

Cold spring weather retards the growth processes. Basalt meal or feather meal (respectively 3 or 6 handfuls per square meter or yard) lightly worked into the soil help to build up warmth more rapidly in the ground. Spraying with Preparation 507 also supports this process.

Bought seeds conform to the legal requirements of purity, true variety and germination, but it must be remembered that they have not been grown by bio-dynamic methods. Experience has shown that it often takes two or three years before plants and seeds from these bought-in seeds can fully benefit from bio-dynamic conditions. When they have become fully adapted this shows in lush growth and considerable resistance to various damaging influences. Much can be done to bring about a balanced situation with the help of compost, spraying the soil and plants with Preparations 500 and 501, and by adhering to Thun's recommendations for particular sowing dates.

A further means is the application to the seeds before germination of the Preparations 502-507. The growth processes are stimulated by the suspensions of the preparations. Kunzel has conducted long-term research on the effects of these seed dressings on various garden and agricultural crops. From her research and that of Hanke the following list of the dressings that favor different seeds has been compiled.

Preparation 500	suitable for all crops, particularly spinach, chard and beetroot
Preparation 502	for rye
Preparation 503	for peas, beans, white radish, brassicas
Preparation 504	for barley
Preparation 505	for all kinds of lettuce, bush beans, oats
Preparation 507	for carrots, chicory, cucumbers, tomatoes, peppers, pumpkin, onions, leek, celery, spinach on peaty soil, fodder and sugar beet (here together with a suspension of cow dung treated with the preparations), wheat, potatoes

---

Seed potatoes are sprayed with a mixture of Preparations 502 and 503.

Preparations 500 and 507 are stirred in the usual way shortly before the seeds are treated with them. Of the solid Preparations 502-506 one teaspoonful is stirred briefly in one liter (2 pints) of hand-warm rainwater until all the particles are well soaked. 20-24 hours later the suspension will be ready to use for seed treatment.

The small amounts of seeds needed for garden sowings are tied in a piece of material and suspended in the seed bath for 10-15 minutes. Then they are placed on a piece of wood or absorbent paper to dry in a shady, airy spot. They can be sown the same day, and should on the whole not be used later than the following day.

Larger amounts of seeds are turned with a shovel while being sprayed with the appropriate preparation. They are then left in a heap covered with sacks on a wooden floor for 12-24 hours. They will then run freely through the seed drilling machine. For 100 kg (3-5 bushels) of seed about 3 liters (5 pints) of liquid are needed, except for beet seeds, which need three times this amount. Here, too, sowing should follow as soon as possible.

In special cases the seed can be sprayed after it has been sown in the drills. This is necessary, for instance, with linseed (which is treated with 503) because treated in a pile the seeds would simply stick together. The observed consequences of treating the seeds are better ground cover by strongly developed plants, increased root formation, resistance to disease, and higher yields. These observations made by Kunzel (1954) have been born out many times by experience in gardens and on farms.

Protection against frost. A plant that has grown harmoniously without any forcing treatment of the soil can withstand a certain amount of frost unless it is surprised just when emerging. Tomatoes, cucumbers, blossoming fruit trees and soft fruit bushes, however, can be badly damaged. Preparation 507 is useful as a protection. 5 ml are stirred for several minutes into 10 liters (2.5 gal) of hand-warm water and then sprayed in a fine mist on the crop the evening before the frost is expected. The plants must not be dripping wet. Especially in spots with cold air currents spraying should be finished before 6 o'clock in the evening. Black currants are extremely sensitive to cold not only while blossoming but also for the subsequent four weeks and should be sprayed with 507 whenever frost is expected. The application is not effective when stronger frost occurs.

Weeds. Between 3000 and 6000 seeds can be found in a square meter or yard of soil. Compost is also sometimes not quite free of weeds, even though proper warmth and dampness in the heap usually causes most seeds to germinate and then rot. Regular hoeing, especially before seeds form, keeps both annual and perennial weeds in check. Once a bed has been worked for resowing or planting, however, resting seeds will start to grow. These newly germinated weeds can be destroyed by making the surface of the bed smooth and then lightly raking it after a few days. This gives the intended crop a head start. This advantage is lost if the top layer of soil is allowed to dry out while it is waiting. If this is likely, a mulch can be laid on the surface straight after flattening. Before sowing the mulch is removed. For planting it is only necessary to remove the mulch where the plants are to be set and once they have rooted a new layer of mulch can be added. By this means a great deal of hoeing is saved and in addition the soil microorganisms are given extra food. There is always something that can be used as a mulch, e.g., fresh nettles, old leaves, bolted garden cress, spinach or Persian clover specially sown for cutting between the rows, wood shavings, etc. Lawn trimmings, especially of young grass, should be allowed to dry out a little before using as a mulch since they can otherwise become rather matted, which is undesirable.

Replanting harvested beds. After shallow working the bed is given a bucket of compost per 4-5 square meters or yards or at least finely sieved compost in the seed drills or plant holes. In summer the soil microorganisms tend to become less active. Compost, watering or the application of liquid manures from plants (see above) help to restimulate them. Preparation 500 also gives excellent help. If the summer is dry some camomile can be added to this preparation. Liquid manure from nettles or a liquid seaweed sprayed in the evening is also a help after such dry days.

Too much damp and coolness can also have its consequences. Apart from improving aeration by working deeper than one usually does with the hoe, Preparation 501 can also be used.

Winter storage of crops. Suitable storage methods are necessary if vegetables are to be kept fresh throughout the winter. Roots are best kept in a cold store or root cellar covered in damp sand. If the store is particularly dry, about a quarter as much peat can be mixed with the sand. Another good method is to bury a box encased in mouse-proof wire netting in the garden so that the top is level with the surface of the earth. The various winter vegetables are placed inside, the box is closed tightly ~ough to keep out the rain and then covered with a thick layer ~lant stalks, bean trailers, leaves or something similar to keep ~he frost. On days without frost the week's supply can be

338

removed to house or cellar from this naturally damp, cool container. The method is particularly useful where there is no cellar or where heating pipes run through the cellar, thus making it unsuitable for winter storage of produce. Fruit can also be kept in this way in a box in the ground. It supplies storage conditions of about 4°C (39°F) and a relative air humidity of 90%. Cabbage can be satisfactorily kept over the winter in spade-deep trenches. The heads are placed root upwards and covered with earth and dead leaves. The root is left protruding, so a cabbage can easily be pulled from the ground when required.

Cold frames in winter. Before the first frost the ground is cleared and mature compost is dug in with a fork. The soil is left in mounds so that it can be well frozen through during the winter. In spring it is then sufficient to smooth it over before planting.

2   Bio-Dynamic Methods in Orchards

Fruit trees have been cultivated since ancient times. The tall trees traditionally grown have their place in the natural distribution of meadow, pasture and field. Recently, the trend has been increasingly toward bush-like trees, and to this has been added the intensification of fruit growing. The modern fruit tree consisting of root stock, trunk and grafting starts producing fruit sooner and also produces more in relation to its woody structure. Intensive fertilizing and pest control have become necessary. Large commercial orchards are replacing the older methods. An orchard run without any connection to the mixed landscape surrounding it cannot be developed as a self-contained farm organism. The fruit farmer is compelled to sell a large yield from his only crop every year, and to achieve this he has to invest heavily. In short, fruit production of this kind has a number of artificial characteristics from the start. In the fifties there were still only a few bio-dynamic growers using bio-dynamic methods. But advisers and the owners of orchards more and more wanted to apply their bio-dynamic experience in this field too and develop fruit growing away from the dead end of increasing chemical usage. Fürst (1976) in Paderborn and Lust (1967, 1962) in Balingen, both in West Germany, have done important pioneering work.

Today it can be stated that adopting bio-dynamic methods of fruit growing is worthwhile not only in private gardens but also in commercial orchards. (This is meant with reference to commercial fruit production in Europe where modern plantations both smaller and larger in size are scattered among farm land. Large continuous areas are rare for fruit production except for grapes, for which this is usual practice.) Experience gathered hitherto is stated here.

1. The habitat itself plays an important part. Under central European conditions the most favorable spots are those with the best deep soils in warm areas. Good results are achieved on northfacing slopes. In cooler areas good results are also achieved, though fruit is not the main income earner. The drier the climate, particularly in summer, the more important is the depth of the soil. Even heavier clays are gradually opened up by the soil life and can thus be included, so long as the groundwater level is not too high.

2. The habitat can only be fully utilized if suitable fruit varieties are selected for growing. Local experts will be able to give advice. Of course the market and the personal wishes of the grower also play a part. But if these last two factors are considered alone, it may not be possible to move away from chemical pest control.

3. Soil analyses are one important tool for management. The figures should be known at the start and any changes noted every three years subsequently. For specialized fruit production, tests should be done every year. It is also a help in elaborating the fertilizer program. More extensive testing programs than the usual have proved useful.

4. Before planting, a strong green manure, which has itself been well-manured, should be given. Later, green manures are grown between the trees. In areas with 700 mm (28 inches) or more of annual rainfall the land is best kept permanently under grass with a high percentage of clover. Relatively young growth

340

is mown regularly and mulched. It can also be raked around the stems of the trees where it helps keep down weeds and stimulates humus formation. Young cut plants are taken in by the soil organisms more rapidly than older ones. Early cutting also reduces the water consumption of the green cover and restricts the water intake to a shallower layer of soil, leaving the remaining ground to the tree roots. In areas that are too dry, every second row should be kept open. Or the whole area is kept without cover from spring until the end of June or beginning of July, or merely mulched, for instance with straw (this achieves by far the lowest degree of water loss through evaporation), and then sown in July. Winter vetch and Persian clover have proved particularly beneficial.

5. Composted cow manure improved with horn-bone meal is the best fertilizer. It is useful to start with 20-30 t/ha (8-12 t/acre) of manure, which contain 80-150 kg/ha (72-135 lbs/acre) of N; plus 0.4-0.7 t/ha (360-630 lbs/acre) of horn-bone meal, which contain 20-50 kg/ha (18-45 lbs/acre) of N; or an equivalent amount of dried chicken manure. Later 10-15 t/ha (4-6 t/acre) plus 0.4 t/ha (360 lbs/acre) of commercial organic material will do. Instead of composted cow manure, fruit pulp compost has also proved useful. A large amount of the nutrients made available by the manure is used first for the growth of the soil cover. The manure may be used in autumn (September) on a green cover that is to remain over the winter. But the best time to apply manure is in early spring before vegetation starts to grow. Length of annual wood growth, amount of blossom, amount of fruit, diseases and pests, as well as soil analysis, are guidelines for the amount of manure and commercial organic fertilizer needed in the following year. Occasionally one also has to decide whether a special material is required in addition. These include basalt and sulphate of potash magnesia.

The application of composted manure and green manure is the prerequisite for starting a healthy orchard. Additional measures are hedges, which protect the air space and provide nesting for birds, and also the rejecting of ordinary pest control. Even where specialization has to a great extent gone well beyond any natural harmony, consistent work can maintain the health of the orchard and produce fruit with plenty of flavor and good keeping qualities.

The following are hints on what can be done in an orchard roughly following the course of the year.

Intensive cultivation calls for intensive measures also in the bio-dynamic method. Preparations 500 and 501 are sprayed quite often. The tree paste made from cow manure and clay gives rise to markedly smooth and healthy bark. Horsetail tea, 1% of blood meal and also calcified seaweed, etc., can be mixed into the paste if required.

In February and March, when clear, sunny days are followed by clear starry nights, considerable temperature differences occur. This can lead to cracks in the rather thin bark of the dwarf trees. Spraying the whole tree early in the morning of a warm day, so that it is quite dry by the time the cold of the evening sets in, gives the tree protection and ameliorates the effects of the temperature changes. For this purpose either the tree paste can be used, or a liquid for spraying can be made. The proportion of materials is 2 kg (4.5 lbs) of clay and 3 kg (6.6 lbs) of calcified seaweed with enough water to make spraying possible. This can be sprayed twice if the first application does not leave a sufficiently thick coating. For autumn applications after normal years with satisfactory development of wood and leaf, diluted cow manure can also be added. This is filtered through a cloth. It prevents the winter eggs of pests from hatching. 5 kg (11 lbs) of waterglass (sodium silicate) are sometimes added.

When budding starts, Preparation 500 is sprayed. If necessary the ground is watered with a horsetail mixture against fungal infestation. 20 g/l (1 oz/2 pints) are used for every 1000 ft². This is either brought to the boil and left to ferment in a warm room for at least 10 or 14 days, or else it is boiled on a low heat for half an hour. Blossoms can be frozen while still in bud. To prevent this, Preparation 501 can be sprayed on to achieve a slight stunting effect and make the bud less sensitive to frost. Preparation 501 should anyway be sprayed when the buds are just about to burst.

The first scab control. As a rule the first long warm rainfall of the year can be expected around blossom time. This releases all the winter spores, which then fill the air and germinate immediately in the damp on young green leaves. This is the moment to start using Bio-S in the orchard; it is subsequently used continuously throughout the spraying season. Before blossoming 1 kg/100 l (2.2 lbs/25 gal) is used; after blossoming 0.5-0.6 kg/100 l (1-1.2 lbs/25 gal). (Bio-S is a formula consisting of wettable sulphur and plant extracts.) If one wants leaves and fruit that are free of scab, the first few sprayings before and after flowering must be carried out carefully. Tables are easily obtainable that indicate how humidity, temperature and the season influence the development of scab. Local warnings about the times for spraying are also given. If the winter spores are prevented from gaining a foothold on leaves and fruit, control of scab later in the year is no longer much of a problem. Bio-S can also be used during flowering without harming the bees.

Frost during flowering can ruin harvest prospects. Between 5 and 6 p.m. when frost is expected Preparation 507 should be sprayed (see p. 337). Temperatures down to -6°C (22°F) have been withstood in this way. If there are also cold air currents, however, the effects of spraying will be uncertain. In this case the usual frost protection methods such as fire, overhead sprinkling, smoke or fans are preferable.

An extract of stinging nettles made by steeping for 24 hours is as effective against aphids in the orchard as it is in the vegetable garden. If left for more than 24 hours it loses its burning properties.

Pyrethrum and derris formulas are also in use against insect pests, e.g., Spruzit-Nova, which also contains a conifer extract. Against codling moth either pyrethrum, or better still Ryania, which does not harm useful predators, are applied according to the instructions given by the maker. A plant extract formula, Artanax, is also available.

When the fruits reach walnut size, cell formation ends and a new growth stage starts in which the cells swell until the fruit reaches its full size. This process is helped by Preparation 501 the use of which is now started.

Once the apple has reached its final size, ripening is also helped by Preparation 501. Round about the end of June until the middle of July the buds for next year's blossom are forming. If the season is not sunny, Preparation 501 should be sprayed again. After picking, 501 is sprayed again, this time in the afternoon, to assist maturing of leaves and woody parts. In the autumn the horsetail mixture is used on the soil for the second time as described above. Fertilizing is carried out as described. Finally, the trunk and crown of the trees are sprayed with tree paste, either clay plus calcified seaweed or loam plus cow manure. If there are only a few trees to be treated, a slightly thicker mixture can be applied to trunk and thicker branches with a paint brush; this results in a thicker coating.

A few hints about pruning in the backyard garden may be useful, since private gardeners usually do this themselves. When the tree starts its winter rest, the sap withdraws from the branches and trunk. Not until the following spring does the new sap start rising from the roots. So pruning can be done any time after winter has started. If freshly cut surfaces are exposed to heavy frost, however, they may be frozen and this leads to damage from drying out. These parts then have to be cut out later. On the whole the tendency is to prune at the end of the winter. Even in this cold season the sap does move, even though only a little, and its rising and falling is linked with the moon. It rises less when the moon is in the declining zodiac (Cancer to Sagittarius). This is the time for winter pruning.

There are many pruning systems. In fact it is not the system as such that leads to success but rather a way of pruning that harmonizes with the growth characteristics of the variety, the shape of the tree and the habitat. Amateur pruning can be quite successful if it takes account of how a tree grows. A tree develops an upright trunk from which the branches spread out more horizontally. At the point where this change of direction takes place, wood formation recedes in favor of fruit formation.

A twig tied downward or pointing horizontally away from the trunk will tend to form fruit. Conversely, a weak twig can be helped to grow stronger if it is tied up more vertically. Thus even without any cutting the formation of wood and fruit can be influenced. If a branch is tied downward it will be noticed in the following year that at the apex of the bend new vertical twigs will have sprouted. A branch tied upward, on the other hand, will rather tend to grow thicker and have only a few lateral twigs. Growth lengthwise is enhanced. This is welcome in the first year but if it is allowed to continue uncontrolled the result will be those long whip-like twigs that only form a few fruits after two or three years; these few fruits will be easily torn off as the branch lashes back and forth in the wind. So it is really best to prune back shoots from the previous growth period. The more the new growth is pruned away, the more the tree will send out new shoots during the following summer. Those shoots that are allowed to remain form buds in the following year. If the shoot is short, then there is less space for fruit to form. If the shoot is longer, more fruit will form, but it may not be quite so large. A further result of cutting shoots rather long is that new shoots growing from them will also be smaller. Cutting is always done just above an eye or bud facing away from the trunk.

If a young tree is pruned and tied in this way from the start, not much cutting will be needed later. The early years are the most important because during this time the foundation is laid for the future shape of the tree's crown. After planting, three or four branches leaving the trunk at roughly the same height and pointing in four different directions are all cut down to roughly a third or a half of their length. They should all be the same length and should be quite clearly shorter than the main shoot, the continuation of the trunk, which is cut correspondingly less. All other branches, if there are any, are not cut but tied down. They will later bear the fruit. In the following year there may be a similar set of branches higher up all at the same height. These are not cut but are destined for fruit bearing. If in the year after that a'third set of such branches appear, three or four of them can once again be cut back to equal lengths. This creates the second level of main branches in the crown structures. As the crown grows, it quite often happens that shoots grow inward. These should be cut out. Larger cutting surfaces should be treated with tree wax.

3 Experience with Bio-Dynamic Methods in the Vineyard

This is a slight abbreviation of a report by Karl Hirsch on the many years of experience he has had in his own vineyard.

I have been using bio-dynamic methods in my vineyard for 35 years. My farm is in Rhineland-Hesse fronting the Rhine. Altogether there are 36 Morgen (approx. 27 acres) of vineyard and some arable land. The soil is mainly loess overlying limestone. The slopes are subject to erosion. Annual precipitation is a little under 500 mm (19.6 inches); the average annual temperature is 9.7°C (49.4°F). The land in this region is used onesidedly and it is thus not surprising that the vines here, where there is no woodland, grassland or cattle (to provide manure), are becoming increasingly weakened in respect of their inner organization of forces. A landscape like this is sick. Old field names show that 150 years ago the tops of the hills were still covered with forest and that there was grassland in the side valleys. Animal and fungal pests are increasing year by year. It would still be possible to introduce healthy methods. But so far industry has again and again introduced new "wonder" remedies that have prevented collapse. This will continue to happen and meanwhile the other vine growers do not see why they should change their methods. Modern weed and pest control make work easy and technical equipment is being constantly improved. Aircraft are used for pest control and it is thought that soon they will also be able to distribute chemical fertilizers. Experiments are under way with hormones that reduce branch growth (CCC), so that less time will be needed for pruning. The grape louse (*Viteus vitifolii*) has been banished by grafting to American root stock. A new chemical will be extensively used next year to combat botrytis on the grapes, thus relieving the vine grower of his worst worry, for recently often as much as half the yield has been lost because of rotting. There is a great shortage of pickers in the autumn so that harvesting often takes until the end of November, thus further increasing losses. At the Vine Growers' Congress at Offenburg, however, a mechanical grape harvester was on show that will be ready for wide distribution in the coming years. After all, the labor market being what it is, the vine grower cannot help having to work as profitably as possible. In conversations with vine growers the view is repeatedly expressed that there is no need to change anything so long as present methods still work and anyway they are not quite as expensive as methods based on humus. But the land cannot possibly cope in the long run with the continued use of chemical sprays and pest control. In addition, a great deal of damage is done by heavy tilling machines that even now dash through the narrow alleys between the rows drawn by their 30 horsepower engines, often when the ground is wet. The general lack of humus in vineyard soils aggravates the situation. In future, there will be problems with residues in the wine. But the wheel of progress cannot be turned back. We cannot do away with machinery and return to the methods of our forefathers. Bio-dynamic methods have become much more profitable in recent years and many farmers have shown that even with these methods work can be done economically so that their farms remain competitive.

It is my opinion that bio-dynamic farms and vineyards must be maintained at all costs. They will be urgently needed in the future both as examples of what can be done and also as training grounds for new recruits. It is my hope that in the future these methods will be taught at agricultural colleges while the farms themselves are given any assistance they might require.

There is a plethora of new literature about growing grapes, but it is most interesting for once to study one of the old books on the subject. In the middle of the last century the apothecary Bronner from Wiesloch in Baden in Southwest Germany journeyed through the vinegrowing districts of Germany. Later, he published a number of books describing his travels. These are no longer in print but can be found in libraries. Bronner reported that cleared vineyards were left fallow for at least seven years and not even used for farm crops. During these years the root excretions of the vines, which eventually exhaust the soil, were broken down while the soil was enriched with valuable physiologically active substances by the many weeds that grew meanwhile. There was also a replacement of humus, nutrients and trace elements. Nowadays vine growers replant without any interval and try to combat exhaustion of fertility with massive applications of fertilizer. Replanting, pruning and tilling were all done with reference to the moon's position in the constellations, something that is today rejected as a superstition. It was particularly important to plant a new vineyard only in a year when the planets were favorably situated. It was known in those days that such vineyards gave better yields and lasted longer. It was quite simply the custom to plant vineyards only in such years. Bronner also described how, when new vineyards were planted in the Palatinate, pebbles were buried in the ground at a certain depth. These collected the effects of the sun like accumulators and later gave them off again. These were dynamic effects, just as was the exact observation of the moon in the constellations, and in those days people still had a delicate sense for such things. The burying of the pebbles was certainly quite justified, for after sunny years the grapes in the following year are always better than expected.

One of the main problems with grape growing is that it is a monoculture. This was not so in the past. The vineyards were hoed only three times a year by hand, so that most of the time the ground was covered with green vegetation. Exhaustion of fertility was thus for the most part eliminated and erosion prevented. Fertilizing was done only every three years, with stable manure that was just as rare in those days as it is today. Diseases existed in those days too, but they were chiefly caused by the weather. Pest control was unknown and in a year with good weather it followed that the harvest would also be good. If we take into account that there was no systematic breeding of vines in those days and that some of the vines in each vineyard bore no or only few grapes, we see that the yields from the good plants were not at all poor. Today only the vines of the best type are used for propagation and success cannot be entirely due to fertilizing. The grape is by nature a light feeder and what it needs above all else is sun and yet more sun. But ever increasing amounts of chemical fertilizer have transformed it into a strong feeder. In the Moselle area amounts of 2.5-3 t/ha (2250-2700 lbs/acre) of a compound NPK fertilizer are not rare. Some plots can still be found that are not sprayed, but the grapes in them have been completely destroyed. This is a sign that in comparison with former days the vines have lost much of their vitality and health. If they are not to deteriorate still further, the cultivation methods that have become customary will have to be entirely reformed.

In 1934 I made the transition to bio-dynamic methods and the results were not at all disappointing. My soil contained only a moderate amount of nutrients and the humus content was meagre. Today I have before me a current report from a Dutch soil laboratory.

Humus content 2.8%; normal humus content with optimum lime content; vigorous biological activity in the soil; good properties for root development and nutrient intake; nitrogen mostly as ammonia; phosphorus content favorable, both in reserve and in available forms; optimum potassium reserves; sodium normal; magnesium sufficient. Conclusion: Favorable general impression; good soil fertility and vigorous biological activity; nutrient content is sufficiently secure if normal amounts of composted manure are given; then follow the figures of 22 different tests.

A compost analysis report says, "Rich in organic substance; rather alkaline reaction; well buffered. Good structure; strong biological activity; relatively few detrimental salts; nitrogen reserves temporarily limited; plentiful reserves of mineral phosphorus of which the major part is readily available; potassium content very high and magnesium content plentiful."

A report from a German soil laboratory: "I have never quite trusted the bio-dynamic method since I assumed that crops would suffer from nutrient deficiencies if fertilized exclusively with compost. The results of the analysis are astounding:

349

Calcium*	23%
pH	7
phosphate	400 ppm = very good
potassium	700 ppm = very high
magnesium	100 ppm = medium
humus	2%
catalase	36 = medium
azotobacter	45 = very good
decomposition of cellulose	= good."

*This is the content in the parent material, soil type: para-rendzina.

These soil analyses show that there has been no impoverishment of the soil but that on the contrary available nutrients and humus contents have doubled or trebled. The high phosphorus and potassium figures are astounding, especially the latter because since the transition my vineyards have never received any potassium salts. In recent years it has been stated frequently in the available specialist literature that particularly in loess soils a great deal of potassium is fixed and high potassium applications are therefore recommended. The soil in most vineyards is dead and my own example shows how important it is to bring it to life. My motto is: "Manure and compost enliven the soil." This is achieved by using exclusively compost or organic fertilizers, also green manure, possibly with straw, all combined with the application of the bio-dynamic preparations. The soil structure must not be damaged and no herbicides must be used. Pesticides are used sparingly indeed. The nitrogen content has always been rather low in my soil and since it is assumed that some nitrogen is lost through composting, the process is not allowed to continue too long. It is best if the manure can be taken straight from the stable, piled and treated with the preparations. Nitrogen losses are smallest with this method. A humus content of 3% should be the aim since this guarantees good biological activity and a constant supply of nitrogen. It is unlikely that a humus content of over 3% will be achieved, since this is hard to reach and maintain on uncovered soil without root residues. My best humus content has so far been 2.8%.

During the early years after the transition I had some diffi-culties with the quality of my wine. Acidity tests showed higher acidity than in comparable wines. This acidity was not reduced until the potassium content in the soil began to rise. The pH figures were also higher, which meant that the wine needed rather longer curing in the cellar. I used to cure my wine for three years. Longer curing improved it greatly and it aged much later, whereas ordinary wines often reach their pinnacle during the first year. Nobody can afford long curing these days and indeed today's fertilizing methods make it unnecessary, for heavy doses of potassium can force maturing. The disadvantage of this is that magnesium is fixed in the soil and so one consequence follows another. Oechsle Test figures are somewhat higher than with comparable wines, while extracts show no great differences. It can be said that the aroma and freshness of bio-dynamic wines are superior to others after longer curing. I have always been satisfied with my yields and I would not have been able to cope with any great losses.

How is fertilizing carried out in a bio-dynamic vineyard? The best fertilizer is stable manure, which must, however, be com-posted. Depending on the soil, about 15 t/ha (6 t/acre) are needed per year. I collect mine in August and pile it, mixed with ¼ of the amount of good earth, if possible from grass with plenty of roots. To this is added about 10% basalt or porphyry meal (whichever is more easily obtainable from the point of view of transport). The whole pile is treated with the bio-dynamic preparations and covered with straw. (The use of the prepara-tions is described in Chapter Four.) It is important that the heap should be neither too moist nor too dry. If necessary some water may have to be added. After harvesting all the available pulp is spread on the heap and then the heap is turned. Then a further treatment with the preparations is given and spreading in the vineyard commences. The distribution of 30 t/ha (12 t/acre) every two years is carried out with a sledge. Usually the job is finished by the middle of December so that the compost can be ploughed under before the first frost. The sledge must be horse-drawn because the pressure from a tractor causes too much damage during the wet season. Unfortunately fewer and fewer horses are used in vineyards. Young farmers in particular no longer want the trouble of feeding and caring for horses. It is regarded as rather old fashioned to be seen walking behind the plough, so horses are being increasingly replaced by tractors.

Soil compaction and poor biological activity are the result. Nutrients and trace elements become fixed and so the grower is forced to resort increasingly to fertilizers. After large harvests and if the soil is rather poor the indicated amounts of compost will not be sufficient. In such cases I apply horn-bone meal or some other organic fertilizer in the interim years. The compost must not be stacked for too long because vines prefer manure that has not fermented for too long. Ash from burning vine wood, and also any foliage removed by pruning are returned to the soil. Wood ash is an excellent fertilizer. Made from one-year shoots it contains 25% potassium, and a large proportion of other minerals and trace elements. If the grapes only are removed while every other residue is returned to the soil, this ensures that two thirds of the nutrient intake by the plants are returned to the soil. Preparation 500 is used on the soil in the spring and Preparation 501 is sprayed on the green leaves in summer. The former stimulates soil life and humus formation; the latter improves the ability of the sun to work on the leaves.

As with every monoculture, also where viticulture is concerned, deficiencies appear in plants and soils. The incidence of so many animal and fungal pests is the consequence of such one-sided cropping. The vine grower therefore has to do whatever he can to ameliorate the effects of monoculture. For this purpose I have started to use a mulch process in one of my vineyards that covers 1 ha (2.5 acres) and has the vine rows planted sufficiently far apart. In this plot the soil is tilled in the usual way until about the middle of June and then all cultivation stops. After about four weeks the result is a thick carpet of weeds. This is then mown with a small mower and left lying as a mulch. This process is repeated two or three times, depending on the weather, until harvest time. I consider mulching with weeds to be the best form of green manure because every weed enriches the soil with effective substances and trace elements and in this way the wild plants can work in a healing way on the soil. The more varied the weeds, the better the effects. For plots lying fallow the same process is used.

Weeds cannot be left to grow in close-planted rows of vines, so another method is appropriate here. In early August, a green manure crop is sown in every second row. The best are legumes or *Phacelia,* frequently also planted for honey bees. The whole is then immediately covered with straw, 1.2-2 t/ha (0.5-0.8 t/acre). The plants germinate quickly under this cover and the method only fails in extremely dry years. The autumn dew supplies enough water and by the time the frosts begin a great mass of plants has grown. The nitrogen in the legumes is needed to break down the straw and by the time growth starts again in spring the straw and the plants have already made a good start with fermentation. Meanwhile the weeds will have started to grow again and shortly before mid-May (when frost is expected) the whole lot is worked in shallowly with a slow rotovator. (In Germany the Saints Bonifaxius, Pancretius and Servatius, whose feast days fall on 11-13 May, are traditionally expected to bring a cold spell). It is important that the working in should be shallow. A green cover then grows once more, and this too has to be superficially worked in. After this, tilling is done as usual but rotating tools are not recommended. While the ground cover is growing, any necessary work is done from the free rows. Permanent ground cover in the vineyard is also possible but only in areas with at least 650 mm (26 inches) rainfall annually. Of course the sowing is done in different rows each year. It is astounding how well the earthworm population develops. Over 100 earthworms per square meter or yard are not unusual. Earthworms are particularly important for maintaining and developing the fertility of the soil. In a neighboring vineyard I could not find a single worm. The ground there is kept utterly clean, so that apart from any other reason there is just no food for the worms.

I had presumed that after a few years of bio-dynamic cultivation there would no longer be any need to spray against fungal disease. This turned out not to be so and my attempts failed. The amounts and concentrations I use, however, are lower than usual. On the other hand, I have not used any pesticides for 15 years. In a Riesling vineyard I have made the following observation over a number of years. In autumn the eggs of the red spider are found at the internodes of the year's growth; the following spring, when growth has started, there are on the young leaves many spider mites; having done nothing about this I return ten days later to find that most of the mites have disappeared while the curled leaves have straightened out and grown larger; the predators have done their work and only a few of the spiders remain to preserve their species. This is as it should be, for were they to be completely wiped out, the beneficial insects would also die for lack of food. Interference with pesticides always decimates the predators more than the pests that always manage to retain the upper hand. I have observed the above process year after year, the only variation being in the numbers both of pests and their predators, which depends on the weather. If other pests appear, the appropriate predators soon follow. The natural community should on no account be disturbed by poisonous substances. Plants fertilized entirely organically have gained from the healthy ground a certain resistance to pests. So I have now reached the stage at which I do nothing against pests and yet I suffer no yield losses.

It would also be possible with time to control fungal diseases, but my slender means are not sufficient to embark on this. What is needed are research laboratories specializing in grape growing. Vines that are almost entirely resistant to fungal diseases are often found in domestic gardens. In the wine-growing regions of the River Saar, in the wasteland between the fortifications of the Second World War, vines have also been found that were almost entirely free of fungal infestation. My recommendation would be to make a new start with such plants on healthy soil.

Chapter Nine

Quality Through Growing Methods

1 Environmental Factors and Plant Characteristics

Ever since man has grown plants he has been able to observe that there is an interrelationship between the characteristics of the plants and those of the environment such as the soil, the climate, etc. Recently, agricultural science has added its discoveries, particularly through chemical investigations. Among these is the fact that certain elements present in the soil, even in small amounts, can obviously limit the growth and development of plants. Moreover, it is known that deficiencies in the soil can also affect animals and man via the plants they use for food, and it is from this that the view arises that deficiencies can be redressed by administering to the soil, plant, animal or man the particular substance that is lacking.

Other influences on plant growth come from onesided conditions (or factors) of the climate or the weather, such as lack of light and warmth, rain, etc. These influences affect particularly the formation of substances such as sugar and vitamins. This kind of "deficiency" is less specific, being more or less pronounced depending on the weather throughout the year.

Another group of "deficiencies" not caused by the soil itself nor by the climate is being increasingly recognized, namely, the

355

influence of management. They are also partly unspecific in the way they manifest, just as are the effects of unfavorable weather. They show, for instance, in reduced resistance to pests, poorer keeping qualities, a deterioration of physiological characteristics and, insofar as this has been registered, a reduced ability of livestock to utilize feeds. An increase of illness among domestic livestock is a consequence of reduced fodder quality.

Once these deficiencies have been overcome through suitable management in the manner attempted and practiced with the bio-dynamic method, the weaknesses mentioned are either reduced or they disappear entirely. People notice the improvement in quality through the better aroma, taste and digestibility. Invalids and convalescents find the food easier to digest. The health both of people and livestock can be expected to improve. Much still remains to be done to demonstrate these facts in biological experiments; the situation is quite clear, of course, in the cases concerned with actual poisoning caused by excessive use of agricultural chemicals. However, the gradual, not sudden, change in foods has raised people's interest in these questions and made it necessary to study and classify the causes. Interest in the health aspect of foods has given the concept of quality an extended significance.

2 Methods of Bio-Dynamic Research

The research that has become a part of bio-dynamic work has developed new and specific methods for studying the characteristics of plants from different viewpoints. Conventional methods are fully made use of parallel with these, though often to seek answers to unconventional questions. Among the first are the morphological and "picture-creating" methods mentioned briefly here. We shall, however, not go into technical details. What interests us here is where these methods are used and how the results are evaluated. (See diagrams, pp. 358-9.)

Observation of growing plants

Normally we always observe the development of the crops we

have planted. In practice, this is often the only way of determining when a particular measure is required. Experienced farmers and gardeners have a well-founded ability to observe and evaluate the development of their crops. For the same kind of observation growth experiments are arranged in which shape, color, stages of development, pests, diseases, etc., are observed.

1. Kolisko (1939) used wheat plants as reactors to light influences. She measured the length of the first two leaves of germinating plants and found that increased light (either from insolation or sandy soil) stimulates the lengthwise growth of the second leaf in relation to that of the first. Her observation method has so far only been applied under laboratory conditions.

2. Klett (1968) observed that light as opposed to shade, compost as opposed to NPK fertilizers, and treatment with Preparation 501 as opposed to no treatment all brought about a more typical root development in spinach (straight, strong tap root) and white radish (spherical thickening of the hypocotyl). Similar differences were noticed in simultaneous observations with regard to the leaf development of the same plants by a method initiated by Grohmann and developed further by Bockemühl (1964-72) and others. Such morphological methods were used for evaluating field experiments that we shall mention again later. Presumably these observation methods could be developed for routine field work.

3. Pettersson (1970) studied the development of side shoots above ground in potato plants. In strong vegetative growth such as can be induced by forcing fertilizers, humus-rich soil or plenty of rain, the number of side shoots increases and the whole plant gains a tendency to grow rather horizontally, often even lying on the ground like lodging cereals. This is accompanied by an inferior tuber quality. In contrast, a more vertical plant with only a few or no side shoots, such as can result from the use of balanced fertilizing, or drier weather, also has a better quality of tuber. This method in conjunction with a knowledge of the characteristics of different varieties has proved useful for the observation and control of potatoes grown to Demeter quality standards.

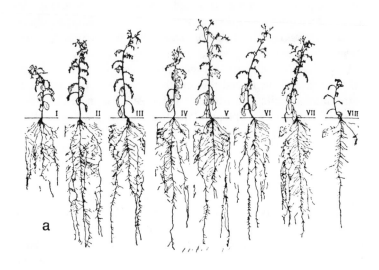

a

Series a and series b show the interaction between organic fertilizer and light in its effect on the development of groundsel (*Senecio vulgaris*), a common weed (Bockemühl 1971). The plants were grown in a greenhouse in vessels that allowed the roots to be observed. Series a was given more and series b less light. These soils used were:

I composted urban garbage

II garden soil + 25% well-fermented cow manure

III garden soil + 25% composted urban garbage

IV garden soil + 25% composted urban garbage containing 10% sewage sludge

V garden soil + 25% composted urban garbage containing 20% sewage sludge

VI garden soil + 25% composted urban garbage containing 30% sewage sludge

358

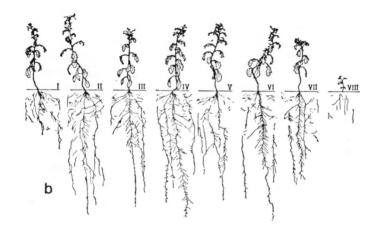

b

VII garden soil + 25% composted urban garbage containing 40%
    sewage sludge
VIII garden soil

The flat containers stood in a slightly slanted position and allowed
for simultaneous observation of both shoot and root development. It
is particularly interesting to note:
a) I and III: Composted garbage with earth gives stronger develop-
    ment than composted garbage alone. With b) I and III the differ-
    ence is less marked.
a) II: Mature composted cow manure gives strong rooting and har-
    monious, strong shoot development.
a) and b): With plenty of light (a) shoot and root are well-developed.
    With less light (b) the plants on cow manure and composted gar-
    bage are stronger compared with those on composted garbage and
    sewage sludge.

Chemical and physiological tests

The chemical analysis of plant substance still plays an important part in quality testing, but it is no longer the sole yardstick. Various kinds of physiological tests are becoming increasingly important. Chemical analysis is used to register changes, but interest is now focussed more on the changes themselves, by which life processes become manifest.

The amount of a particular compound in a plant is usually expressed as a percentage of fresh substance, or better still of dry matter. It is frequently held that a higher content indicates a better quality. In many cases this is so, but taken too far the assumption can lead to some absurd judgments. The desirable chemical composition of a plant is expressed in the ratio of the different substances it contains. This ratio is not constant but changes from one growth phase to the next and even within each period it has a certain variability. If, however, any of the figures fall outside this range of variability, it can then be assumed that the plant is no longer balanced in its composition and disturbances occur. These need not necessarily appear in a prominent way in the plant itself but may bring about strong and sometimes unexpected and undesirable effects in the next organism in the food chain, either animal or man.

Evaluation of the chemical composition can therefore not always be made with reference to the total dry or fresh weight of the plant. This may, for instance, consist mainly of substances like starch or fiber. Rather is it often useful to look into the major groups of compounds. It is, for example, meaningful to study the quality of the proteinaceous compounds and the fertilizing used, or the amount of light given as well as other treatments. Crude protein includes all those substances that contain nitrogen. When proteins are formed, one finds at first nitrate and ammonium ions that are taken up by the plant from the soil. They are (to put it briefly) transformed into amino groups ($-NH_2$) that combine with an organic acid released by the carbohydrate metabolism to form amino acids. These then combine over a number of steps to form true protein. Not until they reach this form do the proteins attain their greatest effectiveness for growth and metabolism and also their highest food value. This protein building process depends above all on the plant's ability to deal with light.

In order to obtain a picture of protein formation in the plant, true protein is reported as a percentage of crude protein. This figure is the relative protein content. High relative protein content denotes good quality. It means that the plant substance has reached a high stage of maturity (for the particular phase of development it is going through). Alternatively, the compounds at a lower level of the protein synthesis may be reported, for instance, the free amino acids or even nitrate, and compared with the crude protein. In these cases a high relative percentage must be evaluated negatively.

One of the physiological testing methods is the observation of decomposition in the crop. This decomposition is the opposite process from the synthesis that takes place in the living plant. It is the only process that continues to function after the plant has been harvested. If the growth processes have been disturbed, decomposition will set in all the more quickly, often accompanied by the formation of obnoxious substances. Decomposition is caused by autolysis, i.e., the effects of enzymes within the plant and bacterial and fungal activity. Such changes can be followed up by chemical analysis, the noting of color changes, etc. Klett (1968) and Klein (1968) measured enzyme activities, e.g., certain carbohydrases, proteinases, dehydrogenases, etc.; Pettersson (1970) has employed the darkening of potatoes as a testing method. The different stages of decomposition can also be followed by the crystallization method (Selawry 1961, Engqvist 1961, Koepf and Selawry 1962).

The decomposition of juices and extracts by bacteria can also be used experimentally. Pettersson (1972) has followed decomposition processes by measuring the electric conductivity in extracts.

There is also a pathology test that uses pathogenic organisms to gauge decomposition. Engqvist, Wiedersheim-Paul and Pettersson have tested the suitability of leaf rot fungi in potatoes (Pettersson 1972).

361

Picture-creating methods

The picture-creating test methods developed in connection with anthroposophical science provide a new source of scientific investigation. However, forty years after the development of the first two methods, the crystallization method after Pfeiffer and the capillary dynamolysis method after Kolisko, they have still not been generally recognized although their usefulness is proven. Their special characteristic is probably one of the reasons for this. They have been elaborated empirically and are thus not speculative. "Reading" the pictures and evaluating the results, however, is difficult and considerable experience must be gained by anyone wanting to master the techniques.

In the crystallization method, copper chloride is allowed to crystallize out on a flat glass plate. If nothing is added to the copper chloride, the resulting pattern of tiny crystals is formed haphazardly. When a solution of plant sap or other organic substance is added, the crystals form an organized picture that is specific for that particular solution.

Thus, this is a picture of the ability of the substance in question to make a coherent pattern out of crystals that without that substance would be arranged incoherently. The situation resembles that of a plant that coordinates inorganic substances from the soil and the air into plant substance and plant shapes. This is a functional test that makes visible the condition of the plant being examined. The methodical basis and progress made with this test method have been reported on by Krüger (1950), Selawry (1957, 1961), Engqvist (1961, 1963, 1970) and Pettersson (1967).

With the capillary dynamolysis method a picture is formed by an organic juice or extract rising in a vertically held cylinder of filter paper (Kolisko 1939). A solution of a metal salt is allowed to rise in the filter paper either before or after the plant juice is applied. The combination of the organic component and the metal results in a pattern of colors and shapes. A modification of this method are the chromatograms developed by Pfeiffer (1959, 1960) in which the filter paper lies horizontally and the

362

solution is allowed to spread outward from a center. This method is relatively simple to carry out and works rapidly. It is suitable for testing soils and composts. It is not as developed as the crystallization method for diagnosing the condition of plants. In agriculture it has been used for testing soils (chromatograms) and also spinach and carrots (Breda 1972). It also plays a part in the selection of plants and picking times for herbs for the production of medicines (Fyfe 1967).

Examples of chromatograms and crystallization pictures will be found facing page 196.

3 Earthly and Cosmic Growth Factors in Food Production

In the Agricultural Course, Rudolf Steiner describes how we may regard plant growth from the viewpoint of earthly unfolding and cosmic moulding. In endeavoring to do this we need not stop short at plant morphology. Morphology and growth progress are related to the quality of substances. We must, however, exercise caution in drawing conclusions about the material composition of the plant from its external appearance, and we should be patient in training ourselves to form such conclusions.

In order to evaluate the qualities of a food plant we need definite points of reference. Chemical analysis gives us some of these. If, however, we carry the stereotyped yardstick that maintains "the more the better" too far, until obvious negative effects appear, then we shall soon fall prey to distorted judgments. The chemical findings must be classified according to a view of the plant's totality and to this end a division into the two polarities of earthly and cosmic growth factors has proved fruitful.

It was thus deemed important in bio-dynamic research to set up and carry out experiments with this in view and the results of some of this research are described in the following.

The plant between light and darkness

Light is probably the most important cosmic factor for plant

363

life. When light is predominant, cosmic laws outweigh the earthly laws and, when light recedes, the earthly laws outweigh the cosmic laws. Between light and darkness lie all the degrees of shade. It is impossible to grow a green plant in complete darkness, but we can study the effects of earthly influences by growing plants in varying degrees of shade. Experiments along these lines were carried out by, among others, Kolisko, Klett and Engqvist (1965).

E. and L. Kolisko (1939) placed germinating wheat plants in a tunnel in which light conditions were graded from sunlight to complete darkness. As mentioned earlier, the comparative lengths of the first and second leaves were measured. As light decreased, the first leaf grew longer and the second shorter. In one experiment the varying growth in different soil mixtures was studied. The diagram shows the relative lengths of the leaves in diagrammatic form. Forty-five pots each containing thirty plants were ranged from the dark to the light end of the tunnel.

The graph referring to garden soil can be used for comparison. We find that the point where the first and second leaves are of equal length falls in pots 32 to 42. If the same garden soil is mixed with sand, the crossing point shifts toward the dark end in pots 29 to 38. This shows that the sand gives an intensified light effect. In the diagram referring to humus soil, the lengths of the first and second leaves do not cross over at all, which shows that here the effect of darkness is intensified. The experiment demonstrates how different soils either intensify or weaken the cosmic light aspect of the plant.

Klett (1968) carried out a three-year field experiment with three grades of light (full light, half shade and deep shade) in combination with two kinds of fertilizer (organic and mineral).

Diagram on facing page.
Growth of first and second leaves of germinating wheat plants in various soils (a, b, c) and in degrees of shade ranging from complete darkness to sunlight (45 pots) (after E. and L. Kolisko 1939).

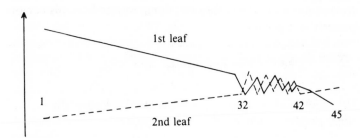

a. garden soil

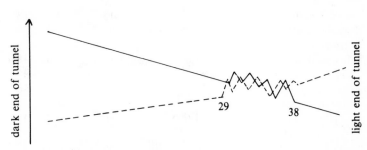

b. garden soil with sand

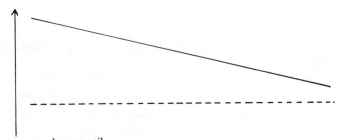

c. humus soil

All combinations were tried with and without bio-dynamic Preparation 501. Seven different varieties of plant were reared, though not all in the same year. Some results from his copious material are given here.

Spinach in a light-shade experiment (Klett, 1968)

Items tested	Treatment					
	light		half shade		deep shade	
	org.	min.	org.	min.	org.	min.
root length, cm	23	18	13	15	13	13
shoot length, cm	89	81	98	97	71	59
crude protein, % in d.m.	12.6	15.6	15.2	18.9	16.0	18.7
*relative protein content, %	97	90	86	81	62	66
nitrate-N, mg in 100 g d.m.	7.6	34.6	13.6	40.4	42.4	47.6
vitamin C, mg in 100 g f.s.	90.4	80.1	78.8	80.2	56.1	47.3
**crystallization standard	95	72	64	51	54	47

f.s. = fresh substance   d.m. = dry matter
* = true protein in % of crude protein
** = visual evaluation of crystallization pictures expressing quality in figures; best quality = 100.

The length of the spinach tap root was increased by light, but the shoot length reached its maximum in the half shade. The crude protein content increased considerably with the degree of shade, while at the same time the true protein content decreased and the amount of nitrate not transformed into protein increased. Vitamin C decreased with increasing shade since it depends on the effect of light. The crystallization standard also

366

of the plant. These can be examined in their polarity to dry conditions. But the combination with warmth, i.e., damp and cold or dry and warm, must also be taken into account.

In bio-dynamic literature and indeed in general agricultural literature there are only a few records of experiments on the influence of moisture and dryness on quality. Some indications arise from a field experiment made by Pettersson (1970). The experiments were located at four sites in Scandinavia with varied rainfall and lasted two years (preceded by two preparatory years). Potatoes were grown in both years. The soil in all cases was sandy though the textures varied. The weather, particularly precipitation, was quite different in the two years and as a result there were large differences in the quality of the potatoes. Quality improvement was taken to entail low crude protein content, high relative protein content, only slight darkening of the tissue and the extract, and a more formed picture in the crystallization test method. These characteristics were summarized in an index figure, the quality index (see p. 373). A comparison between the times of particular meteorological conditions and the quality index showed that the period of 40 days after flowering, i.e., tuber growth and ripening period, was decisive with regard to quality of the tubers.

Potato quality as influenced by meteorological conditions during the 40 days following flowering

Sites:	1	2	3	4
Average temp. °C				
1965	14.6	13.2	14.5	13.5
±1966	+ 0.8	+ 1.4	+ 1.4	+ 2.5
Precipitation, mm				
1965	90	93	86	238
±1966	+69	-46	- 43	- 152
Potato quality in 1966 compared with 1965	unchanged	greatly improved	very greatly improved	greatly improved

Simultaneous increases of temperature and rainfall as in Site 1 produced no change in the quality of the potatoes. An increase in temperature together with a drop in rainfall, which corresponds to a drier weather type (as in Sites 2, 3 and 4), resulted in greatly improved quality. Probably the increased light also had something to do with this.

The soils at the different sites differed somewhat from one another. At Site 4 fine sand with a high capillarity caused on the average more damp conditions than is the case on the other sites. The potato quality was lowest at this site in both years.

This experiment also included five fertilizer dressings. The following tendencies showed clearly:

Organic fertilizers, such as stable manure or, even more so, composted manure, work in the same direction as dry, warm weather.	Mineral fertilizers bring about tendencies similar to those caused by damp, cool weather.

Seasonal influences on quality

Quality differences can also occur in plants planted at different seasons. Fast-growing crops such as spinach can be used to demonstrate this in practice in the open. Breda (1972) used spinach in order to study the difference between planting in spring and autumn. The following table shows some of his results.

In autumn, crude protein and nitrate contents are higher with all fertilizers than in spring. If this is compared with the light-shade series by Klett, it seems that it must be interpreted as lack of light. The increase in fertilizer applications also affects plants like a lack of light, both with mineral and with compost fertilizers. The sugar content also decreases with a decrease in light.

Capillary dynamolysis tests led to parallel findings that can only be described here.

Spring: The pictures from the plots that had received compost differed in their shapes and colors from all plots that had

| Items tested | Treatment | | | | |
| | ammonia sulphate kg N/ha (lbs N/acre) | | composted manure t/ha (t/acre) | | |
	100(90)	200(180)	60(24)	100(40)* 120(48)	180(72)
Crude protein, % in d.m.					
Spring	23.3	29.6	–	22.4	–
Autumn	30.9	34.3	28.7	30.0	31.2
Nitrate-N, mg/100 g d.m.					
Spring	48	207	–	51	–
Autumn	328	444	229	245	403
Total sugar % glucose/100 g d.m.					
Autumn	16.3	13.7	18.9	16.7	14.3

*In spring 100 t/ha (40 t/acre), in autumn 120 t/ha (48 t/acre)

received chemical nitrogen. Composting causes the absence of dark colors in the pictures. One might be inclined to interpret this as a deficiency of the compost variations, but this is not so. When material from the composted plots is subjected to autolysis, the same dark colors appear after three days of "ageing" as in the pictures of the chemical nitrogen. It may be assumed that the spinach grown on the compost variation is different from the chemically treated spinach with respect of enzyme activity and possibly other characteristics of the protein fraction. It keeps longer.

Autumn: The filter pictures showed few differentiations. The pictures from the compost variation resemble those of the mineral fertilizer variations in the spring crop.

The influence of fertilizers

Fertilizers are the most effective means of increasing plant yield. They also influence the plant's qualitative characteristics, in a negative as well as a positive direction. This has long been recognized in bio-dynamic practice, but a great deal of time passed before a wide enough range of experimental work in this field was set up and there is still much to be done. Adequate methods had to be developed for the reliable assessment of the experimental plant material. In the last decade, however, results have been published in Germany and the Scandinavian countries. Some of these have already been mentioned such as the influence of light and shade, humidity and drought, the seasons. Thus, for example, the experiments undertaken by Klett (p.364), Engqvist (p. 368) and Breda (p. 363) show that mineral fertilizers and decreasing amounts of light affect the configuration of substances in the plant in similar ways.

The results of several manuring experiments are shown in the following.

Pettersson (1972) conducted a field experiment on the affects of eight combinations of fertilizer on the quality of potatoes. These were grown in a 4-year rotation of wheat, clover-grass, potatoes, and beetroot, so that all four were grown each year. The following average figures are from the 9th to 12th year, i.e., there had already been 8 years of the experimental sequence. We can assume that the different varieties of fertilizer had had sufficient time to influence the biological state of the soil.

The eight combinations were as follows, the amounts shown being those used for potatoes:

1. Composted stable manure with 1% meat and bone meal, 30 t/ha (12 t/acre), and BD Preparations 500 and 501.
2. As 1. but without the BD preparations.
3. Crude stable manure, 30 t/ha (12 t/acre).
4. Crude stable manure, 15 t/ha (6 t/acre) and ½*

   $NPK = \dfrac{15:13:22 \text{ kg/ha}}{13:12:20 \text{ lbs/a}}$

   *(with reference to the nitrogen content supplied by the manure, see plot 6)
5. Control.
6. NPK = 31:26:43 kg/ha (28:23:39 lbs/acre).

7. 2xNPK = 62:53:86 kg/ha (56:47:78 lbs/acre).

8. 4xN + 2xPK = 124:53:86 kg/ha (112:47:78 lbs/acre)

(the beetroot was fertilized as the potatoes; for wheat in plots 4, 6 and 7 the amount of N was halved; no fertilizer was used for the clover-grass).

The individual results were incorporated in a quality index in the following way:

Items tested	Quality Assessment	
	favorable	unfavorable
crude protein	low	high
relative protein content	high	low
darkening of tissue	slight	considerable
darkening of extract	slight	considerable
speed of extract deterioration	slow	fast
crystallization pattern	better	worse
growth of pathogenic organisms	slight	considerable
QUALITY INDEX	HIGH	LOW

Flavor trials were conducted in autumn and spring: best flavor = 4; poorest flavor = 1. These results are included in the table.

The influence of fertilizers on the qualitative characteristics of potatoes (Pettersson 1972)

Fertilizer	Yield 100=34.1 t/ha 13.6 t/acre	Quality Index	Flavor	
			Autumn	±Spring
1	104	109.3	3.0	±0
2	108	107.1	3.0	- 0.1
3	95	106.1	2.8	±0
4	103	97.5	2.8	- 0.2
5	83	98.5	2.9	- 0.2
6	98	95.9	2.8	- 0.3
7	106	100.0	2.9	- 0.5
8	103	85.6	3.1	- 0.7

Significant variations when:	Yield	Quality Index	Flavor Autumn
P=0.05	12	6.9	0.4
P=0.01	16	9.4	0.6

Yield is significantly lower only in the unfertilized plot. With regard to the quality index there are three quite distinct groups: a) Combinations 1, 2 and 3 in which only organic fertilizers were used; these vary in comparison with one another and the highest grade is achieved by the plot treated in the full bio-dynamic method.

b) The plots with no fertilizer (5), mixed fertilizer (4), low mineral fertilizer (6) and medium mineral fertilizer (7); the favorable position of number 7 is notable.

c) The combination with the greatest nitrogen content (8), which diverges widely from the other plots treated with mineral fertilizers.

The different flavor in autumn and spring is also notable. There was varying durability of flavor and deterioration was greatest in the potatoes grown with mineral fertilizer.

Klein (1968) made field experiments on light sandy soil with graded applications of composted manure and supplementary organic fertilizers such as pig bristles and horn meal. From his extensive material we have selected his records for potatoes in 1965, as this vegetable lends itself well to the detection of qualitative differences. Experiments with potatoes and rye had been in train for two years prior to 1965. Except on the control plot, composted manure and bristles were used in two rates of application and in various combinations with one another, nine treatments in all; each treatment was repeated five times.

In order to show the main results in brief the table shows the values condensed from various of Klein's tables.

Both composted manure and bristles brought about a distinct

The influence on potato quality of various organic fertilizers in graded application (Klein 1968)

Items tested	Control	composted manure		bristles	
		1	2	1	2
fertilizer					
t/ha & kg/ha resp.	–	22.51	45.0	313.0	625.0
t/a & lbs/a resp.	–	9.0	18.0	279.0	556.0
yield, t/ha	11.3	+ 2.8	+ 5.2	+ 1.9	+ 2.6
t/acre	4.6	+ 1.1	+ 2.1	+ 0.8	+ 1.0
crude protein					
% of d.m.	8.5	- 0.5	- 0.4	+ 0.2	+ 0.7
relative protein					
content %	64.3	+ 0.4	- 0.8	- 3.0	+ 0.8
free amino acids					
mg N in 188 g d.m.	219	- 35	- 40	+ 12	+ 16
proteinase activity	58	- 7	- 7	± 0	+ 13
starch, %	13.2	+ 1.4	+ 1.9	- 0.7	- 1.5
vitamin C					
mg/100 g f.s.	23.0	+ 1.2	+ 0.8	+ 5.1	+ 2.4
crystallization					
standard	57	+24	+32	- 17	- 27

increase in yield over the unfertilized plots, but their influence with regard to quality differed as is shown in the table.

With a few exceptions, composted manure clearly tended to have an influence similar to that of light, while bristles worked more like shade. The difference between the fertilized and unfertilized plots, however, was not great. Bristles tend toward the effects of shade less strongly than does mineral nitrogen, though the direction of influence is the same.

Pettersson (1970) tested the effects of the following fertilizer combinations on potatoes:

a) composted stable manure, 60 t/ha (24 t/acre)

Items tested	light tendency affected	by fertilizing with	shade tendency affected	by fertilizing with
crude protein	lower	comp. manure	higher	bristles
relative protein content	higher	fertilizing made no difference in this experiment	lower	fertilizing made no difference in this experiment
free amino acids	lower	comp. manure	higher	bristles
proteinase activity	lower	comp. manure	higher	bristles
starch	higher	comp. manure	lower	bristles
vitamin C	higher	comp. manure	lower	—
crystall-ization standard	higher	comp. manure	lower	bristles

b) meat meal, bone meal, sulphate of potash-magnesium, as NPK = 103:102:145 kg/ha (92:91:129 lbs/acre)

c) ½ a + ½ b

These combinations were also used in the field experiment referred to in the table on page 372.

Yield was approximately the same for each treatment in each year. Quality was distinctly negatively affected only by the com-

The influence of composted manure (a), an organic supplementary fertilizer (b) and a combination of both (c) on the quality of potatoes (Pettersson 1970).

		a	b	c
Yield t/ha(t/acre)	1965	25.4(10.02)	26.7(10.7)	23.8(9.1)
	1966	29.4(11.8)	29.8(11.9)	32.7(13.1)
	average	27.4(11.0)	28.3(11.3)	28.3(11.3)
Quality index	1965	104.4	99.4	102.8
	1966	108.1	96.8	108.6
	average	106.3	98.1	105.7

bination with meat meal, bone meal and sulphate of potash magnesium. When these were composted together with stable manure, no significant negative results occurred.

Fürst (1967) studied the effects of different fertilizers on apple trees and their fruit over a number of years, combined with intercropping of legumes in order to enliven the soil. Detailed reports referring to the years 1964, 1966, 1967 are available on one of these experiments. (The experiments continued until 1972.) A number of organic fertilizers with an annual N-application of 55 kg/ha (50 lbs/acre) were compared with a chemical compound fertilizer (N as ammonium nitrate) in an application of 82 kg N/ha (74 lbs N/acre). It was assumed that the legume crop brought a further annual amount of 85 kg N/ha (76 lbs N/acre).

The apple varieties were Pearmain, Cox, Winston and Boskoop. The trees were intensively planted at 2.5 x 2 m on M-IX root stock. There is no report on yields, but records were kept of the commercial grades achieved and of attack by a number of fungi and insects. The only pest control used against scab was Tecoram (AAteck 0.2%) and from early August on sodium silicate (1.3%).

Commercial grading: the percentage of grade 1a fruits are shown.

Mildew: the following scale was used to record this:

grade 0 = no tips attacked = none
grade 1 = 8 tips attacked = very slight
grade 2 = 16 tips attacked = slight
grade 3 = 24 tips attacked = medium
grade 4 = 32 tips attacked = extensive
grade 5 = 40 tips attacked = very extensive

Scab and codling moth infestation is given in percentage of harvested fruit.

Fertilizer experiment with apple trees showing commercial grades, and infestation with mildew, scab and codling moth. Averages over 3 years (Fürst 1967)

Fertilizer	Commercial grade 1, %	Mildew grade	Scab %	Codling Moth %
Inorganic compound fertilizer	71	3.5	1.6	4.9
Vita-Nova (= blood,horn,bone)	77	0.9	0.4	3.5
Oscorna (= horn,bone)	77	1.4	0.9	3.8
Peru-guano (in 1966 chicken manure)	79	1.6	1.3	4.1
Composted manure	81	2.7	1.8	3.0
½ Peru-guano + ½ Oscorna	81	1.4	1.5	2.9
½ composted manure + ½ Oscorna	84	1.7	0.7	2.4

All the organic combinations compare positively with the inorganic compound fertilizer. The organic combinations differ considerably among each other, particularly in regard to the commercial grade of the fruit. It is striking that the horn/bone meal (Oscorna) is at the bottom of the organic scale when applied alone and at the top when applied together with composted manure. As in the previous example, this implies that a combination of composted manure with commercial organic materials has a favorable affect on quality.

Influence of the bio-dynamic Preparations 500 and 501

A relatively large number of experiments with the bio-dynamic Preparations 500 and 501 have been published as time has gone on. Most describe the influence of the preparations on the growth of the plants, and in particular how treatment influences yield. Few deal with the influence of these two preparations on the quality as determined by analysis. Frequent yield increases of 10-30% and more are reported. In comparison, the qualitative changes registered hitherto are of a lesser order. Some examples are described here.

In the field experiment by Pettersson (1970), from which the climatic influences have already been mentioned (see table on p. 369), various treatments with Preparation 500 and 501 were also given. They were used separately and in combination with one another. As above, the results are given here in units of the quality index. For the sake of comparison some figures for other fertilizer treatments are also given.

During the first year (1965) the influence of the preparations on the quality characteristics examined was positive. In the combination of both 500 and 501 the effect was statistically significant on the 95% level. In the second year (1966) the influence was negative, but this time no significant differences occurred.

The influence of the preparations was lower in comparison with the effects of organic and mineral fertilizers. This means that in practice great care must be taken with the kind and amounts of fertilizer. Once the best possible has been achieved with them, the results can be further improved with the use of the bio-dynamic sprays.

Why did the preparations have positive effects in the first year and negative effects in the second? With the material available this question cannot be answered. The experiment covered a

Fertilizing, treatment with the preparations, and potato quality (Pettersson 1970). Treatment with preparations includes 5 fertilizer variations.

Fertilizer and Preparations	Quality Index	
	1965	1966
Composted stable manure, 60 t/ha (24 t/acre)	104.4	108.1
Crude stable manure, 60 t/ha (24 t/acre)	+ 0.2	- 6.0
Mineral fertilizer, NPK=103:88:144 kg/ha		
(92:78:128 lbs/acre)	- 12.7	- 21.6
Statistical significance, P=0.05	3.9	3.5
P=0.01	5.2	4.7
Untreated	98.7	101.4
Preparation 500	+ 2.4	- 1.7
Preparation 501	·+ 1.4	- 0.1
Preparations 500 and 501	+ 3.8	- 1.8
Statistical significance, P=0.05	3.2	2.0
P=0.01	4.6	2.9

total of 80 variations with 4 sites, 5 fertilizers and 4 preparation combinations. When the experimental results were being calculated, it turned out that with the exception of one of the sites the first year was much wetter than the second. Compared with the untreated plot, the quality indices were positively influenced by Preparation 500 if there was more than 300 mm (12 inches) of rain between May and September and negatively if there was less. It seems from this that this preparation works positively in a moist season. There are no other experimental results that bear this out. Similar consequences for 501 were not derived from this experiment, but there is practical experience pointing in the same direction. It seems possible, therefore, that both preparations have the same tendency in their effects on quality.

Klett (1968) and Klein (1968) have examined the effects of Preparation 501 in a number of experiments. Definite differences emerged when the crystallization method of diagnosis was used. The figures in the following table are from a 4-year experiment by Klett.

Crystallization figures from oats, wheat, potatoes and French beans when experimentally sprayed with Preparation 501 (Klett 1968)

Treatment	Crystallization figure				
	oats	wheat	pota-toes	French beans	average
untreated	61	45	63	67	59
2×500	65	55	68	74	66
— 3x501, middle, morning	76	62	82	82	76
2x500 3x501, middle, morning	95	87	96	99	94
2x500 3x501, middle, afternoon	67	60	76	75	70
2x500 3x501, early, morning	93	80	98	100	93
2x500 3x501, late, morning	78	75	82	79	79

Preparation 500 was applied after sowing. Early, middle and late refer to three stages of growth (with cereals, and analogously with other crops: the 2-4 leaf stage, the 4-leaf stage to heading, and heading to the milky stage).

The figures must not be evaluated as absolutes but as relative to one another. They indicate a series from qualitatively better (high figure) to qualitatively lower (low figure). The calculated averages must be seen in the same light.

The table shows that both preparations had favorable effects on the crystallization figures. Treatment with 501 in the morning turned out to be more favorable than applications made in the afternoon. Treatment at the early or middle growth stages was more effective than at the later stages. Moreover the experiment included up to six treatments with 501 that were extended over a longer period of time. It became particularly clear that evening treatment is less effective than morning treatment. Klett summarizes the results:

According to the crystallization method, the time during the growth period when Preparation 501 is most effective is between the first leaf development and flowering, or with cereals until the elongation of the stalk commences. Judging by yields and partly by the results of analyses, optimum effects are achieved by spraying when the leaves have reached full development. With regard to yields, spraying during the earlier growth phase is not quite as effective as spraying at maturity. The analytical results suggest opposite directions of effects resulting from early spraying and spraying during the ripening phase respectively. Apart from a number of opposing tendencies (free amino acids, electric conductivity, lipase activity) it is also significant that spraying during the latest phase (particularly with potatoes and wheat) brings about a decrease of crude protein and an increase of relative protein content, and also a decrease in proteinase activity and an increase in the total vitamin C content. We can conclude from this that Preparation 501 does not simply have one specific effect but rather affects those physiological processes that are typical for the particular phase of development the plant has reached. This is not contradicted by the fact that the crystallization figures, which are indicative of quality, showed more favorable results after early than after late spraying. The result of diagnosis by the crystallization method is not a function of certain physiological processes or substances but a correlate of vitality and/or the degree of its typical expression in the plant.

4   The Creation of Quality in the Plant Between the Earthly Unfolding Forces and the Cosmic Forming Forces

The examples from bio-dynamic research discussed above have shown how different kinds of environmental influence can bring about similar effects. The plant lives between the two main poles of its environment, the earthly and the cosmic. In the interplay between the two it has to unfold and mould its form and its substances. The working of the two poles is mediated to the plant by various substances. In accordance with Steiner's agricultural lectures this situation may be described as follows:

The cosmic light pole works in a formative way in the realm of substance and form. This working is mediated by air and warmth, in the soil and also partly in the air also by the silica substances.

The life processes take place through protein, which consists of five basic elements.

The earthly dark pole works in the plant to produce more plant matter so that the form is filled with matter. This working is strengthened by water, humus, calcium and kindred substances (Ca, K, Na, Mg, etc.). On the whole, soluble salts strengthen the working of water.

The ratio between the various components of protein mirrors the influence of the two poles. Four of the five elements of protein come from the soil (N, S, O, H). Nitrogen is more than any other the bearer of the earthly influences. Too little and also too much of this element can be a disadvantage for the plant. The amount must be in proportion to the plant's ability from its cosmic side to work on the elements it takes in. If the plant contains many untransformed or only partly transformed nitrogen compounds such as nitrate, amides and free amino acids, this indicates an imbalance towards the earthly pole. A high relative protein content, on the other hand, must be regarded as having been brought about by the cosmic pole.

The main bulk of plant substance consists of carbohydrates: sugar, starch, cellulose. They are formed by sunlight from the carbon in the air (CO_2) and the water in the green leaves. The cosmic forces stimulate the ripening process in the plant. Carbohydrates and fats are produced and from these derive aroma, flavor and color causing substances. If the plant has sufficient light it completes its ripening process undisturbed and this will also bring about better keeping qualities. If this process is accelerated, premature ripening takes place that in turn impairs the quality. If ripening takes place under the onesided influence of the earthly processes, it is incomplete and among other things there is insufficient inhibiting of the hydrolyzing enzymes. This also impairs keeping qualities.

The simultaneous influence of several factors

Since positive as well as negative effects are brought about by the working of several factors, we must ask how the combination of various factors affects the quality characteristics of the plant. This question cannot be entirely answered as yet. In polyfactorial experiments, however, cumulative effects as shown in the following table were found to occur.

The cosmic, light influence is made stronger by:	The earthly, dark influence is made stronger by:
sandy soil	very rich humus soil
dryness	moisture
warmth	lack of warmth
organic fertilizers	forcing with easily soluble N fertilizers
compost (more than unfermented manure)	P and K fertilizers (less than N fertilizers)
Preparation 500	
Preparation 501	

In this table, Preparations 500 and 501 are in the left-hand column. Organic fertilizers are also in this column. This is not a contradiction of the table on p. 209, although there Preparation 500 appears on the "terrestrial" side and Preparation 501 on the "cosmic' side. It must be remembered that between the two poles there is a sliding scale. This particular table lists the more extreme earthly factors on the right, all of which are ameliorated by both preparations and by organic fertilizers.

All these factors can be taken into account by the grower of crops. Adapting to the environmental conditions is to a great degree possible by choosing suitable varieties. Once plants are sown, of course, it is less easy to make adjustments. The weather is always an incalculable factor. But even this can be partly balanced by measures such as irrigation and additional fertilizers. Preparations 500 and 501 help to improve quality when the weather is damp and cold.

One of the phenomena often observed in practice is the gradual improvement in quality that takes place during the changeover of a farm or garden from ordinary to bio-dynamic methods. This period of gradual improvement usually lasts for 3-4 years, after which a quality level is established that does not vary much from year to year. Put simply, this phenomenon is caused by the progressive accumulation of positive influences. In fact, though, the process is more complicated because a sick organism grows well again according to other laws apart from effects that are purely additive.

Since the different factors contributing to quality work at varying intensities, there are in practice different quality levels. There is therefore no uniform bio-dynamic quality. The examples quoted show that bio-dynamic measures in planting, fertilizing and treatment with the preparations enable considerable improvements in quality to come about. Soil conditions, the climate and other given factors, on the other hand, change from place to place. The fertilizing methods and the preparations have to work on these differing foundations. It is quite possible for a comparison of bio-dynamic methods with conventional (NPK) methods in different conditions to yield results like those shown in a simplified form in the table below, which are taken from an

Quality improvement	Site 1	Site 2	Site 3
	BD		
		BD	
	NPK		
			BD
		NPK	
			NPK

experiment by Pettersson (1970). The indications refer to the quality characteristics measured, which needless to say did not include the whole spectrum.

The conditions of soil and climate decreased from Site 1 to Site 3 in their suitability for a particular crop. Throughout there was a difference between BD and NPK treatments. On Site 2 the BD quality was still better than the NPK quality on Site 1. But on Site 3 the initial conditions were so bad that the BD quality was lower than the NPK quality on Site 1. If we add to this the point that during the transition period of 3-4 years the BD quality gradually improves until it has reached the maximum possible for a particular locality, it becomes obvious that there can, in fact, be variations of bio-dynamic quality. This is easily forgotten when products from highly different areas are considered, and it has even been used in attempts to prove that bio-dynamic methods are not effective. This is of course unfair. If bio-dynamic measures are properly applied and the results compared with ordinary measures on similar habitats, the bio-dynamic products are of a higher quality. The above example shows, however, that fundamental amelioration of a habitat is as important for high-quality production as is the improvement of fertilizing measures.

For the sake of completeness it should once more be pointed out that in the above discussion the term "quality" applies to the present stage of development of the quality concept and of the testing methods that have become available so far. Work both on the thought content of the concept of quality and on methods to determine quality is unceasing.

Can quantity and quality be combined?

It is the aim of bio-dynamic research to point the way to the production of high quality foods. In the practical situation there are usually more ways than one of achieving this. With the aim of the product he wants constantly before him, the farmer or gardener has to decide which way to choose.

One of the recurring questions is: Quantity or quality? The bio-dynamic grower is interested in the first place in quality. The economic viability of his work, however, also depends on quantity. Since bio-dynamic work started in the twenties, the demands on field and stable in terms of yields have approximately doubled. Bio-dynamic producers are faced with the problem of remaining competitive despite their different attitudes and aims. Yields from bio-dynamic farms have risen at approximately the same rate as those on other farms. The ever-recurring question is: Can we increase the yields still further and yet combine this with the quality achieved hitherto? In essentials, research over the last fifteen years and also on-going analysis of products in practice have found that this question can be answered in the affirmative. The measures that work positively toward quality are today better known than before, particularly in the degree to which they work. To apply them at the right time and in the right way in different places is one of the most important tasks of farm management.

5 Quality Grading

There has been no lack of attempts to classify the characteristics of agricultural products according to the most varied viewpoints, e.g., purpose, commercial grade, etc. In the following we outline a system based on Schuphan's scheme (1961), which included the properties of the products as foods and also points regarding growing and storage.

Suitability of the cultivar. This comprises above all those characteristics that play a part in the growing process, e.g., suitability as seed, cuttings or tubers, etc., resistance to viruses, bacterial and fungal infection and also pests.

Keeping properties. This comprises the ability of a product to be stored and transported with the least possible losses of weight, shape, consistency, freshness, food value, etc., as well as resistance to damage by various parasites. These properties have been rather neglected hitherto because they have been regarded as above all technical problems. They are so only to a certain extent. If the product is insufficiently ripe physiologically, which can easily be the case with N fertilizers, keeping qualities are poor and this soon shows in transport and storage.

Commercial grading. This classification is well worked out and more clearly defined. It denotes the properties that will make it easier to conduct trade with the products. It comprises a number of external qualities such as size, shape, color, freshness, ripeness, consistency, volume weight, absence of damage and foreign substances. The latter include residues from herbicides and pesticides that can considerably diminish the usability of a product or even be poisonous. Commercial grading has only recently become really interested in residues. They are among the negative properties of a product.

Technological quality. The qualities required for processing are in many ways the same as those taken into account for commercial grading. There are in addition a number of inner characteristics that ease processing and make the produce more valuable as a food.

Nutritional quality. In none of the sectors of quality grading mentioned so far is it so difficult to find criteria acceptable to all as in the matter of nutritional value. The same applies to fodder though it is somewhat simpler here because certain definite aims are pursued. The final criterion for the food value of a product does not lie in the product itself but in another organism. It is the reaction of this other organism that indicates the food value. All the standards of modern nutritional science have been gained by exmining the relationship between the composition of a food and the reaction of the organism that eats it. But many questions remain open. One to be put to a modern science of nutrition is whether it is right never to regard a food as a totality. On the whole, nutritionists do not regard a carrot as a carrot but as a bearer of nutrients such as sugar, carotin, etc. In a similar way Schuphan himself defines the nutritional quality of a product as "the sum of all its positive attributes minus the negative ones." So far, no mention has been made in the field of nutrition of effects other than those that can be detected by chemical analysis. Nor is there any discussion regarding the totality of a product that is expressed among other things in the ratio of different substances contained in it, a matter that is, after all, not without consequences for the animal or human organism eating it. In fact, people do choose their foods as totalities rather than according to the definitions made by nutritionists. Nutritional science needs to take fully into account the concept of wholeness. Every food plant represents in its shape and in its ratio of substances a certain type that can be only partly described chemically. Understanding how this type is formed under the various influences of the environment is one of the tasks of bio-dynamic quality research.

This is also the background against which to judge the efforts to develop new methods of research and evaluation. It is not a matter of doing without the results of modern nutritional research. On the contrary, if the details discovered in this research can be viewed in conjunction with the totality, they can themselves also be better understood.

It is not always easy to meet in full all the quality expectations demanded of a particular product. Indeed, it is not unusual that different quality expectations conflict with each other. Most frequently this conflict probably lies between the demands for certain external characteristics (size, uniformity, color, etc.) and demands for the highest food value. With regard to the latter there need be no difference between a tomato that is perfect in shape and size and one that is perhaps smaller or irregular in shape. The perfection of one quality characteristic often means that another has to be neglected. In such a situation, which requirement is the more important? This can only be decided on the basis of the use to which the product is to be put. Characteristics that have only an indirect bearing on this are of secondary importance.

For the grower it is important to aim for high biological value in the quality of his products. The methods he uses must be planned to this end. If necessary he can discard any individual pieces that do not correspond to the usual sizes required. But the biological qualities can hardly be improved by means of sorting! Among these biological properties are, for instance, resistance to parasites, also in storage, a high content of flavor-enhancing substances, and also high food value. Experience and experiments have shown that if the biological quality in one of these characteristics is good, this is often the result of a desirable general biological quality. The product will thus also possess a number of other biological characteristics that are desirable. This is frequently the case, though not always. Our knowledge of the way all these things are linked together is constantly being added to, which means in practice that quality research has to follow up a great many details. To assess quality and find the criteria for cultivation methods is less complicated. The foundations for a number of important qualities in the product can be laid with relatively few measures so long as they are the right ones.

Chapter Ten

Bio-Dynamic Production and the Consumer

As mentioned in Chapter One, bio-dynamic products were brought onto the market in Germany in 1928, only four years after the work began, by the Demeter Cooperative (Verwertungsgenossenschaft Demeter), which had its headquarters in Bad Saarow in Mark Brandenburg. In 1930 the organization was reconstituted as the Demeter Trading Association (Demeterwirtschaftsbund), which continued to function until it was banned in 1941. The adoption of the trade mark *Demeter* was an important step. This was in all probability the first attempt to market basic foodstuffs that offered more than the conventional market qualities encompassing in the main only appearance and certain cooking characteristics. The name Demeter indicated the biological value of the products, arising out of a particular production method. In those days the number of people interested in such foodstuffs was limited but today there is a wide demand for foods that are qualitatively good.

These consumer requirements apply not only to products that travel straight from the producer to the kitchen table but also to those that have been processed or stored for longer or shorter periods. Therefore, not only the producer but also the processor and distributor influence the quality. So if high quality products are to come onto the market, the services and interests of grower, processer, distributor and consumer must all be taken into consideration.

In the Federal Republic of Germany it is the task of the Demeter Association (Demeterbund e.V.) to develop this area and create secure legal foundations. In 1954 this Association took up the work that had had to be abandoned for awhile in 1941. Its work will be briefly described here, since it is both in commercial and idealistic terms directly linked to the endeavors of bio-dynamic farmers and market gardeners. The producers expect to receive fair prices for their products. These can nowadays, however, not be high enough to make economics their sole motivation. It must be recognized that they are also motivated out of devotion to their work. Experience has shown repeatedly that the agricultural code of ethics is a reality provided that the way in which the work is carried out is able to foster this ethic. The consumer, on the other hand, must be able to rely on the quality offered being as high as he wishes. So the interests of producers and consumers must be coordinated. Based on the work achieved so far, the Demeter movement provides a possible pilot for such coordination. The necessary discernment on which the consumer can base and increase his confidence in what he buys still needs further developing.

Environmental conservation is a matter much in the public eye at present. It has many points of contact with agricultural and horticultural production methods and also with processing. Environmental problems have stimulated the interest of wider circles than ever before in residue-free and also "organically" grown food. Words like "natural," "pure," "organic," and many others are on everybody's tongue. There is a danger in this. Over a number of years interests of the most divergent kinds, based both on understanding and on the crassest lack of understanding, have been competing for attention. To date, this applies more to some countries than others. We must now see to it that a justified striving for high quality can win through. Otherwise it could become discredited. Availability of information about production methods on the one hand, and production according to properly supervised quality guide-lines on the other, can contribute to the generation of confidence in the quality of the produce offered.

Bio-dynamically grown products must reach the consumer either directly or after processing without any chemical preservatives or cosmetics. Marketing under the brand name and trade mark of "Demeter" guarantees this. The owner of this trade mark in the Federal Republic of Germany is the Experimental Circle for Bio-Dynamic Agriculture and Horticulture (Forschungsring für biologisch-dynamische Wirtschaftsweise). This circle encompasses farmers, scientists and consultants who want to take responsibility for the further development and dissemination of the bio-dynamic method. Its members develop the guidelines for production and processing of Demeter products. Consequently, these guidelines stem from on-going knowledge and quality research and not from a group consisting only of producers or distributors.

The only legal entity in the Federal Republic of Germany entitled to put into effect Demeter guidelines and standards is the Demeter Association. This fulfills its tasks through the collaboration of its members who are producers, processors, distributors and consumers. The executive committee consists of representatives of all four groups. The Demeter Association itself does no trading. Its activities are borne by support contributions that are incorporated in the final price of products. The income from these support contributions is used not only for administration and the regular issue of the Demeter News Letter, but also for quality control, research and advisory services to member farms and market gardens.

The above, as described for the Federal Republic of Germany, also applies with minor modifications to other countries in which the Demeter name and trade mark are protected. Regular conferences of all groups in European countries serve to bring about standardization of guidelines and labeling.

Protection of Demeter standards is enforced by covering contracts between the Demeter Association and producers, processors and distributors. Contract partners are entitled to use the protected Demeter name and trademark and the words "Produced by bio-dynamic methods" or "Produced by bio-dynamic cultivations." There is an annual harvest registration regarding proper application of the methods. A contract can be entered into once a representative of the Experimental Circle has pronounced the products in question to be up to the required Demeter standards. Similar contracts are entered into with processers and wholesale distributors. The production program is enlarged and controlled in collaboration with the Working Group for the Processing and Distribution of Demeter Products. We have therefore what is probably a unique situation. The protection bestowed by the Demeter name extends from the soil to the kitchen table.

The Institute for Bio-Dynamic Research in Darmstadt runs a testing service that carries out systematic quality control tests. Moreover, random samples of primary and secondary products are sent to the experimental stations to be tested for chemical residues. The work of the testing service also assists the Experimental Circle's advisers in their task of helping the producers to improve and maintain high standards. The samples sent in by the producers are accompanied by a report on the growing method used. In this way a body of experience on varieties, crop management and quality characteristics is gradually built up that the consultants can combine with their own soil analysis and knowledge of the individual farms.

The above points have been mentioned for the sake of completeness. The central question we are concerned with here is: How can one determine quality in a reliable fashion? This is certainly not easy. Yield can be measured, and so, to a sufficient degree of accuracy, can pesticide residues, lead or other poisons. The question of cost determines how many samples one can have tested. But it is not easy to determine the nutritional quality of foods in their physiological and health aspects. Here again cost prevents the routine laboratory investigation of numerous individual factors. There are, of course, methods of testing protein content, relative protein content, sugar, fat, unsaturated fatty acids, vitamins, minerals, keeping properties, enzyme activities and much else. Apart from the cost of such tests, the evaluation of the results is a problem in itself. Each characteristic is influenced by a variety of growth conditions. In one year weather conditions might be dominant, and in another the manures used. Rational quality control must take the totality of growth conditions into account. These are determined in the first place by the fertility and vitality of the soil and are influenced by manuring and crop management in general. The quality characteristics, too, must be seen in their totality. The manner in which orthodox analytical methods and the more recent "picture-creating" methods are applied to this end is described in the previous chapter.

In practice the totality of growth conditions is taken into account in the definition of standards (to which the above-mentioned quality and residue tests are added). Crop management takes place along lines that experience and the results of present research show can guarantee balanced growth conditions. Without going into detail, the main features of the guidelines as they are applied at present are set out in the following.

The first condition for recognition of Demeter quality is that, in the main, bio-dynamic production methods should have been in use for not less than two years. Fertilizer use must be aimed at building up soil productivity. This must be achieved through the application of farm manures with bio-dynamic preparations in such a way that neither organic nor inorganic forcing manures are required. Commercial organic fertilizers should, as far as possible, be applied via compost, and their manurial value must not exceed that of the manure derived from the farm itself. Synthetic nitrogen compounds, soluble phosphates, and chlorine containing potash salts are out of the question. Sewage sludge, physiologically questionable dressings, other pesticides and weed killers are not used. The purchase of fertilizers and soils must be undertaken only in agreement with the representative of the Experimental Circle who is also a trustee of the Demeter Association. Similar conditions apply to horticultural products. Here, too, if recognition is sought, the manurial value of bought-in organic fertilizers must not exceed that of the garden's own compost, green manure and stable manure. Bought-in young plants must be bio-dynamically cultivated. If necessary, any additional bought-in materials may have to be analyzed for residues. With regard to livestock, the farm must carry out its own breeding program or buy in animals only from other bio-dynamic farms (except male animals, of course). The proportion of bought-in feedstuffs in the total ration is laid down, and feeding is aimed at lasting good performance, fertility and health rather than at record yields, which anyway tend to be short-lived.

The system of annual reconfirmation of recognition presupposes a close collaboration between the farms and the advisory service. As will have been seen in earlier chapters, importance is attached to the development of each farm as a biological unit. Variety of cropping and the combatting of weeds by crop rotation and tillage means there is extra work. The carefully considered and progressive use of modern equipment can ease this in a number of ways. The price structure can also redress the balance to a certain extent. The overall state of agriculture, however, also has its consequences for bio-dynamic farms. The general aims of agriculture and its extension services at the present time with regard to agrarian structure and productivity are in many ways opposed to those of Demeter farms. Whether the Demeter aims, as they stand at the moment, are the final ones, must in every respect remain open for discussion. The bio-dynamic farms and Demeter organizations are in their own way and within their own possibilities developing models for the realization of an agriculture that is guided by ecological principles, that is favorable for the environment, that encourages people to remain on the land, and that achieves the production of high-quality foods as free as possible from undesirable residues.

Chapter Eleven

Conclusion and Future Prospects

I

When an existing situation has become questionable and new goals are called for, we usually speak of a crisis. A crisis is insecurity experienced. Agriculture and horticulture are now going through a severe crisis and this shows signs of becoming a prolonged one. Also, it is one that has many facets. In seeking for a common denominator for its numerous symptoms we find that the interplay between the parts and the whole is disturbed. Here are only a few examples of this; more could be mentioned.

Production schemes have achieved high yields per acre and man hour, but they are questionable with regard to their affects on the ecosystem, the raw material and energy input, and the nutritional value of feed and food. There are limits to the possibilities of achieving high profits or—as is also said—income parity with industry by means of specialization in profitable enterprises or by increasing farm sizes. In the rich countries the rural districts are becoming depopulated, while the cities and their problems grow. "Market and income oriented" production often results in grain or butter or beef "mountains" that confuse the market and make consistent farm planning impossible. Government interference creates untrue prices and aggravates the problem. In short, in the industrialized countries the

397

approach to agriculture is beginning to defeat some of the goals initially envisaged, not to mention its inapplicability to the problems of the Third World. The symptoms now obvious include not only the excessive use of agricultural chemicals. The crisis, rather, is a structural one, and includes the single farm and the position of farming in the social fabric. In the rich countries the rifts between rural, urban and recreational districts are widening. Finally, the crisis includes man. When his interest is absorbed in a onesided way by economics and technology, important human values are lost. The ethical foundations of the farming profession remain undernourished.

At the beginning of this book the first paragraph of von Thaer's "Principles of a Rational Agriculture" was quoted. In the second paragraph he says: "The higher these profits continue to be, the more is the purpose of agriculture fulfilled. The most perfect farm is the one that makes the highest possible continuous gains in relationship to its potential, strength, and circumstances." In other words, production must not be achieved at the expense of the production potential. Living in the early days of modern economic thinking, Thaer stresses profitability, but he does not mean exploitation. Today it is a matter of putting his principle into practice in all its numerous and varied implications. Indeed, there is a growing awareness of the fact that ecological and economic interests are linked in both the private and the public sectors. This is one of the reasons why regional planning authorities for rural, residential and industrial development are set up in many states and countries. But one should admit that mere analytical-technical thinking is bound to arrive at faulty answers. Even the most careful analyses—which are also needed of course—will not allow for the correct weighting of all parameters. Only a way of thinking oriented towards biological categories can cope with this complex situation. This kind of thinking conceives the idea of the farm as an organism.

The mixed, largely self-sustaining farm is the basic unit that in turn sustains a healthy continuously productive landscape. This holds for the greater part of all agricultural land the world over. If this archetypal cell is destroyed, sooner or later ecological, economic and sociological damage is bound to occur. Some of it is irreversible. On the whole the number of bio-dynamic and also of good organic farms is small. But their importance exceeds by far their relative number. The latest developments in general agriculture underline their function as models. Bio-dynamic farms have become pilot projects. They demonstrate—and many of them have been doing so for 20, 30, 40 or more years—the answers that work, based on a deepened understanding of the life processes, can provide to many of the grave problems plaguing contemporary agriculture. They prove that the spiritual insight, on which Steiner's lectures in 1924 were based, contains solutions that are needed now. Not merely theories are offered, but rather an approach that has stood the test in practice. If a farm or garden is organized according to bio-dynamic principles, a number of negative side effects and weaknesses of present-day production methods give way to more positive results:

- Starting from the soil, the quality of products is improved; these are as free of residues as possible since biocides are not used.
- The farm or garden does not add to the load of pollution that penetrates the ecosystem.
- The biological cycles of substances and forces are closed and put to their optimum use.
- The communities of flora and fauna within the farm are stabilized.
- The processes of self-renewing productivity are strengthened.
- The present wastage of natural resources and energy is reduced to a reasonable level.
- The production becomes varied. This encourages a more regionally oriented provisioning of the population that leads to contacts and mutual understanding.

This regional pattern of providing for the people benefits the rich and the developing countries. There remains enough space for trade and relief shipments. But production will show fewer fluctuations. According to available figures, yields on bio-dynamic farms compare favorably with those of their neighbors.

Bio-dynamic farms do not have all the answers. The method needs further development, but they have worked out important alternatives in practice, not in theory.

II

The future development of agriculture is a concern of society as a whole, not just of farmers and gardeners. Emancipated agriculture, privately run, is a rather recent new development in Europe but also in large parts of North America. Land laws, tradition and the farmers' sense for what is biologically right maintained order in the past. Mistakes leading to the destruction of soils and forests were made in earlier times too. The damage is permanent, though it was easier to cope with it when the population was less dense. Today three factors coincide that have hitherto not arisen with such intensity. They are:

1) The demand for food is increasing rapidly;
2) Economic individualism;
3) Technical and chemical possibilities to manipulate life processes have grown to unprecedented dimensions.

These factors have set dangerous developments in motion. Therefore it will be necessary:

1) To focus attention on the welfare of the agricultural means of production: soils, animals, plants. This can be achieved by organizing ecologically sound farms, and nurturing an image of farming that is based on a deepened understanding of life as a prerequisite for responsible action.
2) To harmonize the interplay between agriculture and the other sections of the economy, so that it becomes conducive to sound farming. More than in times past, the fate of farms and their members is at the mercy of the rest of the economy, i.e., does not depend on the farmers' image of farming but on that of consumers, businessmen, proces-

sors, politicians, scientists, industrialists and journalists.

Agriculture is expected to produce enough food for all while at the same time preserving the landscape, its water reserves, its flora and fauna, and its aesthetic value. The national heritage of the land and all that grows on it has been entrusted to the care of farmers, gardeners and foresters. Apart from the fact that these people have to earn their living, they thus also have a public function to fulfill in maintaining the self-renewing production capacity of the land. This "task for the common weal" must be emphasized all the more strongly the more the three factors mentioned above are at work. It cannot be the task of the farmer alone to meet this situation. If economic pressures force him into increasing specialization, which damages the environment, there is no point in demanding that he should do his ecological duty. Contributions to the task must also be made by training plans at all levels, agricultural advisory services, and also economic and structural measures taken by the government. It will be essential that agricultural colleges include alternative approaches in their programs.

III

On the one hand the depopulation of the farms is progressing. (This need not necessarily be the same as the depopulation of the flat land.) On the other hand there is a growing movement toward the land. In Europe this often takes the form of part-time farming (Priebe 1972). The longing for a secure investment sometimes plays a part when people acquire a piece of land, though this is by no means the sole motive. There is a spirit of search coming from deeper levels.

To encourage the depopulation of farms by increasing rationalization and mechanization beyond a reasonable level is a fallacy. This holds for rich countries. In developing countries an intermediate approach offers the best chances for employing people and depending largely on local resources. Bio-dynamic farms have for many years been unable to accommodate all the potential apprentices. Varied mixed farms, although they have to make full use of modern implements, need more workers. Various examples exist that show how this can be economically feasible. Such farms, some of which also do a limited amount of processing, have been found to develop well when linked to cultural and charitable institutions. Both members enjoy economic benefits. Similar forms of cooperation could develop in connection with industries. The tendency toward mixed farming is encouraged wherever there are direct sales outlets for at least part of the production. This situation stirs interest in the quality of products and leads to closer contacts between the producers and consumers. On a much larger scale this occurs when products are marketed under a trade mark. The Demeter organization provides a functioning model for this approach. The products come from mixed bio-dynamic farms and gardens. Finally, there exist on larger commercial farms examples for partnership farming, which also leads to program diversity. The demand that more rather than fewer people should find a livelihood and meaningful work on the land is not illusory but backed by practical experience.

These matters reveal still another much deeper layer. Many people from all walks of life are today seeking work on the land. They often have only a vague feeling of what they want. But they reject or have no means of relating to the piecemeal kind of life offered in modern industrial and bureaucratic society. Seen from one aspect, this is a marvellous fact; material goods or social prospects provide no answer to these souls for their questions about life. Where can people find meaningful work? They hope that caring for the land, working with plants and animals, the rhythms of the seasons and the stars, they will be dealing with entities that will interest them in quite a different way from the functioning of dead objects. The work to be done is ordered by the processes of life and growth through the seasons. It is a realm that can be viewed comprehensively and understood and to which one can feel personally related.

The relationship of the parts to the whole needs to be put in order not only at the level of the means and processes of production. Man, too, belongs to the totality of agriculture, which would not exist without him. From time immemorial his work has transformed natural wilderness into cultivated land. For doing this work in the past he developed adequate forms of social life some of which were better and some not so good. But they were rooted in the wisdom that underlies the processes of life. Nowadays the relationship between man and nature has grown shallow and onesided; it does not nourish depth of inner life. We have today the astonishing fact that the ecological crisis and the human crisis are being recognized simultaneously, though the latter, which relates to the inner substance of work on the land, is experienced only by a minority. This is, however, a minority that cannot be overlooked. Meaning cannot be given to work by extracting more goods out of shorter and shorter working hours. The manner in which mere technological skills and attitudes can cope with the realms of nature is more than a matter of survival. It stirs deeper levels of man's being. For some it has become an existential question. Answers can be found when one searches for the spirit in nature and man. This quest helps create the determination to embark on a form of land cultivation that is biologically right, that improves the quality of the environment and of the produce, and also gives real value to the farming profession. This, too, is borne out by the remark of Rudolf Steiner mentioned in the first chapter of this book: "The course (of lectures) will show us how intimately the interests of agriculture are bound up, in all directions, with the widest spheres of human life."

Bibliography

ABELE, U.: Vergleichende Untersuchungen zum konventionellen und biologisch-dynamischen Pflanzenbau unter besonderer Berücksichtigung von Saatzeit und Entitaten. Diss. Giessen 1973.

ADVISORY LEAFLET 320 Ministry of Agriculture, Fisheries and Food: Poultry Manure. HMSO, London 1969.

AEHNELT, E., und HAHN, J.: Beobachtungen über die Fruchtbarkeit von Besamungsbullen bei unterschiedlicher Grünlandbewirtschaftung. Experimentelle Pflanzensoziologie, Den Haag 1969.

AGRIC. RES., Stubble Mulching. USDA, Washington, D. C., July 1964.

ALBRECHT, W. A.: Soil Fertility and the Human Species. Amer. Chem. Soc., Chem. Eng. News 21, 1943, 221-227.

Put the Cow Ahead of the Plow. Guernsey Breeders Journal 84, 1173-1177, 1952.

Physical, chemical and biochemical changes in the soil community. In: Man's Role in Changing the Face of the Earth. Univ. Chicago Press, 648-673, 1956.

Soils - their effects on the nutritional values of food. Consumer Bull. 44 (1), 20-23, 1961.

ALDRICH, S. R.: Supplemental Statement in the Matter of Plant Nutrients. Illinois Pollution Control Board, R 71-15, March 28, 1972.

ALLISSON, F. E.: The enigma of soil nitrogen balance sheets. Adv. Agr. 7, 213-250, 1955.

The fate of nitrogen applied to soils. Adv. Agron. 18, 219-258, 1966.

APPELL, H. R. and FU, Y. C.: Converting organic Wastes to Oil. Agric. Eng. (ETHZ) 53, No. 3, 17-20, 1972.

BALFOUR, E. B.: The living soil. Faber and Faber Ltd., London 1943.

BARTHOLOMEV, W. V.: Maintaining Organic Matter. Yearbook of Agric. (USA), Yearb. Sep. No. 2814, 245-252, 1957.

BIELENBERG, H.: Wer baut richtig?. Leb. Erde 2, 1959.

BOCKEMÜHL, J.: Gartenkresse, Kamille, Baldrian. Elemente Naturwiss. 11 (2), 13-28, 1969.

Entwicklungsbilder zur Charakterisierung von Löwenzahn und Brennessel. Elemente Naturwiss. 12 (1), 1-14, 1970.

Beobachtungen am Pflanzenwachstum auf Erden mit Kompostzusatzen aus Stadtmüll und Klärschlamm. Elemente Naturwiss. 15 (2), 21-32, 1971.

et al., Elemente der Naturwissenschaft, Hefte 1-18, Dornach (Schweiz), 1964-1973.

BOGUSLAWSKI, E. v.: Zur Entwicklung des Begriffes Bodenfruchtbarkeit. Z. Pflanzenernahr. Düng. Bodenkunde 108 (2), 97-115, 1965.

BOOS, R.: see Steiner 1957.

BORNEFF, J., KUNTE, H., FARKADSI, G., und GLATHE, H.: Krebs durch Benzpyren in natürlichem Dünger. Umschau 73 (20), 626-628, 15. Okt. 1973.

BOSSE, J.: Ein Versuch zur Bekampfung der Bodenerosion in Hanglagen des Weinbaus durch Müllkompost. Weinberg und Keller 15, 385-397, 1968.

BREDA, E.: Unterschiedliche Ausbildung von Spinat im Frühjahrsanbau und Herbstanbau. Bericht über Arbeiten aus dem Institut fur biologisch-dynamische Forschung. Leb. Erde 3, 1972.

BRIEJER, C. J.: Silberne Schleier. Verlag Biederstein, Munich 1970.

BRONNER, H. und JANIK, V.: Bodenkundliche Untersuchungen bei rinderhaltenden und rinderlosen Zuckerrübenbaubetrieben in Oberösterreich. Die Bodenkultur, 25, 223-251, 1974.

BRUGGER, G.: Die Aufbereitung von städtischen Siedlungsabfällen und ihre Verwendung im Weinbau. Rebe und Wein, 344-348. 1966.

BÜNNING. E.: Die physiologische Uhr. Verlag Springer, Berlin 1963.

CARLGREN, F.: Rudolf Steiner, 1861-1925. Philosophisch-Anthroposophischer Verlag am Goetheanum, Dornach, Schweiz, 1972.

CHRISTEWA, L. A.: On the effect of humic acids on plants. Rep. All. Union Acad. of Agri. Sci. im. Lenin, 7, 29, 1948. Quot. from Krasilnikov (1961).

CIRCULAR 777: The Morrow-Plots, University of Illinois 1960.

CLARK, F.: Soil microorganisms and plant roots. Advan. Agron. 1, 241, 1949.

CLOUDSLEY-THOMPSON: Rhythmic Activity in Animal Physiology and Behaviour. Academic Press, New York, London 1961.

CORRIN, G.: Handbook on Composting and the Bio-Dynamic Preparations. Bio-Dynamic Agric. Assoc., London 1960.

DAVY, J.: et al: Work Arising from the Life of Rudolf Steiner. Rudolf Steiner Press, London 1975.

DAWES, J. H., LARSON, Th. E. and HARMESON, R. H.: Nitrate Pollution of Water. Intern Report, University of Illinois 1967.

DEBRUCK, J.: Ewiger Weizenanbau ein Geschäft? Landw. Zeitschr. Rheinl. Bonn, 36, 1640-1642, 1973 (Quot. from "Kurz und Bündig").

DLG, Legehennen, Aufzucht und Haltung. DLG-Verlag, Frankfurt 1966.

DOMSCH, K. H.: Microbial Stimulation and Inhibition of Plant Growth. Transact. 9th Internat. Congr. Soil Sci. III, 455-463, 1968.

ECKART, W. v.: The Challenge of Megalopolis. Washington 1964.

EDMOND, J.B., SENN, T. L., and ANDRESS, F. S.: Fundamentals of Horticulture. McGraw Hill, New York 1964.

EGNER, H., and ERIKSSON, E.: Current Data on the Chemical Composition of Air and Precipitation. Tellus 1, 134-139, 1955.

ELLENBERG, H., SCHREIBER, K.-F., SILBEREISEN, R., WELLER, F., und WINTER, F.: Grundlagen und Methoden der Obstbau-Standortskartierung. Der Obstbau 75, 3-12, 1956.

ENGQUIST, M.: Strukturveränderungen im Kupferchloridkristallisationsbild von Pflanzensubstanzen durch Alterung und Düngung. Leb. Erde 3, 1961.

Pflanzenwachstum in Licht und Schatten. Leb. Erde 2, 1963.

Gestaltkräfte des Lebendigen. Vittorio Klostermann, Frankfurt am Main 1970.

FARKADSI, G., GOLWER, A., KNOLL, K.-H., MATTHESS, G., und SCHNEIDER, W.: Mikrobiologische und hygienische Untersuchungen von Grundwasserverunreinigungen im Unterstrom von Abfallplätzen. Städtehygiene 2, 1969.

FERRIS, V. R. and FERRIS, J. M.: Inter-relationships between nematode and plant communities in agricultural ecosystems. Agro-Ecosystems, 1, 275-299, 1974.

FLEIGE, H. und CAPELLE, A.: Feldversuche über den Verbleib von markiertem Düngerstickstoff in Boden und Pflanze. Mitteilgn. Dtsch. Bodenkundl. Gesellsch. (Proc. German Soil Sci. Soc.) 20, 400-408, 1974.

FRANCK, G.: Gesundheit durch Mischkultur. Verlag Boden und Gesundheit, Langenburg o.J.
Frühling im Kraütergarten. Gartenrundbrief aus der biologisch-dynamischen Arbeit, Nr. 79, 1972, Pforzheim, Mathystr. 34.

FRIED, M., and BROESHART, H.: The Soil-Plant System. Academic Press, New York 1967.

FÜRST, L.: Untersuchuggen zur Erzeugung von Qualitätsobst. Arbeitsgemeinschaft zur Erzeugung von Qualitätsobst, Paderborn 1967.

FYFE, A.: Die Signatur des Mondes im Pflanzenreich. Verlag Freies Geistesleben, Stuttgart 1967.

GEIGER, R.: Das Klima der bodennahen Luftschicht. Verlag Vieweg u. Sohn, Braunschweig 1950, 1965.

GERMAN COMMISSION FOR UNESCO: Problems of the Rational Use and Conservation of the Resources of the Biosphere, Cologne, 1969.

GEUTTER, M.: Herbs in Nutrition. Bio-Dynamic Agricultural Association, London 1962.

GRIGORAKIS, Chr.: Erosionsprobleme in Griechenland. Z. Kulturtechnik und Flurbereinigung 8 (1), 1-12, 1967.

GROTZKE, H.: Mulching. Bio-Dynamics 79, 17-28, 1966.

HABER, W.: Principles of the Development and the Formation of the Entire Living Space. German Commission for UNESCO, Cologne 1969, 44-50.

HANCOCK B., and ESCHER, P.: Studies of litter management for growing and for laying Leghorn Fowl. Bio-Dynamics 74, 7-12, 1965.

HAURI, P.: Vergleich von Kompost mit verschiedenen Eisenpräparaten bei der Prophylaxe der Ferkelanämie. Diss. Zürich 1968.

HEINZE, H.: Vom biologisch-dynamischen Prinzip. Leb. Erde 45, 105, 151, 1965.
Vom ökologischen Aspekt des landwirtschaftlichen Kurses. Leb. Erde 45, 1968.

HEINZE, H., und BREDA, E.: Versuche über Stallmistkompostierung. Leb. Erde 2, 3-10, 1962.

HEMLEBEN, H.: Rudolf Steiner. Rowohlt Monographien, Hamburg 1963.

HERWEIJER, S., and DENIG, E.: Developing and Shaping the Entire Space. In: Problems of the Rational Use and Conservation of the Resources of the Biosphere, German Commission for UNESCO, 148-156, Köln 1969.

HESS, D.: Pflanzenphysiologie. Verlag Ulmer, Stuttgart 1972.

HEYNITZ, B. v.: Betriebsergebnisse 1936. Manuskriptdruck Heynitz über Meissen 1937.

Betriebsergebnisse 1937, Manuskriptdruck Heynitz über Meissen 1938.

Die biologisch-dynamische Wirtschaftsweise im Lande Sachsen, Privatdruck, Gromitz 1951.

HEYNITZ, K. v.: Wird das Grünland der Höfe richtig gepflegt und genutzt? Leb. Erde, 203, 1966.

HOWARD, Sir A.: An Agricultural Testament. O. U. P. 1940.

JAHN-HELD, W.: Über Herstellung, Eigenschaften, Wirkung pflanzenverfügbarer Kaliverbindungen mit "langsamer Losungsgeschwindigkeit" unter besonderer Berucksichtigung von Kalium-Magnesium-Phosphat. Diss. Giessen 1971.

IBN AL AWAM: Kitab al Felahah, 12th century (excerpts by A. Glyn-Jones) 1969.

ILLINOIS POLLUTION CONTROL BOARD: In the Matter of Plant Nutrients. R 71-15. March 28, 1972.

JONES, L. H. P., and HANDRECK, K. A.: Silica in Soils, Plants and Animals. Adv. Agron. 19, 107-149, 1967.

JUNG, L.: Beobachtungen und Untersuchungen über Bodenerosion in Thrazien und Westanatolien. Z. Kulturtechnik und Furbereinigung 3, 337-355, 1962.

KABISCH, H.: Die Anwendung der Präparate. Forschungsring für Biologisch-Dynamische Wirtschaftsweise, Darmstadt, o. J., 8. Edit.

KELLNER, O., und BECKER, M.: Grundzüge der Fütterungslehre. Verlag Parey, Hamburg und Berlin 1971.

KLAUSEWITZ, W., SCHÄFER, W., und TOBIAS, W.: Umwelt 2000. Kleine Senckenbergreihe, Frankfurt 1971.

KLEIN, J.: Der Einfluss verschiedener Düngungsarten in gestaffelter Dosierung auf Qualität und Haltbarkeit pflanzlicher Produkte. Inst. f. Biol.-Dyn. Forschung, Darmstadt 1968.

408

KLETT, M.: Die boden- und gesteinsbürtige Stofffracht von Oberflächengewassern. Arb. d. Landw. Hochschule Hohenheim, Bd. 35, 1965.

Untersuchungen über Licht- und Schattenqualität in Relation zum Anbau und Test von Kieselpräparaten zur Qualitätshebung. Inst. f. Biol.-Dyn. Forschung, Darmstadt 1968.

KOCH, W.: Unkrautbekämpfung. Verlag Ulmer, Stuttgart 1970.

KONONOWA, M.M.: Soil Organic Matter. Pergamon Press, Oxford, 1961.

MISHUSTIN, Y. N., and SHTINA, E. A.: Microorganisms and the transformation of soil organic matter, research results and tasks. Sov. Soil Sci. 4 (2), 202-212, 1972.

KOEPF, H. H.: Soil tests and chromatograms. Bio-Dynamics 69, 1-13, 1964.

Evaluation of soil productivity for management and other basic decisions. Bio-Dynamics 71, 3-14, 1964.

Experiment in treating liquid manure. Bio-Dynamics 79, 1-12, 1966.

The matter that matters. Bio-Dynamics 80, 2-17, 1966.

Die Bodenluft. Handbuch der Pflanzenernahrung und Düngung, Bd. II, Boden und Düngemittel. Verlag Springer, Wien, New York 1966a, 745-774.

Die Bodentemperatur. Handbuch der Pflanzenernährung und Düngung, Bd. II, Boden und Dungemittel. Verlag Springer, Wien, New York 1966b, 717-744.

Relations between soil management and the quality of surface and groundwater supplies. Qual. Plant 17, 45-65, 1968.

Agriculture and the Quality of Water Supplies. J. Soil Assn. 15 (8), 442-449, 1969a.

Bodennutzung und Wasserqualität. Mitt. DLG 84, 242-246, 1969b.

Die Stoffverfrachtung durch ursprungsnahe Oberflächengewässer. Wasser und Boden, 11, 326-328, 1970.

Der Einfluss der Landwirtschaft auf den Eutrophierungsprozess stehender Gewässer. Gewässerschutz, Wasser, Abwasser, 4, II, Aachen 1971, 397-415.

Bio-Dynamic Sprays. Reprint from Bio-Dynamics 97, Bio-Dynamic Farming and Gardening Assn. Inc. Duxbury, Mass. (USA) 1971.

Responsible Dealing with the Kingdoms of Nature. Bio-Dynamics 102, 1-10, 1972.

Organic management reduces leaching of nitrate, Bio-Dynamics,

108, 20-30, 1973.
Die Bedeutung der Forschung auf dem Gebiet der biologischen Landwirtschaft. Swiss Foundation for the Advancement of Biological Farming, Winterthur, No. 2, Oct. 1974.
and SELAWRY, A.: Application of the Diagnostic Crystallization Method for the Investigation of Quality of Food and Fodder. Bio-Dynamics 64 and 65, 9-24 and 1-12, 1962 and 1963.

KOHL, D. H.: Testimony before Pollution Control Board Hearings. Illinois Pollution Control Board, Dec. 10, 1971.

KOHNKE, H., and BERTRAND, A. R.: Soil Conservation. McGraw Hill, Inc., New York 1959.

KOLISKO, E. and L.: Agriculture of Tomorrow. Kolisko Archive, Stroud (England), 1939.

KRASILNIKOV, N. A.: Soil Microorganisms and Higher Plants. Acad. Sci. USSR, Moscow, 1958; Nat. Sci. Found. USDA, The Israel Program for Scientific Translations, 1961.

KRÜBER, H.: Kupferchloridkristallisation, ein Reagens auf Gestaltungskräfte des Lebendigen. Weleda Schriftenreihe 1, Schwäb. Gmünd 1950.

KTL-Flugschrift Nr. 15: Flüssigmistverfahren in der Rindvieh- und Schweinehaltung, 1966.

KÜNZEL, M.: Von der Saatgutbehandlung. Leb. Erde, 3/4, 1954.

LEIHENSEDER, W.: Kieselspritzungen bei Stangenbohnen. Leb. Erde 2, 1961.

LINDER, M.: Selling Biodynamic Ideas to the Public. Bio-Dynamics 97, 17-24, 1971.

LIPPERT, F.: Vom Nutzen der Kräuter im Landbau. Forschungsring für Biologisch-Dynamische Wirtschaftsweise. Stuttgart 1953.

LUST, V.: Naturgesteigerter Obstbau. Leb. Erde, 153, 1967; Der Obstbau 5-6, 1967.
10 Jahre ''Naturgesteigerter Obstbau'' — ein praktischer Beitrag zum Umweltschutz. Obst und Garten 2, 37, 1972.

MATILE, Ph.: Wege der Biologie. Symp. Umweltprobleme und Landwirtschaft, Bern 13./14. Okt. 1971. Abdruck in Leb. Erde 5, 1616-170, 1972.

McCALLA u. Mitarb., in: Science 108, 163, 1948.

MERCKENS, G.: Erfahrungen mit der Drilldüngung. Leb. Erde, 149, 1963.

MILLER, C. E.: Soil Fertility. John Wiley & Sons, New York 1955.

NATIONAL ACADEMY OF SCIENCES, Publ. 1400: Waste Management and Control. National Research Council, Washington D. C. 1966.

NICHOLSON, M.: The Environmental Revolution. Hodder and Stoughton, London 1970.

NN: Abfallbeseitigung. Schweiz. Z. Hydrologie 31, Fasc. 2, 185-783, 1969.

NN: The Faithful Thinker, a symposium, Hodder and Stoughton, London 1961.

NÖRING, F., FARKADSI, G., BOLWER, A., KNOLL, K. H., MATTHESS, G., und SCHNEIDER, W.: Über Abbauvorgange von Grundwasserverunreinigungen im Unterstrom von Abfalldeponien. GWF, 109, 6, 137-142, 1968.

OBER, J.: Stallbelichtung und Stallfenster. Grub 1954.

PELIKAN, W.: Heilpflanzenkunde, Bd. II. Philosophisch-Anthroposophischer Verlag, Dornach 1962.

PETTERSSON, B. D.: Betriebsbericht eines biologisch-dynamisch arbeitenden Spitzenbetriebes in Dänemark. Leb. Erde 5/6, 1951.

Die Produktivität biologischdynamischer Wirtschaftsweise im Nor. den. Leb. Erde, 3-15, 1963.

Resultat af danske bedriftsuntersogelser, 1957-1961. Tidskrift for bio-dynamisk Jordbrug, Nr. 4, 1964.

Beiträge zur Entwicklung der Kristallisationsmethode mit Kupferchlorid nach Pfeiffer. Leb. Erde 1, 1957.

Verkan av växtplats, gödsling och tillvaxtreglerande substanser pa matpotatisens kvalitetsegenskaper. Nordisk forskningsring, Järna, nr. 23, 1970.

Die Einwirkung von Standort, Düngung und wachstumsbeeinflussenden Stoffen auf die Qualitätseigenschaften von Speisekartoffeln. Leb. Erde 3/4, 1970.

Gödslingens inverkan pa matpotatisens kvalitetsegenskaper II. Nordisk forskningsring, Jarna, nr. 25, 1972.

PFEIFFER, E. E.: Bio-Dynamic Farming and Gardening, Anthroposophic Press New York, 1938.

The Earth's Face and Human Destiny. Faber and Faber Ltd., London 1947.

The Art and Science of Composting. Bio-Dynamics 49, 1959.

Chromatography Applied to Quality Testing. Bio-Dyn. Farm & Gard. Assn., Stroudsburg, Pa., 1960.

411

Die Fruchtbarkeit der Erde. 5. Aufl., Verlag R. Geering, Dornach 1969.

Rudolf Steiners landwirtschaftlicher Impuls. In: Wir erlebten Rudolf Steiner. Verlag Freies Geistesleben, Stuttgart 1970.

und RIESE, E.: The Fair Garden Plot. Philosophisch-Anthroposophicher Verlag, Dornach 1970.

PHILBRICK, H., and GREGG, R.: Companion Plants and How to Use Them. Devin-Adair Co., New York 1966.

PIMENTEL, D., HURD, L. E., BELLOTTI, A. C., FORSTER, M. J., OKA, J. H., SHOLES, O. D., WHITMAN, R. H.: Food Production and the Energy Crisis. Science 182, no. 4 III, 2. Nov. 1973, 443-449.

POPKEMA, B.: Die Herstellung und Bewertung von Stadtmüllkompost. Wageningen (Niederlande), 1. 1. 1967.

PORTMANN, A.: Die Gestalt als Selbstdarstellung des Lebendigen. Neue Wege der Biologie 1960.

PREUSCHEN, G.: Unkrautbekämpfung durch Abflammen. Mitt. DLG, A. 22 Mai 1968.

PRIEBE, H.: Landwirtschaft als Nebenberuf. Leb. Erde 3, 89-95, 1972.

REISCH, E.: Entwicklungslinien der Organisation und Produktionsstruktur landwirtschaftlicher Betriebe. Hohenheimer Arbeiten 57, 34-73, 1971.

REMER, N.: Lebensgesetze im Landbau. Philosophisch-Anthroposophischer Verlag, Dornach 1968.

Kurze praktische Anleitung. Forschungsring für Biologisch-Dynamische Wirtschaftsweise, Stuttgart 1949.

RICHTER, G.: Bodenerosion. Gutachten im Auftrage des Bundesministeriums für Ernährung, Landwirtschaft und Forsten. Bundesanstalt für Landeskunde und Raumforschung, Selbstverlag, Bad Godesberg 1965.

SANTON, J.: Trees and Shrubs for Shelterbelts. Bio-Dynamics 81, 23-35, 1967.

SATTLER, F.: Von der Entwicklung des Talhofes. Leb. Erde 5, 52-63, 1968.

Erfahrungen bei Fütterung von Milchkuhen. Leb. Erde 5, 181, 1969.

Hohe Milchleistung aus wirtschaftseigenem Futter. Leb. Erde, 129, 1971.

SAUERLANDT, W., und TRAPPMANN, M.: Untersuchungen über

Stallmist- und Strohkompost. Z. Acker- u. Pflanzenbau 98, 233-251, 1954.

SCHAUMANN, W.: Gesichtspunkte zur Schwemmentmistung. Leb. Erde, 153, 1963a.

Gesundheit und Krankheit als Lebensprozesse der Landwirtschaft. Sonderdr. Leb. Erde, 1963b.

Über die Behandlung der Jauche. Leb. Erde 1970.

SCHEERER, G.: Fruchttragende Hecken. Siebeneicher Verlag, Berlin, 1965.

SCHEFFER, F.: Erhaltung und Mehrung der Bodenfruchtbarkeit. Verlag Landwirtschaftskammer, Hannover 1946.

und ULRICH, B.: Humus und Humusdüngung. Verlag F. Enke, Stuttgart 1962.

SCHÖNBECK, F.: Untersuchungen über Vorkommen und Bedeutung von Hemmstoffen in Getreideruckständen innerhalb der Fruchtfolge. Z. Pflanzenkrankh. u. Pflanzensch. 63 (9), 513-545, 1956.

und BREISEWITZ, G.: Untersuchungen über den Abbau phytotoxischer Substanzen durch *Lumbricus terrestris*. Naturwiss. 44 (2), 42, 1957.

SCHREIBER, K.-F.: Landschaftsökologische und standortskundliche Untersuchungen im nördlichen Waadtland als Grundlage für die Ort- und Regionalplanung. Arbeiten der Universität Hohenheim, Bd. 45, Stuttgart 1969.

SCHROEDER, D.: Bodenkunde in Stickworten. Verlag F. Hirt, Kiel 1969.

SCHULTZ, J.: Samenjahre bei Waldbäumen und Planetenperioden. Sternkalender, Math. Astr. Sekt., Goetheanum, Dornach 1948, 46-60.

Kosmische Perioden bei den Samenjahren der Waldbäume. Sternkalender Forsch.-Ring, Stuttgart 1951.

SCHUMACHER, E. F.: Small Is Beautiful. A study of economics as if people mattered. Blond Briggs, London 1973.

SCHUPAN, W.: Zur Qualität der Nahrungspflanzen. BLV-Verlagsgesellschaft Munich 1961.

SCHWERDT, K.: Unerwartete Fruchtfolgewirkungen im Getreidebau. Landw. Wochenbl. Kurhessen, Waldeck, Kassel 32, 2111-2113, 1974 (Quot. from "Kurz und Bündig").

SCHWILLE, F.: Nitrate im Grundwasser. Dt. Gewasserkde. Mitt. 6 (2), 25-32, 1962.

Vortrag in Geisenheim am 25. 10. 1967.

SEKERA, F.: Gesunder und kranker Boden. Verlag Parey, Berlin und Hamburg 1951, 3. Aufl.

SELAWRY, A.: Neue Einblicke in die Samenkeimung. Dornach 1961. und SELAWRY, O: Die Kupferchloridkristallisation in Naturwissenschaft und Medizin. Verlag G. Fischer, Stuggart 1957.

SHEPHERD, A. P.: A Scientist of the Invisible. Hodder and Stoughton, London 1975.

SMITH, G. E., in: Farm Journal, Philadelphia (USA), Febr. 1967.

SPOHN, E.: Wohin mit dem Klärschlamm. Städtehygiene 8, 1968.
Wege und Irrwege des Düngens. Inst. f. Bodenhygiene, Blaubeuren 1971.
Das Atemverfahren zur Kompostierung. Blaubeuren 1971.

STEINER, R.: The Course of My Life. Anthroposophic Press, New York 1950.
An Outline of Occult Science. Anthroposophic Press, Spring Valley, N. Y. 1972.
Knowledge of the Higher Worlds and its Attainment. Anthroposophic Press, Spring Valley, N. Y. 1947.
Goethe's Conception of the World. Anthroposophical Publishing Company, London 1928.
Agriculture. A course of eight lectures. Bio-Dynamic Agricultural Association, London 1972.
Neuordnung des Bodenrechts als soziale Forderung der Gegenwart. Stuttgart 1957.
Geisteswissenschaftliche Grundlagen zum Gedeihen der Landwirtschaft. Verlag R. Steiner Nachlassverw., Dornach 1963.

STRUTT, N.: Modern Farming and the Soil. Ministry of Agriculture. Fisheries and Food, Agricultural Advisory Council. HMSO, London 1970.

THIESS, H.: Die Verwendung von Tonerde im Zusammenhang mit der organischen Düngung. Leb. Erde 1/2, 1958.

THUN, M.: Bewährte Fruchtfolgen im Kleingarten. Leb. Erde 3, 1963.
Aussaatzeit und Kieselversuch bei Möhren und Rote Beete 1966. Leb. Erde 1, 9, 1967.
Mehrjähriger Weiteranbau von Kartoffeln im siderischen Mondrhythmus sowie Nachbau unter gleichen Bedingungen und verschiedenen Kieselbehandlungen. Leb. Erde 1, 19-34, 1969.
Mond-Tierkreisversuche mit Kieselanwendung bei Gurken. Leb. Erde 1, 20-21, 1971.
und HEINZE, H.: Weitere Nachbauversuche mit Stangenbohnen 1968. Leb. Erde 1, 14-20, 1968.

und HEINZE, H.: Anbauversuche über Zusammenhange zwischen Mondstellungen im Tierkreis und Kulturpflanzen, B.1, Darmstadt 1973.

TIETJEN, C.: Aufbereitung und Unterbringung von Kot und anderen Abfällen der Hühnerhaltung. Mitt. DLG 42, 1965.

USDA Agric. Inform. Bull., No. 30: Guide to Agriculture, Superint. of Documents, Washington, D. C. (USA) 1955.

UTERMÖHLEN, W.: Basaltsteinmehldüngung 1892-1937, ihre Auswirkung und deren praktische Anwendung. Selbstverlag, Heimgarten (Schweiz).

VOGTMANN, H.: Proc. 23rd Alberta Poultry Ind. Conf. Calgary Alta, Can. 1974.

VOISIN, A.: Grundgesetze der Düngung. BLV-Veriagsgesellschaft, Munich 1966.

Boden und Pflanze. BLV Munich 1959.

VOITH-MÜLLEX: Schriftenreihe Aufbereitung und Nutzbarmachung von Siedlungs abfällen i.J.

WARD, J. F.: The Haughley Experiment 1952-1965. Mimeogr. Manuscr., Haughley, Suffolk.

WARING, J. J. and SHANNON, D. W. F.: Investigations with poultry. Brit. Poult. Sci. 10, 331-336, 1969.

WEBER, F., und HEYNITZ, K. v.: Die Möglichkeiten der mechanischen Unkrautbekämpfung. Leb. Erde 43, 1972.

WEINSCHENCK, G.: Alternativen der EWG-Agrarpolitik. Hohenheimer Arbeiten, Nr. 57, 9-33, 1971.

WEST, S. H.: Growth Promotion and Control in Plants. Proc. Soil and Crop Sci. Soc., Fla. 22, 44-48, 1962.

WESTERMANN, L. T., and HANCK, R. D.: Recovery of 15N-labeled Fertilizers in Field Experiments. Soil Sci. Soc. Proc. America 36, 82-86, 1972.

WINTER, A. G., und BUBLITZ, W.: Untersuchungen über antibakterielle Wirkungen im Bodenwasser der Fichtenstreu. Naturwiss. 40, 345-346, 1953.

WINTER, A. G., und SCHÖNBECK, F.: Untersuchungen über wasserlösliche Hemmstoffe aus Getreideboden. Naturwiss. 41, 145-146, 1954.

WINTER, A. G., und SIEVERS, E.: Untersuchungen über die Beeinflussung der Samenkeimung durch Kaltwasserextrakte aus der Blattstreu verschiedener Gramineen. 39, 191, 1952.

WHITE HOUSE REPORT: Restoring the Quality of our Environment. Superint. of Documents, Washington, D. C., 1965.

WILSON, B. J. and McNAB, J. M.: Brit. Poult. Sci. 13, 67-73, 1972.

WIRTH, A. G.: Höchsterträge durch Mischkultur wahlverwandter Gemüse- und Obstarten. Verlag Ulmer, Stuttgart 1962.

WISTINGHAUSEN, E. v.: Die Verlagerung von Nitrat- und anderen Ionen in Böden und die Wirkung der Bewirtschaftung auf diesen Vorgang. Diss. Hohenheim 1971.

WITZEL, S. A.: Nitrogen Cycle in Surface and Subsurface Waters. Techn. Completion Report DWRRB - 004 - WIS, The University of Wisconsin, 1969.

WOLBER, G., und VETTER, S.: Samenjahre der Rotbuche und Planetenstellung im Tierkreis. Sternkalender. Philosophisch-Anthroposophischer Verlag, Dornach 1973, 94-99.

WOODWELL, G. M.: Toxic Substances and Ecological Cycles. March 1967 in Man and the Ecosphere. Readings from Sci. Amer., W. H. Freeman, San Francisco.

WORLD HEALTH ORGANIZATION: Survey of Solid Wastes Management Practices. Internat. Reference Center, Dübendorf (Schweiz) 1971.

Index

419

Clay, 134, 135f., 183, 186f., 190-1; in compost, 331; content in soil, 133, 140f.; as fertilizer, 333
Clay minerals, in soil, 48
Click beetle (*Agriotes sp.*), 129
Climate, 140-1
Climatic zones, 58-9
Clover-grass-herb mixture, 86
Clubroot, 68
Cobalt, 167, 177
Cock-chafer, 237
Coddling moth, 343, 378
Cold frames, 321, 327; in winter, 339
Colorado beetle, 90
Colostrum, 299
Companion plants, 17, 30, 202, 234f.; with field crops, 235-6; in the garden, 236-8; with trees, 235
Compost/Composting, 17, 30, 57, 134, 149, 161f., 163, 164f., 173, 220f., 229, 230, 234; of fruit pulp, 341; garbage, 115, 162, 166, 174f.; garden, 162, 164, 166, 321, 327f.; large scale, 165, 166; leaf, 330; orchard, 341; sheet, 44, 172, 230, 243
Compost pile, 164f, 328f. (*see* Compost/Composting)
Compost starter, 208, 251
Compound preparations, 169, 170, 173, 196, 219, 266
Concentrates, 18, 80, 91, 256, 272, 293, 294-5, 298, 308
Coniferous trees, 125, 140, 142, 227, 233
Constipation, in cattle, 318
Copper, 177, 180, 208; in manure, 167; in pesticides, 5; in plant, 3; in soil, 139
Copper chloride, 362, 367
Coriander (*Coriandrum sativum*), 91
Corn, 159-60
Cornbelt (USA), 156, 160
Corn root worm, 68
Cosmic forces, *see* Formative forces
Cotton wilt, 144
Couch grass (*Agropyron repens*), 35, 82, 89, 248
Coumarin, 227

Covered yard, *see* Loafing yard
Cowhouse, *see* Livestock, housing
Cowpox, 317
Crop residues, 226, 229-30
Crystallization method, 361f., 368, 380f.
Cultivation, 141 (*see* Tillage)
"Cultivation fertility", 78
Cybernetic models, 7, 26
Cycle of substances and forces, 47f.
"Cylinder Experiments", 97
Cystin, 241

D

Dairy farming, *see* Livestock
Damping off, 232
Dandelion (*Taraxacum officinale*), 207, 225; dwarf (*Taraxacum laevigatum*), 218
Darkness, 363f. (*see* Shade)
DDT, 5, 6
Decomposition, of plant, 361
Dehydrogenase, 361, 368
Demeter Association (*Demeterbund*), 21, 392f.; founding of, 20; new Demeter Association, 22
Demeter, Cooperative (*Verwertungs-genossenschaft Demeter*), 391; products, 239, 393f.; quality control, 394-5; standards, 357, 395f.; trademark, 22, 31, 182, 391f.; Trading Association (*Demeterwirtschaftsbund*), 391 (*see* Quality, of produce)
Denitrification, 153, 156, 157
Derris, 343
Deschampsia cespitosa, see Tufted hair grass
Diabase, 137
Diarrhea, in cattle, 274, 292, 297, 317; in calves, 318; treatment of, 317
Diastase, 368
Dieldrin, 5
Dill (*Anethum graveolens*), 92
Diorite, 137
Disease control, 116-17
Diseases, animal, 312f.; in calves, 318-19; in cattle, 294, 315f.; medicine

420

chest for, 314; in piglets, 319-20
Diversified (mixed) farming, 67f., 105, 112, 401-2
Divining rod, 310
Dock (*Rumex obtusifolius*), 275
Dolomite, 9, 136, 139, 179, 279
Drainage, 61, 63, 103, 104, 113, 133, 191, 240
Drifting, 104
Dry period, before calving, 298
Dung channel, 168, 174, 309

E

Earthly forces, *see* Formative forces
Earthworms, 134, 140, 187, 194, 215, 219, 228, 234, 329, 353
Ecological system (eco-system), 12, 31, 106f.
Ecology, 12, 44, 59; and farm organization, 65
Edaphon, 140
"Efficiency principle", of nutrient utilization, 54f.
Ellenberg, H., 61
Energy, requirements for modern agriculture, 41f., 57
Engqvist, N., 361, 362, 364, 368, 372
Enteric fever, 176
Environment, 105f., 111, 116, 355; cosmic and earthly, 24, 31
Equisetum arvense, see Horsetail
Erosion, *see* Soil
Erucic acid, 240, 241
Erysipelas, *see* Swine fever
Etheric forces, 16
Ethylene, *see* Apple gas
"Exchange capacity", 221
Experimental Circle for Bio-Dynamic Horticulture and Agriculture (West Germany), 22, 393f.
Experimental Circle of Anthroposophical Farmers, 17, 18
Eyespot, 34, 68

F

Factory farming, 114, 161, 266, 268, 300

Family farms, 43
Farm individuality, 46, 47 (*see* Organism)
Farm organization and management, 47, 64f., 111f., 145f., 178, 195f.
Farm reorganization, 189f., 269, 278, 295, 385
Fats, in plants, 49
Fatty acids, 271, 289
Fauna, 109
Feather meal, 181, 332, 335
Feces, 177
Feed, for livestock, 269f., 270f.; herbs in, 294-5; minerals in, 296f.; in spring and summer, 272f.; in winter, 289f. (*see* Fodder, Silage)
Feeding, *see* Livestock, feeding
Feldspar, 137f., 186, 207
Fencing, electrical, 274, 276
Fennel (*Foeniculum vulgare*), 91
Fermentation, 193, 266; in compost pile, 162f., 329; in digestion, 286; in silage, 282, 287
Fertilizer, mineral (NPK), 3, 4, 9, 54, 94, 114f., 134, 149f., 158, 159, 161, 175, 179, 244, 357, 367-8, 370, 386; organic, 80, 103, 178f., 180, 182, 244f., 358-9, 367, 370
Fertilizing, 148f., 178f.; in garden, 331f.; influence on quality, 371f.; in orchard, 341f.; in vineyard, 351
Field bean (*Vicia faba*), 240, 241, 293, 300
Field experiments, 94
Fish waste, 164
Flea beetle, 237
Fluorine, 108
Fodder, 17, 18, 79-80, 269f., 289f.
Fodder beet, 91, 270, 288, 291f.
Foeniculum vulgare, see Fennel
Folic acid, 231
Food chain, 6, 360
Food production, 40, 42
Forests, 62, 120
Forestry, 66
Form creation, 48
Form, of plant, 382f.

Persian clover (*Trifolium resupinatum*), 88, 248, 338, 341
Persistence, 5
Pest control, biological, 116, 117
Pesticides, 5, 6, 9, 10, 14, 117
Pettersson, B.D., 70, 357, 361, 362, 369, 372, 379
Pfeiffer, E.E., 9, 15, 16, 21, 188, 208, 219, 362
pH, 134, 183, 185, 186, 188, 192, 242, 279; in rumen, 286f.; in silage, 283f.
Phacelia, 353
Phaseolus vulgaris, see French beans
Phosphate, 57, 107, 114, 115, 134, 140, 167f., 175, 176, 191, 244f., 278, 293; in cattle feed, 274, 296f.
Phosphorus, 50, 139, 147, 152, 178, 183, 185, 208, 231, 234, 251, 265, 278, 288
Photosynthesis, 49, 50, 227, 280
Phthium, 232
Phyllite, 138
"Picture-creating" methods, 356, 362f.
Pig rearing, 264, 265, 267, 300f.
Piglets, diseases of, 319f.
Plagioclase, 137
Plant forms, 48, 49
Plant growth, accelerators of, 228, 233; influences on, 355; inhibitors of, 227, 228, 232, 233; observations of, 356f.
Plant hormones, *see* Hormones
Ploughing, *see* Tillage
Podsol, 148, 231-2
Polio, 176
Pollution, 41, 107, 109, 111, 116, 174-5, 177
Population growth, 36f., 57, 60, 101-2, 106, 176
Porphyry, 137; meal, 351
Potash, 3, 13, 38, 54, 114, 167
Potassium, 50, 53, 54, 135, 137f., 147, 150, 151, 152, 168f., 175, 176, 178, 186, 244f., 273, 278, 297, 352
Potato blight, 237
Potato scab, 4

Potatoes, 20, 369, 372f.
Poultry, raising, 304f.; housing, 169; at Talhof, 93
Preparations, *see* Bio-Dynamic compost preparations
Propionic acid, 271, 285, 286
Protein, 360, 368f., 383; in cattle feed, 91, 272f., 287f., 293; crude, 91, 157, 241, 360, 361; digestible, 91, 289; formation of, 361; in pig feed, 303; in plants, 47, 49; in poultry feed, 305; relative content, 361; in rumen, 287, 289
Protein shortage, 40, 41
Proteinase, 361
Protozoa, 232
Pruning, 344
Pseudogley, 148
Pyrethrum, 343
Pyrolysis, 177

Q

Quality, of produce, 20-21, 356f., 373-4, 386f., 394f.; control, 394-6; guarantee, 178; grading, 387f.; index, 373-4; influence on, 363f.; testing, 360f.
Quartz, 135f., 140, 183, 185, 207
Quercus robur, see Oak bark
Quicklime, 328, 329

R

Radiation (sun), 119, 120
Radish, 204, 229
Ranunculus acer, see Buttercup
Rape seed, 240, 241, 243, 244; cakes, 240
Recycling, of organic matter, 145f., 175
Red clover, 40, 69, 173, 242
"Red earth", 58
Remer, N., 266, 302
Rendzina, 148
Reproduction, cattle, 296, 298f.
Residue formation, 5
Rhizosphere, 54, 144, 231
Rhythms, cosmic, 31, 48, 49, 119f., 126, 202f.
Riboflavin, 231

Richter, G., 101
Rickets, 265, 319
Ripening, process in plant, 383
Robinia (*Robinia pseudoacacia*), 53
Rock, ground, 181
Rock crystal, 507
Rock phosphate, 9, 13, 96, 162, 170,
    171, 179, 181, 244, 279
Rodale, I.J., 11
Root crops, 78-9, 147, 180; planting
    times for, 206
Root, forms, 49; growth, 357f.; life,
    149; system, 228f.
Root rot, 232
Root wilt, 232
Rotation, 39, 69, 145, 147, 156, 192,
    196, 239f., 248; in garden, 323f.;
    on Morrow Plots, 95-6, at
    Talhof, 86f.
Roughage, in feed, 271f., 279f., 289,
    293, 295
Roundworms, 176
Rue (*Ruta graveolens*), 91
Rumen, 270f., 285f., 289
Rumex obtusifolius, see Dock
Run-off, 100, 102, 103, 104, 112, 113,
    116, 133
Ryania, 343

S

Sage (*Salvia officinalis*), 92
Salmonella, 176, 266
Salt, 91, 273-4, 284, 297
"Saltation", 104
Sanborn field, 97
Sandstone, 136
Scab, 343, 378
Scheffer, F., 38, 39, 56, 94, 147
Schist, 138
Schutz-Lupitz, 241
Schumacher, E., 40
Schuphan, W., 5, 387, 388
Seaweed, calcified, 245, 278, 279, 304,
    328, 329, 330, 332-3, 342, 344;
    ground, 245; liquid, 331
Sedementary rocks, 135-6
Seed, baths, 90, 336; boxes, 335f;
    dressings, 215, 219; treatment,
    173

Selawry, A., 361, 362
Selenium, 208
Sewage, 114; sludge, 174, 175f.
Shade, 375; and mineral fertilizer,
    367-8
Shale, 136
Shelter belts, 123f.
Shepherd's purse (*Capsella bursa-
    pastoris*), 331
Silage, 271, 273, 282f., 288, 290f., 303
Silage, maize, 291, 297
Silo, 284f., 290
Silviculture, 18
Silica, 134, 135f., 183f., 209, 213, 383;
    in cattle fodder, 298; as formative
    element, 184; in grasses, 183; in
    soil, 138, 140, 178
Silica cycle, 184
"Silica plants", 183
Silicic acid, 135, 138, 184, 223, 305
Silicon, 3, 137
Site, improvement of, 63
Slopes, 61, 63, 119, 276; in orchard,
    340
Slugs, 237
Slurry, 168, 169, 171, 172, 173, 219,
    222, 230, 246, 266, 273, 275, 277
Sodium, 135, 186; in clay, 135; in
    plant, 3
Sodium hydrocarbonate, 271, 286
Sodium silicate, 377 (*see* Waterglass)
Soil, 56, 103-4, 119, 132f., 187f.;
    aeration, 61, 129, 133-4, 140, 145,
    182; analysis, 60; classification,
    138; degradation, 101-2; erosion,
    43, 57, 99f., 101, 102f., 124, 127,
    228; fertility, 56, 57, 61, 67,
    139f.; formation, 133, 135f.;
    horizons, 148; management, 141;
    moisture, 61; profiles, 148; respi-
    ration, 126-7, structure, 67, 103,
    133f.; testing, 187f., 244f., 340;
    types, 148; water-holding
    capacity, 113, 133, 139
Soil and Water Conservation Plan
    (USA), 60
Soil Association (Great Britain), 10,
    11, 40
Soil Conservation Service (USA), 9,

427

428